T0329636

Ultra-Reliable and Low-Latency Communications (URLLC) Theory and Practice

Ultra-Reliable and Low-Latency Communications (URLLC) Theory and Practice

Advances in 5G and Beyond

Edited by Trung Q. Duong, Saeed R. Khosravirad, Changyang She, Petar Popovski, Mehdi Bennis and Tony Q.S. Quek

WILEY

This edition first published 2023
© 2023 John Wiley & Sons Ltd

All rights reserved. No part of this publication may be reproduced, stored in a retrieval system, or transmitted, in any form or by any means, electronic, mechanical, photocopying, recording or otherwise, except as permitted by law. Advice on how to obtain permission to reuse material from this title is available at http://www.wiley.com/go/permissions.

The right of Trung Q. Duong, Saeed R. Khosravirad, Changyang She, Petar Popovski, Mehdi Bennis and Tony Q.S. Quek to be identified as the authors of the editorial material in this work has been asserted in accordance with law.

Registered Offices
John Wiley & Sons, Inc., 111 River Street, Hoboken, NJ 07030, USA
John Wiley & Sons Ltd, The Atrium, Southern Gate, Chichester, West Sussex, PO19 8SQ, UK

For details of our global editorial offices, customer services, and more information about Wiley products visit us at www.wiley.com.

Wiley also publishes its books in a variety of electronic formats and by print-on-demand. Some content that appears in standard print versions of this book may not be available in other formats.

Trademarks: Wiley and the Wiley logo are trademarks or registered trademarks of John Wiley & Sons, Inc. and/or its affiliates in the United States and other countries and may not be used without written permission. All other trademarks are the property of their respective owners. John Wiley & Sons, Inc. is not associated with any product or vendor mentioned in this book.

Limit of Liability/Disclaimer of Warranty
While the publisher and authors have used their best efforts in preparing this work, they make no representations or warranties with respect to the accuracy or completeness of the contents of this work and specifically disclaim all warranties, including without limitation any implied warranties of merchantability or fitness for a particular purpose. No warranty may be created or extended by sales representatives, written sales materials or promotional statements for this work.This work is sold with the understanding that the publisher is not engaged in rendering professional services. The advice and strategies contained herein may not be suitable for your situation. You should consult with a specialist where appropriate. The fact that an organization, website, or product is referred to in this work as a citation and/or potential source of further information does not mean that the publisher and authors endorse the information or services the organization, website, or product may provide or recommendations it may make. Further, readers should be aware that websites listed in this work may have changed or disappeared between when this work was written and when it is read. Neither the publisher nor authors shall be liable for any loss of profit or any other commercial damages, including but not limited to special, incidental, consequential, or other damages.

A catalogue record for this book is available from the Library of Congress

Hardback ISBN: 9781119818304; ePub ISBN: 9781119818335; ePDF ISBN: 9781119818311;
oBook ISBN: 9781119818366

Cover image: © Krunja/Shutterstock
Cover design by Wiley

Set in 9.5/12.5pt STIXTwoText by Integra Software Services Pvt. Ltd, Pondicherry, India
Printed and bound by CPI Group (UK) Ltd, Croydon, CR0 4YY

C9781119818304_010323

Contents

Preface

Pursuing ever higher data rates has been the central design goal in all the previous generations of mobile communications. This has been changed in the 5th generation (5G) mobile communications that aims to support various new emerging services with diverse and stringent quality-of-service requirements. The most formidable challenge in 5G is to achieve Ultra-Reliable Low-Latency Communications (URLLC) for many mission-critical services including autonomous vehicles, industry automation, and tele-robotic surgery, i.e. the roundtrip delay of 1 millisecond and less than 1 out of a million in packet loss. In the 4th generation (4G) systems, the average latency is usually a few hundred milliseconds, and the packet loss probability is around 1%. 5G systems need to significantly improve the latency and reliability by several orders of magnitude compared to 4G systems. This presents unprecedented challenges.

This book covers a range of topics from fundamental theories to practical solutions in URLLC.

In Chapters 2 and 3, the authors analyze the statistical features and tail distributions of wireless channels and provide useful insights on the performance of URLLC. *Chapter 2* presents the statistical aspects of URLLC in both frequentist and Bayesian approaches. The authors have analyzed the statistical features and guarantees for outage probability in a narrowband wireless channel. *Chapter 3* considers various metrics of URLLC including tail distribution, higher-order statistics, extreme events with very low occurrence probabilities, worst-case metrics, and reliability/latency. The authors have introduced readers to the entropic risk measure in financial mathematics, generalized extreme value distribution, and generalized Pareto distribution to investigate these metrics.

In Chapters 4–7, the authors have introduced several techniques to guarantee the reliability and latency of URLLC, including machine learning, candidate channel codes, sparse vector coding, and network slicing. Two problems of resource allocation in URLLC are addressed in *Chapter 4* with an unsupervised learning approach. The results have shown that bandwidth utilization efficiency of URLLC can be improved more significantly by exploiting frequency diversity than by multi-user diversity. *Chapter 5* discusses the channel coding and decoding schemes for URLLC. This chapter reviews state-of-the-art channel codes for URLLC and analyzes them in terms of performance and complexity. Furthermore, the Ordered Statistics Decoding (OSD) is promoted as one of the potential universal decoding algorithms for URLLC. In *Chapter 6*, a new approach to support short packet transmissions, referred as Sparse Vector Coding (SVC) is introduced. The numerical evaluations and performance analysis which validate the proposed SVC technique is highly effective in URLLC transmissions. *Chapter 7* studies a CoMP-enabled RAN slicing system simultaneously supporting URLLC and eMBB services. The authors address a joint bandwidth and CoMP beamforming optimization problem to maximize the long-term total slice utility.

In Chapters 8 and 9, downlink Orthogonal Frequency Division Multiple Access systems (OFDMA) and full-duplex relay system are optimized for URLLC, respectively. *Chapter 8* investigates the beamforming design for downlink ODFMA to enable the stringent delay requirement. In particular, the authors address a non-convex optimization problem to maximize the weighted system sum throughput subject to Quality-of-Service (QoS) of URLLC users. *Chapter 9* presents an up-to-date overview of the end-to-end latency for a Full-Duplex (FD) relay system in the context of URLLC. The authors not only provide an insightful investigation of reliability and latency together for FD relay assisted URLLC but also discuss possible relaying latency reduction solutions in the chapter.

In Chapters 10 and 11, the authors investigate URLLC in vertical industries: Tactile Internet and Industrial Internet-of-Things. More specifically, *Chapter 10* addresses an optimization problem that maximizes the number of URLLC services by jointly optimizing time and frequency resources and the prediction horizon. The numerical results clearly demonstrate the effectiveness of the proposed solution. In addition, a proof-of-concept experiment with the remote control in a virtual factory is also provided to illustrate a typical application of Tactile Internet. Finally, *Chapter 11* considers relay robots-aided URLLC in 5G factory automation, which consists of multiple relay robot deployment and decoding error probability minimization problems. There are two different approaches introduced for relay robot deployment, including Deep Neural Networks (DNN) and the K-means clustering algorithm. A low-complexity iterative algorithm is also provided to deal with the joint blocklength and power allocation problem to minimize the decoding error probability.

List of Contributors

Mehdi Bennis
Centre for Wireless Communications
University of Oulu
Oulu, Finland

Xianbin Cao
School of Electronics and
Information Engineering
Beihang University
Beijing, China

Jingxuan Chen
School of Electronics and
Information Engineering
Beihang University
Beijing, China

Hanjun Duan
The School of Electronic and
Information Engineering
Harbin Institute of Technology
Shenzhen, China

Trung Q. Duong
School of Electronics Electrical
Engineering and Computer Science
Queen's University
Belfast, UK

Walid R. Ghanem
Friedrich-Alexander-University
Erlangen-Nuremberg (FAU)
Erlangen, Germany

Zhanwei Hou
The School of Electrical and
Information Engineering
The University of Sydney
Sydney, Australia

Yung-Lin Hsu
Graduate Institute of
Communication Engineering
National Taiwan University
Taipei, Taiwan

Dang Van Huynh
School of Electronics Electrical
Engineering and Computer Science
Queen's University
Belfast, UK

Vahid Jamali
Technical University Darmstadt
Darmstadt, Germany

Yufei Jiang
The School of Electronic and
Information Engineering
Harbin Institute of Technology
Shenzhen, China

Tobias Kallehauge
Connectivity Section at the
Department of Electronic Systems
Aalborg University
Aalborg, Denmark

Anders E. Kalør
Connectivity Section at the
Department of Electronic Systems
Aalborg University
Aalborg, Denmark

Saeed R. Khosravirad
Nokia Bell Labs
Murray Hill
New Jersey, US

Yonghui Li
School of Electrical and Information
Engineering
University of Sydney
Sydney, Australia

Chen-Feng Liu
Technology Innovation Institute
Masdar City
Abu Dhabi, UAE

Antonino Masaracchia
School of Electronics Electrical
Engineering and Computer Science
Queen's University
Belfast, UK

Yuexing Peng
Beijing University of Posts and
Telecommunications
Beijing, China

Petar Popovski
Connectivity Section at the
Department of Electronic Systems
Aalborg University
Aalborg, Denmark

Tony Q.S. Quek
Information Systems Technology
and Design
Singapore University of Technology
and Design
Singapore

Pablo Ramirez-Espinosa
Connectivity Section at the
Department of Electronic Systems
Aalborg University
Aalborg, Denmark

Robert Schober
Friedrich-Alexander-University
Erlangen-Nuremberg (FAU)
Erlangen, Germany

Changyang She
The School of Electrical and
Information Engineering
The University of Sydney
Sydney, Australia

Byonghyo Shim
Institute of New Media and
Communications and Department of

Electrical and Computer Engineering
Seoul National University
Seoul, Korea

Mahyar Shirvanimoghaddam
The School of Electrical and
Information Engineering
The University of Sydney
Sydney, Australia

Chengjian Sun
School of Electronics and
Information Engineering
Beihang University
Beijing, China

Branka Vucetic
The School of Electrical and
Information Engineering
The University of Sydney
Sydney, Australia

Hung-Yu Wei
Department of Electrical
Engineering
National Taiwan University
Taipei, Taiwan

Dapeng Wu
Department of Electrical and
Computer Engineering
University of Florida
Gainesville, USA

Xing Xi
School of Electronics and
Information Engineering
Beihang University
Beijing, China

Chenyang Yang
School of Electronics and
Information Engineering
Beihang University
Beijing, China

Peng Yang
Information Systems Technology
and Design
Singapore University of Technology
and Design
Singapore

Chentao Yue
The School of Electrical and
Information Engineering
The University of Sydney
Sydney, Australia

Fu-Chun Zheng
The School of Electronic and
Information Engineering
Harbin Institute of Technology
Shenzhen, China

Xu Zhu
The School of Electronic and
Information Engineering
Harbin Institute of Technology
Shenzhen, China

1

URLLC: Faster, Higher, Stronger, and Together

Changyang She[1,], Trung Q. Duong[2], Saeed R. Khosravirad[3], Petar Popovski[4], Mehdi Bennis[5], and Tony Q.S. Quek[6]*

[1] School of Electrical and Information Engineering, University of Sydney, 2006, NSW, Australia
[2] School of Electronics Electrical Engineering and Computer Science, Queen's University Belfast, BT7 1NN, Belfast, UK
[3] Nokia Bell Labs, NJ 07974-0636, Murray Hill, USA
[4] Connectivity Section at the Department of Electronic Systems, Aalborg University, 9220, Aalborg, Fredrik Bajers Vej 7A, Denmark
[5] Centre for Wireless Communications, University of Oulu, Oulu, Finland
[6] Information Systems Technology and Design, Singapore University of Technology and Design, 487372, Singapore
* Corresponding Author

As one of the new communication scenarios in the 5th Generation (5G) mobile communications, Ultra-Reliable and Low-Latency Communications (URLLC) are crucial for enabling a wide range of emerging applications, including industry automation, intelligent transportation, telemedicine, Tactile Internet, and Virtual/Augmented Reality (VR/AR). According to the requirements in 5G standards, to support emerging mission-critical applications, the End-to-End (E2E) delay cannot exceed 1 ms and the packet loss probability should be 10^{-5}–10^{-7}. Compared with the existing cellular networks, the delay and reliability require significant improvements by at least two orders of magnitude for 5G networks. This capability gap cannot be fully resolved by the 5G New Radio (NR), i.e. the physical-layer technology for 5G, even though the transmission delay in Radio Access Networks (RANs) achieves the 1 ms target. Transmission delay contributes only a small fraction of the E2E delay, as the stochastic delays in upper networking layers, such as queuing delay, processing delay, and access delay, are key bottlenecks for achieving URLLC. Beyond 5G systems or so-called 6th Generation (6G) systems should guarantee the E2E delay bound with high reliability.

In addition to the latency and reliability requirements, some other Key Performance Indicators (KPIs) should also be taken into account, including Spectrum

Ultra-Reliable and Low-Latency Communications (URLLC) Theory and Practice: Advances in 5G and Beyond, First Edition. Edited by Trung Q. Duong, Saeed R. Khosravirad, Changyang She, Petar Popovski, Mehdi Bennis and Tony Q.S. Quek.
© 2023 John Wiley & Sons Ltd. Published 2023 by John Wiley & Sons Ltd.

Efficiency (SE), throughput, Energy Efficiency (EE), Age of Information (AoI), jitter of latency, round-trip delay, network availability, and security (shown in Table 1.1). These requirements will pose unprecedented challenges in terms of design methodologies and enabling technologies in the 6th Generation (6G) mobile communications. To fill the gap between 5G URLLC and the diverse KPI requirements, we shall investigate novel methodologies and innovative technologies for the next generation URLLC (xURLLC), also known as eXtreme URLLC, [23]. This book will cover various methods and technologies to achieve URLLC from the physical layer, link layer, and network layer, to diverse applications in vertical industries of 5G/6G communications.

Table 1.1 KPIs and research challenges. ©IEEE 2021. Reprinted with permission from [28].

Indoor large-scale scenarios		
Applications	**KPIs**	**Research Challenges**
Factory automation	SE, EE, and AoI	Scalability and network congestions
VR/AR applications	SE and throughput	Processing/transmission 3D videos
Indoor wide-area scenarios		
Applications	**KPIs**	**Research Challenges**
Tele-surgery	Round-trip delay, throughput, and jitter	Propagation delay and high data rate
eHealth monitoring	EE and network availability	Propagation delay and localization
Outdoor large-scale scenarios		
Applications	**KPIs**	**Research Challenges**
Vehicle safety	AoI, SE, security, and network availability	High mobility and scalability
Outdoor wide-area scenarios		
Applications	**KPIs**	**Research Challenges**
Smart grid	SE	Propagation delay and scalability
Tele-robotic control	SE, security, network availability, and jitter	Propagation delay and high data rate
UAV control	EE, security, network availability, and AoI	Propagation delay and high mobility

1.1 Requirements of URLLC: Faster, Higher, Stronger, and Together

The next generation URLLC is expected to be "faster, higher, stronger - together". The specific requirements and research challenges are discussed in the sequel.

1.1.1 Faster Responses and Movement

In factory automation and autonomous vehicles, mobile devices need to make decisions according to their local observation in a real-time manner. Given the fact that the energy budget and the computing capability of each device are limited, it may need the help of Mobile Edge Computing system (MEC). Unlike centralized mobile cloud computing with routing delay and propagation delay in backhauls and core networks, the E2E delay in MEC systems consists of Uplink (UL) and Downlink (DL) transmission delays, queuing delays in the buffers of users and Base Stations (BSs), and the processing delay in the MEC [26]. Although MEC helps reduce latency in communication systems, there are two bottlenecks for providing fast responses to mobile devices. First, optimization problems in MEC systems are generally non-convex. To find the optimal solution, such as user association and task offloading. The computing complexity for executing searching algorithms is too high to be implemented in real-time. Second, exchanging information among different edge servers will lead to high overheads and latency. To avoid this issue, edge servers need to make decisions in a distributed manner. As a result, a local decision may not be optimal for all the devices.

Supporting high mobility URLLC is critical for some outdoor applications, e.g. Unmanned Aerial Vehicle (UAV) control. Since the mobile devices are moving fast, the Doppler frequency shift is large. Thus, the inter-symbol interference is strong and the receiver needs to adjust the carrier frequency according to the Doppler shift. Besides, frequent handovers in high mobility URLLC will result in service interruption. The BS with good channel quality may not have sufficient radio resources due to the dynamic traffic load in high mobility scenarios. Therefore, how to serve high mobility URLLC remains an open issue in 6G.

1.1.2 Higher Throughput and Density

As one of the killer applications in 5G networks, VR/AR applications require ultra-reliable and low-latency tactile feedback and high data rate 360° videos [8]. Meanwhile, as the sizes of devices shrinks, battery lifetime will become a bottleneck for enabling high data rate URLLC [24]. To implement VR/AR applications in future wireless networks, we need to investigate the fundamental trade-offs among throughput, energy efficiency, reliability, and latency in communications, caching, and computing systems [34], as well as enabling technologies such as touch user interface and haptic codecs [4, 31].

Due to the explosive growth of the numbers of autonomous vehicles and mission-critical IoT devices [9], future wireless networks are expected to support massive URLLC. To support massive URLLC, novel communication and learning techniques are needed. With orthogonal multiple access technologies, the required bandwidth increases linearly with the number of devices. To achieve better trade-offs among delay, reliability, and scalability, other multiple access technologies should be used, such as non-orthogonal multiple access and contention-based multiple access technologies [29, 30]. Meanwhile, we may need to exploit the above 6 GHz spectrum including mmWave [16] and the Terahertz band [36].

1.1.3 Stronger Connectivity and Security

Multi-connectivity is a promising approach to provide seamless services to URLLC users [17]. As illustrated in [22], one way to improve network availability without sacrificing spectrum efficiency is to serve each user with multiple BSs over the same subchannel (or subcarrier). The disadvantage of this intra-frequency multi-connectivity is that the failures of different links are highly correlated. For example, if there is a strong interference on a subchannel, then the signal to interference plus noise ratios of all the links are low. To alleviate cross-correlation among different links, different nodes can connect to one user with different subchannels or even with different communication interfaces [21]. Further considering that terrestrial networks may not cover rural areas and marine areas, we need to use non-terrestrial networks to provide global connectivity for long-distance URLLC services, e.g. Tactile Internet.

Future URLLC systems will suffer from different kinds of attacks that result in inefficient communications [19]. The widely used cryptography algorithms require high-complexity signal processing, and may not be suitable for URLLC, especially for IoT devices with low computing capacities. To defend against eavesdropping attacks in URLLC, physical layer security is a viable solution [3]. The maximal secret communication rate in the short blocklength regime over a wiretap channel was derived in [37]. The results show that there are trade-offs among delay, reliability, and security. Based on this fundamental result, we can further investigate the technologies for improving physical-layer security.

1.1.4 Human Intelligence Together with Artificial Intelligence in URLLC

As illustrated in Figure 1.1, a new trend of developing communication networks is to integrating human intelligence (expert knowledge in wireless

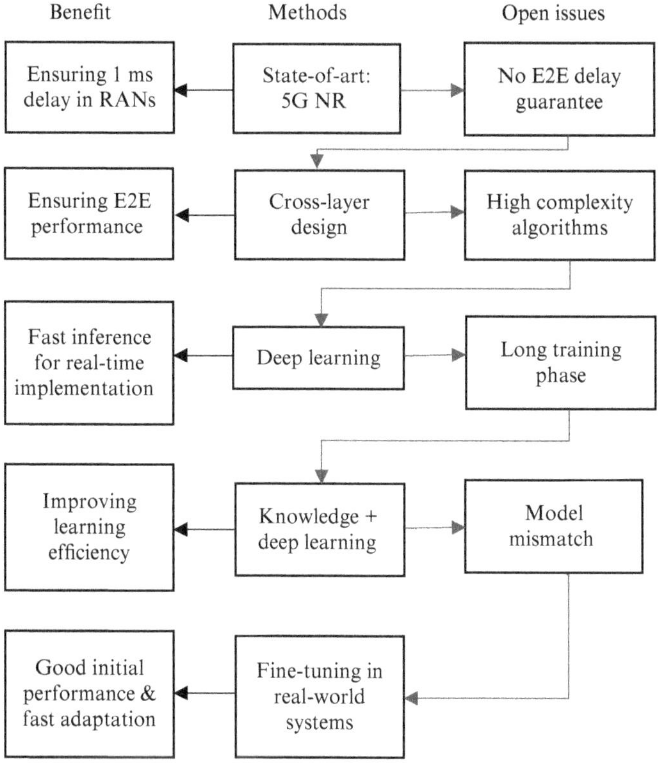

Figure 1.1 Wireless AI for developing URLLC systems. ©IEEE 2021. Reprinted with permission from [28].

communications) into artificial intelligence (deep learning) for optimizing communication systems.

Cross-layer Design Existing design methods divide communication networks into multiple layers according to the Open Systems Interconnection model [18]. Communication technologies in each layer are often developed without considering the impacts on other layers, despite the fact that the interactions across different layers are known to significantly impact on the E2E delay and reliability. Most existing approaches do not reflect such interactions; this leads to suboptimal solutions and thus we are yet to be able to meet the stringent requirements of URLLC. To guarantee the E2E delay and the reliability of the communication system, we need accurate and analytically tractable cross-layer models to reflect the interactions across different layers.

Deep Learning With 5G NR, the radio resources are allocated in each Transmission Time Interval (TTI) with a duration of $0.125 \sim 1$ ms [1]. To implement optimization algorithms in 5G systems, the processing delay should be less than the duration of one TTI. Since the crosslayer models are complex, related optimization problems are non-convex in general. Most of the existing optimization algorithms incur high computing overheads, and hence can hardly be implemented in real-world systems. Deep learning has significant potential to address the above issue in beyond 5G/6G networks. The basic idea is to approximate the optimal policy with a Deep Neural Network (DNN). After the training phase, a near-optimal solution of an optimization problem can be obtained from the output of the DNN in each TTI. Essentially, by using deep learning, we are trading off the online processing time with the computing resource for off-line training.

Integrating Knowledge into Learning Algorithms Although deep learning algorithms have shown significant potential, the application of deep learning in URLLC is not straightforward. As shown in [27], deep learning algorithms converge slowly in the training phase and need a large number of training samples to evaluate or improve the E2E delay and reliability. If some knowledge of the environment is available, such as the estimated packet loss probability of a certain decision, the system can exploit this knowledge to improve the learning efficiency [13]. Domain knowledge of communications and networking including models, analytical tools, and optimization frameworks have been extensively studied in the existing literature [12, 14]. How to exploit them to improve deep learning algorithms for URLLC has drawn significant attention as well, including in [15, 32, 38].

Fine-tuning in Real-world Systems Communication environments in wireless networks are non-stationary in general. Theoretical models used in off-line training may not match this non-stationary nature of practical networks. As a result, a DNN trained off-line cannot guarantee the Quality-of-Service (QoS) constraints of URLLC. Such an issue is referred to as the model mismatch problem in [5]. To handle the model mismatch, wireless networks should be intelligent to adjust themselves in dynamic environments, explore unknown optimal policies, and transfer knowledge to practical networks.

1.2 Scope of This Book

We invited leading researchers in both academia and industry from diverse backgrounds to share their recent studies in URLLC.

In Chapter 2, T. Kallehauge et al. focus on the physical layer and present the statistical aspects of URLLC, detailing both frequentist and Bayesian approaches. Specifically, the authors analyze the statistical features and guarantees for outage probability in a narrowband wireless channel. As a motivating example, they treat the practical case in which a Base Station (BS) collects channel statistics for users at different locations and attempts to predict the performance of a user at a new location. Their results show that the BS can obtain high-quality predictions of the reliability performance even for locations that are not in proximity.

In Chapter 3, instead of analyzing the average latency and the delay outage probability, C.-F. Liu et al. investigate the statistical information and/or metrics, rooted in the tail behavior of probability distributions to gain insights in URLLC systems. Specifically, the authors analyzed the tail distribution of the delay, channel fading, or packet inter-arrival time; variance and higher-order statistics; threshold deviation with a very low occurrence probability; worst-case metrics; age of information. Useful methodologies and extensive numerical results are provided in this chapter.

In Chapter 4, C. Sun et al. establish a unified framework of using unsupervised deep learning to solve both kinds of problem with both instantaneous and statistic constraints. For a constrained variable optimization, the authors first convert it into an equivalent functional optimization problem with instantaneous constraints. Then, to ensure the instantaneous constraints in the functional optimization problems, the authors use DNN to approximate the Lagrange multiplier functions, which is trained together with a DNN to approximate the policy. By taking resource allocation problems in URLLC as examples, the authors show that unsupervised learning outperforms supervised learning in terms of quality-of-service violation probability and approximation accuracy of the optimal policy.

In Chapter 5, C. Yue et al. overview candidate channel codes for URLLC, and compare them in terms of performance and complexity. Their respective strengths and weaknesses are investigated in terms of the performance gap to theoretical limits and the computational complexity of practical decoding algorithms. Furthermore, Ordered Statistics Decoding (OSD) is introduced as one of the potential universal decoding algorithms for URLLC, which can achieve near-optimal performance for any block code. The error performance and computational complexity of OSD are investigated in this chapter. Finally, recent improvements on OSD, including decoding rules and complete decoder design, are studied.

In Chapter 6, B. Shim et al. introduce a new type of short packet transmission framework named the Sparse Vector Coding (SVC) technique. The key idea behind SVC is to transform an information vector into the sparse vector in the transmitter and to exploit the sparse recovery algorithm in the receiver. Metaphorically, SVC can be thought as marking dots on the empty table. As long as the number of dots is small enough and the measurements contains enough information to figure out the marked cell positions, accurate decoding of the SVC packet can be guaranteed. The numerical results demonstrate that SVC is very effective in the short packet transmission for URLLC scenarios.

In Chapter 7, P. Yang et al. consider a CoMP-enabled RAN slicing system simultaneously supporting URLLC and eMBB traffic transmission. In the presence of eMBB traffic, the authors orchestrate the shared network resources of the system to guarantee a more reliable bursty URLLC service provision from the perspectives of lowering both URLLC packet blocking probability and codeword decoding error probability. The authors formulate the problem of RAN slicing for bursty URLLC and eMBB service multiplexing as a resource optimization problem and develop a joint bandwidth and CoMP beamforming optimization algorithm to maximize the long-term total slice utility. Several bandwidth allocation and beamforming algorithms are evaluated in the RAN slicing system.

In Chapter 8, W. R. Ghanem et al. investigate the beamforming design for downlink Orthogonal Frequency Division Multiple Access (OFDMA) URLLC systems. To enable the stringent URLLC delay requirements, finite blocklength transmission is adopted for the beamforming algorithm design. The authors formulate the beamforming algorithm design as a non-convex optimization problem for maximization of the weighted system sum throughput subject to constraints on the Quality of Service (QoS) of the URLLC users. A sub-optimal algorithm is proposed based on Sequential Convex Approximation (SCA). Numerical results reveal that the proposed design can achieve a considerable gain compared to several baseline schemes.

In Chapter 9, H. Duan et al. first present an up-to-date overview of the end-to-end latency for a Full-Duplex (FD) relay system. The authors investigate the possible solutions in the literature to achieve the goal of URLLC. The efficient solution is to allow a simple Amplify-and-Forward (AF) FD relay mode with low-complexity SI radio frequency and analog cancellations, and process the residual SI alongside the desired signal at base station in an adaptive manner, rather than being canceled at relay in digital domain. Their results show that the FD relay assisted system with adaptive SI utilization or cancellation enables extended network coverage, enhanced reliability, and reduced latency, compared to the existing overview work.

In Chapter 10, Z. Hou et al. aim to reduce the user experienced delay through prediction and communication co-design, where each mobile device predicts its future states and sends them to a data center in advance. Since predictions are not error-free, the authors consider prediction errors and packet losses in communications when evaluating the reliability of the system. Then, the authors formulate an optimization problem that maximizes the number of URLLC services supported by the system by optimizing time and frequency resources and the prediction horizon. Simulation and experiment results verify the effectiveness of the proposed method, and show that the trade-off between user experienced delay and reliability can be improved significantly via prediction and communication co-design.

In Chapter 11, D. V. Huynh et al. investigate the URLLC supported Industrial Internet-of-Things (IIoT) devices in industry automation. To enhance the URLLC performance, the authors propose two approaches to optimize the deployment of multiple relay robots in assisting URLLC system, namely DNN-based deployment, and the K-means clustering algorithm. An optimal resource allocation scheme is proposed to minimize the error probability at the IIoT devices. To solve the highly non-covex optimization problems of URLLC, the authors propose an effective iterative algorithm for solving the reliability maximization. Representative numerical results demonstrate the proposed scheme can significantly improve the reliability over various conventional approaches.

1.3 Future Directions

Considering the diverse application scenarios and KPIs of URLLC in 6G, the design of future communication systems requires considerable additional research efforts beyond what the community has done so far. We discuss some promising research directions in this section.

1.3.1 Constrained Deep Learning for URLLC

When applying deep learning for URLLC applications, the reward is usually defined as a weighted sum of different KPIs. With different weighting coefficients, the final achieved KPIs are different. Thus, we need to select these weighting coefficients manually to achieve satisfactory KPIs. A potential way to overcome this difficulty is to formulate the problem as a constrained optimization problem and use constrained unsupervised deep learning to find the optimal policy [7]. If the problem turns out to be a sequential-decision making problem with constraints, constrained deep reinforcement learning algorithms can be applied [20]. Nevertheless, the reliability of URLLC is extremely high and the closed-form results may not be available. Therefore, the required number of training samples is extremely large. How to achieve the target KPIs with a reasonable amount of training samples remains an open issue.

1.3.2 Distributed Learning for URLLC

In IIoT, collecting the status of all the devices in the central server will bring considerable overheads to wireless networks, and hence the devices should be controlled in a distributed manner. To achieve this goal, distributed learning with partial observation is a promising framework. With this framework, edge computing servers, VR glasses, and IoT devices can take actions based on local observations. In this way, the overheads for exchanging control information and updating global status can be reduced remarkably [25]. The major issues of distributed learning include long convergence time, poor performance with partial observation, and limited computing and storage resources of mobile devices.

1.3.3 Graph Neural Networks for Network Management of URLLC

Since the dimensions of the input and the output of a Fully-connected Neural Network (FNN) grows with the number of devices, we need to adjust the hyper-parameters of the FNN and retrain the parameters whenever the number of devices varies. Thus, FNNs are not flexible in managing dynamic networks. To address this issue, a promising approach is to use GNNs to represent the topology of wireless networks [11]. As indicated in [35], GNN is a very general structure that can be applied to solve large-scale problems with non-Euclid data structure. Since the number of parameters of a GNN does not increase with the dimension of the input, GNNs are suitable for resource management in dynamic networks [6]. In most of the existing literature, the network status is assumed to be perfectly known at the central control plane. How to find optimal policy with inaccurate/outdated/partial network status remain open problems.

1.3.4 Few-shot Learning for URLLC

The real-world data samples from practical systems are limited and could be non-stationary in wireless networks. A promising learning framework for fast adaptation is known as few-shot learning [33]. The basic idea is to use meta learning to optimize hyper-parameters, including initial parameters, learning rates, and the structures of neural networks [2, 10]. After off-line training in existing tasks, the neural networks can be transferred to new tasks with limited data samples. Most of the existing few-short learning algorithms for image classification cannot achieve high reliability. Whether it is possible to meet the reliability requirement of URLLC with few-shot learning remains open.

Bibliography

[1] 3GPP. Study on new radio (NR) access technology; Physical layer aspects (release 14). Stockholm, Sweden, Jun.2017. TR 38.802 V2.0.0.

[2] Marcin Andrychowicz, Misha Denil, Sergio Gomez, Matthew W Hoffman, David Pfau, Tom Schaul, Brendan Shillingford, and Nando De Freitas. Learning to learn by gradient descent by gradient descent. In *Proc. Neural information processing systems (NIPS)*, pages 3981–3989, 2016.

[3] Riqing Chen, Chunhui Li, Shihao Yan, Robert Malaney, and Jinhong Yuan. Physical layer security for ultra-reliable and low-latency communications. *IEEE Wireless Commun.*, 26 (5): 6–11, 2019.

[4] Victor Adriel de Jesus Oliveira, Luciana Nedel, and Anderson Maciel. Assessment of an articulatory interface for tactile intercommunication in immersive virtual environments. *Computers & Graphics*, 76: 18–28, 2018.

[5] Rui Dong, Changyang She, Wibowo Hardjawana, Yonghui Li, and Branka Vucetic. Deep learning for radio resource allocation with diverse quality-of-service requirements in 5G. *IEEE Trans. Wireless Commun.*, 20 (4): 2309–2324, 2020.

[6] Mark Eisen and Alejandro Ribeiro. Optimal wireless resource allocation with random edge graph neural networks. *IEEE Trans. Signal Process.*, 68: 2977–2991, 2020.

[7] Mark Eisen, Clark Zhang, Luiz FO Chamon, Daniel D Lee, and Alejandro Ribeiro. Learning optimal resource allocations in wireless systems. *IEEE Trans. Signal Process.*, 67 (10): 2775–2790, 2019.

[8] Mohammed S Elbamby, Cristina Perfecto, Mehdi Bennis, and Klaus Doppler. Toward low-latency and ultra-reliable virtual reality. *IEEE Netw.*, 32 (2): 78–84, 2018.

[9] Ericsson. Internet of things forecast. *Ericsson Mobility Report*, 2019. URL https://www.ericsson.com/en/mobility-report/internet-of-things-forecast.

[10] Chelsea Finn, Pieter Abbeel, and Sergey Levine. Model-agnostic meta-learning for fast adaptation of deep networks. In *Proc. International Conference on Machine Learning (ICML)*, pages 1126–1135, 2017.

[11] Fernando Gama, Joan Bruna, and Alejandro Ribeiro. Stability properties of graph neural networks. *IEEE Trans. Signal Process.*, 68: 5680–5695, 2020.

[12] Andrea Goldsmith. *Wireless Communications*. Cambridge university press, 2005.

[13] Shixiang Gu, Timothy Lillicrap, Ilya Sutskever, and Sergey Levine. Continuous deep Q-learning with model-based acceleration. In *Proc. International conference on machine learning (ICML)*, pages 2829–2838. PMLR, 2016.

[14] Mor Harchol-Balter. *Performance modeling and design of computer systems: Queueing Theory in Action*. Cambridge University Press, 2013.

[15] Hengtao He, Shi Jin, Chao-Kai Wen, Feifei Gao, Geoffrey Ye Li, and Zongben Xu. Model-driven deep learning for physical layer communications. *IEEE Wireless Commun.*, 26 (5): 77–83, 2019.

[16] Ibrahim A Hemadeh, Katla Satyanarayana, Mohammed El-Hajjar, and Lajos Hanzo. Millimeter-wave communications: Physical channel models, design considerations, antenna constructions, and link-budget. *IEEE Commun. Surveys & Tuts.*, 20 (2): 870–913, 2017.

[17] Tom Hobler, Lucas Scheuvens, Norman Franchi, Meryem Simsek, and Gerhard P. Fettweis. Applying reliability theory for future wireless communication networks. In *Proc. IEEE Personal, Indoor and Mobile Radio Communications (PIMRC)*, 2017.

[18] Xiaolin Jiang, Hossein Shokri-Ghadikolaei, Gabor Fodor, Eytan Modiano, Zhibo Pang, Michele Zorzi, and Carlo Fischione. Low-latency networking: Where latency lurks and how to tame it. *Proc. IEEE*, 107 (2): 280–306, 2018.

[19] Chong Li, Chih-Ping Li, Kianoush Hosseini, Soo Bum Lee, Jing Jiang, Wanshi Chen, Gavin Horn, Tingfang Ji, John E Smee, and Junyi Li. 5G-based systems design for tactile internet. *Proc. IEEE*, 107 (2): 307–324, 2018.

[20] Qingkai Liang, Fanyu Que, and Eytan Modiano. Accelerated primal-dual policy optimization for safe reinforcement learning. *Proc. Neural Information Processing Systems (NIPS)*, 2019.

[21] Jimmy J. Nielsen, Rongkuan Liu, and Petar Popovski. Ultra-reliable low latency communication (URLLC) using interface diversity. *IEEE Trans. Commun.*, 66 (3): 1322–1334, Mar. 2018.

[22] David Ohmann, Ahmad Awada, Ingo Viering, Meryem Simsek, and Gerhard P. Fettweis. Modeling and analysis of intra-frequency multi-connectivity for high availability in 5G. In *Proc. IEEE Vehicular Technology Conference (VTC) Spring*, 2018.

[23] Jihong Park, Sumudu Samarakoon, Hamid Shiri, Mohamed K Abdel-Aziz, Takayuki Nishio, Anis Elgabli, and Mehdi Bennis. Extreme URLLC: Vision, challenges, and key enablers. *arXiv preprint arXiv:2001.09683*, 2020.

[24] Walid Saad, Mehdi Bennis, and Mingzhe Chen. A vision of 6G wireless systems: Applications, trends, technologies, and open research problems. *IEEE Netw.*, 34 (3): 134–142, 2019.

[25] Mohit K Sharma, Alessio Zappone, Mohamad Assaad, Mérouane Debbah, and Spyridon Vassilaras. Distributed power control for large energy harvesting networks: A multi-agent deep reinforcement learning approach. *IEEE Trans. Cog. Commun. and Netw.*, 5 (4): 1140–1154, 2019.

[26] Changyang She, Yifan Duan, Guodong Zhao, Tony Q. S. Quek, Yonghui Li, and Branka Vucetic. Cross-layer design for mission-critical IoT in mobile edge computing systems. *IEEE Internet of Things J.*, 6 (6): 9360–9374, 2019.

[27] Changyang She, Rui Dong, Zhouyou Gu, Zhanwei Hou, Yonghui Li, Wibowo Hardjawana, Chenyang Yang, Lingyang Song, and Branka Vucetic. Deep learning for ultra-reliable and low-latency communications in 6G networks. *IEEE Netw.*, 34 (5): 219–225, 2020.

[28] Changyang She, Chengjian Sun, Zhouyou Gu, Yonghui Li, Chenyang Yang, H Vincent Poor, and Branka Vucetic. A tutorial on ultra-reliable and low-latency communications in 6G: Integrating domain knowledge into deep learning. *Proc. IEEE*, 109 (3): 204–246, Mar. 2021.

[29] Mahyar Shirvanimoghaddam, Mischa Dohler, and Sarah J Johnson. Massive non-orthogonal multiple access for cellular IoT: Potentials and limitations. *IEEE Commun. Mag.*, 55 (9): 55–61, 2017.

[30] Bikramjit Singh, Olav Tirkkonen, Zexian Li, and Mikko A Uusitalo. Contention-based access for ultra-reliable low latency uplink transmissions. *IEEE Wireless Commun. Lett.*, 7 (2): 182–185, 2017.

[31] Eckehard Steinbach, Matti Strese, Mohamad Eid, Xun Liu, Amit Bhardwaj, Qian Liu, Mohammad Al-Ja'afreh, Toktam Mahmoodi, Rania Hassen, Abdulmotaleb El Saddik, et al. Haptic codecs for the tactile internet. *Proc. IEEE*, 107 (2): 447–470, 2018.

[32] Chengjian Sun and Chenyang Yang. Learning to optimize with unsupervised learning: Training deep neural networks for URLLC. In *Proc. IEEE Personal, Indoor and Mobile Radio Communications (PIMRC)*, pages 1–7. IEEE, 2019.

[33] Qianru Sun, Yaoyao Liu, Tat-Seng Chua, and Bernt Schiele. Meta-transfer learning for few-shot learning. In *Proc. IEEE Computer Vision and Pattern Recognition (CVPR)*, pages 403–412, 2019.

[34] Yaping Sun, Zhiyong Chen, Meixia Tao, and Hui Liu. Communications, caching, and computing for mobile virtual reality: Modeling and tradeoff. *IEEE Trans.Commun.*, 67 (11): 7573–7586, 2019.

[35] Zonghan Wu, Shirui Pan, Fengwen Chen, Guodong Long, Chengqi Zhang, and S Yu Philip. A comprehensive survey on graph neural networks. *IEEE Trans. Neural Netw. learning Syst.*, 32 (1): 4–24, 2020.

[36] Yunchou Xing and Theodore S Rappaport. Propagation measurement system and approach at 140 GHz-moving to 6G and above 100 GHz. In *Proc. IEEE Global Communications Conference (GLOBECOM)*, pages 1–6. IEEE, 2018.

[37] Wei Yang, Rafael F Schaefer, and H Vincent Poor. Wiretap channels: Nonasymptotic fundamental limits. *IEEE Trans. Inf. Theory*, 65 (7): 4069–4093, 2019.

[38] Alessio Zappone, Marco Di Renzo, and Mérouane Debbah. Wireless networks design in the era of deep learning: Model-based, AI-based, or both? *IEEE Trans. Commun.*, 67 (10): 7331–7376, 2019.

2

Statistical Characterization of URLLC: Frequentist and Bayesian Approaches

*Tobias Kallehauge, Pablo Ramirez-Espinosa, Anders E. Kalør, and Petar Popovski**

Connectivity Section at the Department of Electronic Systems, Aalborg University, 9220, Aalborg, Fredrik Bajers Vej 7A, Denmark
** Corresponding Author*

2.1 Introduction

URLLC is one of the most significant novelties brought by 5G, aiming to support wireless connections with very stringent requirements in terms of latency and reliability, as specified, for example, in 3rd Generation Partnership Project (3GPP) [1]. Meeting these requirements is highly non-trivial, as the strict latency constraint limits the degrees of freedom that can be used to ensure reliable communication, and thus high reliability comes at a very high cost in terms of spectral efficiency. Low latency and high reliability are contradicting requirements [9, 42], but both can be boosted by investing in more bandwidth [34], such that the problem of choosing the necessary resources (e.g., bandwidth) is of central importance. From a physical layer viewpoint, special attention must be paid to channel modeling, since the required amount of resources and the actual reliability are closely related to the wireless propagation characteristics [7, 44]. In this context, a bad channel model may force the operators to use an overly conservative provisioning scheme, which will decimate the performance of the system. Of course, investing more in the estimation of channel statistics will ensure better models, hence, more suitable provisioning for the channel, but the excessive number of samples required to characterize the ultra-rare events that URLLC concerns, may leave very little to no time for data transmission [7]. This is particularly detrimental for non-stationary

Ultra-Reliable and Low-Latency Communications (URLLC) Theory and Practice: Advances in 5G and Beyond, First Edition. Edited by Trung Q. Duong, Saeed R. Khosravirad, Changyang She, Petar Popovski, Mehdi Bennis and Tony Q.S. Quek.
© 2023 John Wiley & Sons Ltd. Published 2023 by John Wiley & Sons Ltd.

systems since the time required to estimate, say, the probability that the signal-to-noise ratio (SNR) is below a certain threshold with high confidence may exceed the time that channel statistics remain constant.

To tackle the difficulties of *assuring* and *ensuring* that the strict requirements for URLLC are met, this chapter applies a rigorous statistical approach to the problem. Hence, we define statistical measures that evaluate if a URLLC system fulfills its requirements (i.e., assurance) and then uses the insight gained from the statistical analysis to choose an appropriate transmission scheme to fulfill them (i.e., ensurance). Central to the statistical analysis of URLLC is the characterization of ultra-rare events which necessitates special statistical measures and careful analysis. For example, the average SNR is not very informative about the rare event when the SNR is outside its 95% lower bound, therefore another statistic is required to characterize this event.

The amount of literature on statistical guarantees for URLLC and adjacent topics since the introduction of URLLC has been modest although some relevant articles have been published on the issue. One of the early works in the area is [7], which discusses the fundamental concepts of assuring communication reliability, defines statistical measures for this purpose, proposes different resource allocation schemes tailored to the statistical measures and highlights the issues of model mismatch and the number of samples required to estimate rare-event statistics. Other promising directions include the conditional value at risk (CVaR) as a statistical measure to characterize worst-case events [6, 26] and extreme value theory (EVT) which offers a more direct way of characterizing rare events (see Section 2.3.5 for a brief introduction to EVT).

This chapter will introduce a fundamental approach to fulfilling service requirements for URLLC from a statistical perspective. We focus on reliability at the physical communication layer with narrowband transmission to limit the scope and simplify system models. As the chapter title suggests, we will explore both *frequentist* and *Bayesian* approaches. For this, consider the following motivating example. An industrial port services large container ships where cranes lift shipping containers onto autonomous guided vehicles known as *shuttle carriers* that place the containers in the port for temporary storage — see Figure 2.1 for an illustration. On the route between the cranes and storage locations, the shuttle carriers are in constant communication with a central control unit that schedules tasks for each shuttle, determines container locations and manages traffic in the port to optimize flow and avoid collisions. The wireless communication link between the control unit and each shuttle is therefore critical to performance and safety. In 3GPP, this type of communication can be classified under *mobile robots* with strict service requirements, such as a mean time between communication outages of the order of 10 years [2, p. 15]. The quality of the wireless channel varies in both

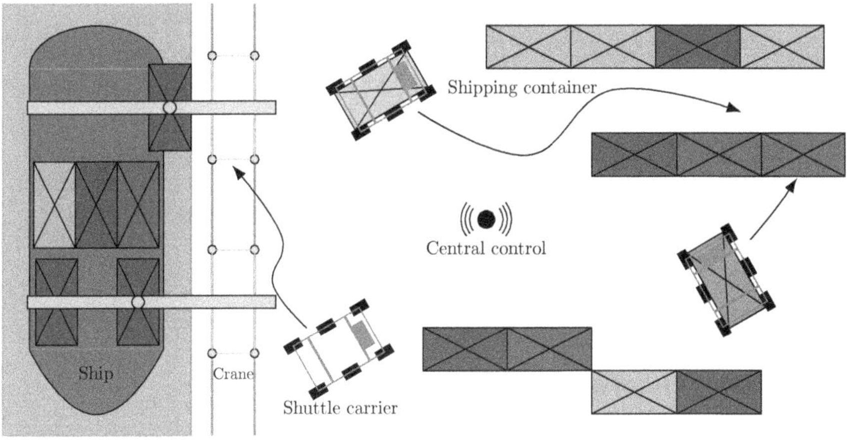

Figure 2.1 Illustration of industrial port with automatic guided shuttle carriers controlled wirelessly.

space and time, so channel state information (CSI) must continuously be updated to meet these requirements. A frequentist approach in this context would acquire CSI by estimating the channel every so often using pilot signals and then possibly assigning confidence intervals to the estimated CSI values. A Bayesian approach on the other hand would also rely on prior information to obtain a posterior belief about the likely CSI given the information at hand. The distinction between the frequentist and Bayesian approaches is somewhat subtle[1], but for the purposes here, the Bayesian approach is especially attractive since it directly allows prior information to be incorporated into CSI acquisition. Prior information is readily available in this scenario, e.g., by using previous CSI estimates from other shuttle carriers in proximity to the location where a new transmission takes place. By relying on prior information, the CSI can be estimated with higher precision, or less time can be dedicated to pilot transmission. For URLLC, a Bayesian approach, therefore, has the potential to solve the difficulties of estimating rare-event statistics under strict latency constraints assuming that accurate prior information is available. Despite this, there has been a rather limited utilization of Bayesian statistics for URLLC in the recent literature, which is therefore explored here along with the frequentist approach.

The remainder of the chapter is organized as follows. Preliminaries for statistical channel modeling and reliability at the physical layer are given in Section 2.2.

1 In essence, the frequentist approach views CSI as a *constant* but unknown value, while the Bayesian approach assumes that CSI is a *random variable* given the available information.

Section 2.3 introduces different statistical guarantees and shows how to fulfill them in a frequentist context. We then formally introduce the concept of Bayesian statistics and how it can be used in resource allocation for URLLC. Two illustrative examples are given. The first example (Section 2.4.2) shows the benefits of using Bayesian statistic in the ideal case where prior information about channel parameters are perfectly known. The second example (Section 2.5) shows how such prior information could be obtained in practice through *statistical radio maps*.

2.2 Preliminaries

2.2.1 Channel Models

The reliability of a communication system is inherently determined by the random wireless channel that alters the signal from the transmitter to the destination. Consequently, modeling the channel is central to describing the reliability of a wireless system. However, characterizing the channel in a wireless communication link is by no means a trivial task. The electromagnetic waves propagating from the transmitter to the receiver antenna are subjected to multiple physical phenomena, such as scattering and diffraction, giving rise to signal changes and fluctuations. Considering the propagation environment as a linear time-invariant (LTI) system — the reader is referred to standard textbooks, e.g., [20, 43, 46], for a more detailed description — these propagation effects are encapsulated in the baseband equivalent channel impulse response

$$h(t) = \sum_{i=1}^{p} g_i \delta(t - \tau_i), \tag{2.1}$$

where p is the number of resolvable paths, $g_i \in \mathbb{C}$ is the channel coefficient associated with the i-th path, δ is the Dirac delta function and τ_i is the corresponding delay. The ability of a system to resolve the different paths is related to the bandwidth B and the delays τ_i. If $\tau_i \ll 1/B \;\forall\, i$, then only one path can be resolved with aggregated channel coefficient

$$h = \sum_{i=1}^{p} g_i. \tag{2.2}$$

We say therefore that the transmission is *narrowband*, and we will pay attention to this particular case in the rest of the chapter.

As stated before, the channel coefficient h (or simply, *the channel*), captures the variations in the received signal, usually referred to as *fading*. Depending on the spatial scale of these variations, propagation effects are usually classified as [46]:

1. Large-scale fading, associated with the attenuation due to shadowing and distance (pathloss). It occurs at a scale of tens of meters.
2. Small-scale fading, due to constructive and destructive interference of the scattered waves arriving at the receiver, noticeable at a wavelength scale.

Pathloss is a consequence of the propagation of the electromagnetic waves throughout the medium, and is characterized by Friis' widely-used transmission formula [20, Eq. (2.7)]

$$P_L = \frac{\overline{P}_R}{\overline{P}_T} = G\left(\frac{\lambda}{4\pi d}\right)^2, \tag{2.3}$$

where $\overline{P}_R$ and $\overline{P}_T$ are the averaged received and transmitted powers, respectively, d is the line-of-sight (LoS) distance, λ is the wavelength and G is a generic term accounting for the antenna gains, polarization losses, etc.

Whilst pathloss can be seen as the averaged loss in power due to the distance, slow variations in space are produced by large obstacles like trees or buildings, giving rise to the so-called *shadowing*. This slow fluctuation, here represented by a random variable β, is usually assumed to follow a lognormal distribution [43]. Due to the mathematical complexity of the lognormal formulation, other distributions have been used to characterize shadowing, such as Gamma [4, 5] or inverse Gamma [36].

Finally, the small-scale fading (or simply, fading) represents the interference of multiple paths that cannot be resolved at the receiver. In its simplest form, it gives rise to a single complex coefficient representing the sum of multiple homogeneous planar waves [12, 13]:

$$\alpha = \sum_{n=1}^{N} V_n e^{j\phi_n}, \tag{2.4}$$

where $V_n \in \mathbb{R}$ is the amplitude of the n-th planar wave and $\phi_n \in \mathbb{R}$ its phase. Since small differences in distance render noticeable changes in the phase of the incoming waves, the different ϕ_n's are usually assumed to be independent and, in most cases, uniformly distributed. Moreover, if $N \to \infty$, then α can be approximated as Gaussian by the central limit theorem, i.e., $\alpha \sim \mathcal{CN}(\mu_\alpha, \sigma_\alpha^2)$, leading to the most common fading distribution:

- *Rayleigh* fading, in which $\mu_\alpha = 0$, used to characterize non-line-of-sight (NLoS) environments with probability density function (PDF)

$$f_{|\alpha|^2}(w) = \frac{1}{\sigma_\alpha^2} e^{-w/\sigma_\alpha^2}, \tag{2.5}$$

which is also known as the *exponential* distribution for $|\alpha|^2$ with scale σ_α^2.

- *Rician* fading, in which $\mu_\alpha \neq 0$ represents some dominant path, used to characterize LoS environments with PDF

$$f_{|\alpha|^2}(w) = \frac{(1+K)e^{-K}}{\overline{W}} e^{-(1+K)w/\overline{W}} I_0\left(2\sqrt{\frac{K(1+K)w}{\overline{W}}}\right) \tag{2.6}$$

  and parameters $K = |\mu_\alpha|^2/\sigma_\alpha^2$ and $\overline{W} = E[|\alpha|^2] = |\mu_\alpha|^2 + \sigma_\alpha^2$ where I_0 is the modified Bessel function of the first kind of order 0.

Aiming to generalize both Rayleigh and Rician distributions, several fading models have been proposed over the years [40, 49], although for the purpose of this chapter we stick to the simpler aforementioned models.

Combining pathloss, shadowing, and fading gives rise to the widely-used signal model

$$y = \sqrt{P_L}\,\beta\,\alpha\,s + n = h\,s + n, \tag{2.7}$$

where s is the transmitted complex symbol and the noise n is white Gaussian with power spectral density N_0, i.e., $n \sim \mathcal{CN}(0, N_0)$. In this simplified model, $E[|\beta|^2] = E[|\alpha|^2] = 1$, since the average power is captured by P_L.

2.2.2 Outage Probability

From the last section, and specifically from the random variable model in (2.7), we have seen that the transmitted symbol s is directly affected by the channel h, and then corrupted by noise. Naturally, more terms may be added to this model representing, e.g., interference coming from other users in the same system or even other systems operating at the same frequency band. However, for the sake of simplicity, we here stick to the simpler form in (2.7). It should be noted that the resulting received symbol is also a random variable, allowing a statistical analysis of the communication link. Along this line, one of the most extended metrics to characterize the performance of a system is the SNR, defined as the ratio between the received signal power P_R and the noise power. Since the noise is assumed to be normalized, the SNR reads

$$\gamma = \frac{P_R}{N_0 B}, \tag{2.8}$$

with B as the bandwidth.

A common assumption in wireless communications analyses is block-fading; that is, we consider the transmission of long blocks composed of several symbols s, and for all of them the channel coefficient h remains constant (albeit random) while the noise takes independent realizations. Under block-fading, the SNR can be rewritten as

$$\gamma = |h|^2 \frac{E[|s|^2]}{E[|n|^2]} = |\alpha|^2 \frac{P_L |\beta|^2 E[|s|^2]}{N_0}. \tag{2.9}$$

The last equality in (2.9) is useful to characterize the system in a local area where shadowing and pathloss remains constant, and thus we can rewrite $\gamma = |\alpha|^2 \overline{\gamma}$, with $\overline{\gamma} = E[\gamma] = \frac{P_L |\beta|^2 E[|s|^2]}{N_0}$ being the SNR in the case of no fading. Consequently, we observe that the SNR is also a random variable, whose distribution is a scaled version of that of $|h|^2$ (or directly $|\alpha|^2$ in a local area).

Once the SNR is presented in terms of random variables, we can introduce an important metric in performance analysis: the *outage probability*. It is defined as the probability of the SNR to fall below a given threshold γ_{th} [40], i.e.,

$$P_{\text{out}}(\gamma_{\text{th}}) = \Pr(\gamma < \gamma_{\text{th}}) = \int_0^{\gamma_{\text{th}}} f_\gamma(t) dt = F_\gamma(\gamma_{\text{th}}), \tag{2.10}$$

where f_γ and F_γ denote, respectively, the PDF and cumulative distribution function (CDF) of γ. If γ_{th} is the minimum SNR required to successfully decode the received symbol, this outage event can be used to characterize the impact of the channel in the communication reliability.

2.2.3 The Rate Selection Problem

We have seen that the outage probability, one of the key metrics in reliability analysis, completely depends on the SNR threshold γ_{th}. In practice, this threshold is the result of several signal processing and communication techniques such as the modulation scheme or the channel coding (again, the reader is referred to [20, 33, 35, 41] for detailed explanations). However, from a theoretical point of view, the concept of channel *capacity* is used. The channel capacity is the maximal transmission rate that could be used for which the probability of error can go to zero as the number of channel uses (symbols) goes to infinity. The capacity of the additive white Gaussian noise (AWGN) channel is given as: [39]

$$C_{\text{AWGN}} = B \log_2 \left(1 + \frac{P_R}{N_0 B}\right), \tag{2.11}$$

although usually the normalized capacity C_{AWGN}/B is used. Then, if we transmit at a rate $R < C_{\text{AWGN}}$, we are theoretically able to decode the signal with a vanishing error probability[2].

It should be noted that the important assumption in deriving the capacity (2.11) of the AWGN channel is that the SNR is constant during all channel uses *and*

2 Shannon's capacity is generally used as the upper bound of the system performance. However, the assumptions under which this expression is valid imply, among others, a theoretically infinite block length which contradicts the low latency requirement in URLLC. For a comprehensive analysis of the finite block length regime, see [32].

both the transmitter and the receiver know the SNR value. Relating (2.11) with the previously introduced random variable model (2.9), we can consider that each realization of γ leads to an AWGN channel; however, now the SNR of that channel is not known to the sender. For convenience, we can express the *instantaneous* normalized capacity as

$$C = \log_2(1 + \gamma), \tag{2.12}$$

Since the sender does not know γ for the given channel realization, it fixes a transmission rate to R and if the instantaneous realization of the channel cannot support this rate, an outage occurs. In analytical form: the outage probability in (2.10) in terms of the transmission rate R is expressed as:

$$P_{\text{out}}(R) = \Pr(R > \log_2(1 + \gamma)) = \Pr(2^R - 1 > \gamma) = F_\gamma(2^R - 1), \tag{2.13}$$

which directly relates the statistics of the SNR with the transmission rate R. Using this relation, we can deterministically select a transmission rate such that an outage probability constraint is met. In other words, if we want to ensure $P_{\text{out}}(R) \leq \epsilon$, then the maximum transmission rate is

$$R_\epsilon = \sup_R \{R \geq 0 : P_{\text{out}}(R) \leq \epsilon\}, \tag{2.14}$$

which, from (2.13) easily follows

$$R_\epsilon = \log_2(1 + F_\gamma^{-1}(\epsilon)). \tag{2.15}$$

This expression is known as the *outage capacity* given ϵ. However, in order to operate at the outage capacity, the distribution of the SNR must be *perfectly* known, which is extremely complicated in a real scenario.

2.2.4 How to Define Reliability?

The outage probability is a key indicator of performance for wireless communication systems. However, it is arguably too simplistic since it only accounts for the physical channel in a communication system with many layers. In practice, the overall performance of a system is determined by the aggregation of all protocol layers. For instance, two systems can have two different latencies and packet error probabilities when using different automated repeat request (ARQ) techniques, even though their channel conditions are identical.

Ideally, the definition of reliability should account for the whole system, providing an end-to-end and non-ambiguous requirement. However, this is by no means a simple task, and consequently several definitions of reliability are available in the literature [9]. An often cited definition of URLLC is from the 3GPP, which states that systems require a success probability of 99.999% for a 32 bytes-length packet with user plane latency of 1 ms [1], although the specific requirements vary between applications. 3GPP has another definition for reliability in [3], where reliability is the percentage value of successfully transmitted packets within a certain time frame.

Along with reliability, 3GPP also introduces some *characteristic parameters* that define the performance of URLLC systems. For instance, *communication service availability* is measured as the percentage of time the network is available, where availability is the ability for a network to perform as required when required [2]. On the other hand, *communication service reliability* is how long a system is available until it becomes unavailable (i.e., the time between failures), and *survival time* measures the time that an application consuming a communication service may continue without an anticipated message [3].

The reader may notice that a common factor in all these definitions is the lack of a proper statistical context. In other words, what are the conditions under which these requirements should be met? In this chapter, we focus on providing statistical guarantees to URLLC, not only introducing a framework to statistically characterize reliability but also proposing approaches on how to achieve these guarantees. Throughout the chapter we use the outage probability as a measure of reliability, since despite not being complete, it is an analytically tractable measure that characterizes the physical layer behavior of the system. Moreover, given a specific system with given ARQ, modulation and coding schemes, one should be able to translate the required performance into the outage probability.

2.3 Statistical Approaches to URLLC

The fundamental challenge in the rate selection problem lies in the uncertainty about the channel. We have seen that, if the channel distribution is perfectly known, then the rate can be maximized simply by selecting the ϵ-quantile as in (2.15). However, in practice the problem is more involved, since learning the channel distribution is not trivial. Even assuming a parametric model for the channel, the estimation of the distribution parameters, such as the mean in case of Rayleigh fading, is not perfect and thus introduces some error in the rate selection. Ultimately, this implies that the outage probability requirement is not met all the time, and characterizing this probability is of capital importance for a thorough analysis of URLLC reliability.

2.3.1 Revisiting the Rate Selection Problem

The standard way to proceed is assuming that the channel follows a parametric distribution, e.g., Rayleigh, and then obtaining the maximum likelihood estimate for the parameters using samples drawn from the channel, e.g., using pilot symbols. Although this introduces an estimation error as any other estimator, this error is often neglected in the literature as the distribution is assumed to remain the same over time, and thus needs to be estimated only once. Nevertheless, in practice the channel distribution needs to be re-estimated every time there are significant changes in the environment or when a user moves significantly. In this regard, the number of samples used to estimate the distribution parameters determines the accuracy of the resulting estimates. In turn, this can have significant impact on the performance of the rate selection mechanism in a URLLC system.

In order to take the parameter uncertainty into account, it is convenient to pose the rate selection as a decision theoretic problem. Specifically, suppose that the true (but unknown) channel has the CDF $F_W(W)$, from which we have drawn a set $W_1, W_2, \dots, W_n$ of n independent samples collectively denoted as W^n. Here, note that we are considering as channel power $W = |h|^2$, with h as in (2.7), and assume without loss of generality that $N_0 = 1$. The channel distribution F_W is assumed to belong to a family of distributions $\mathcal{F}$, also referred to as the hypothesis class, which is often assumed to be a set of parametric distributions parameterized by θ within the parameter space Θ, i.e., $\mathcal{F} = \{F_W(W\,;\,\theta) : \theta \in \Theta\}$. For instance, if $\mathcal{F}$ is the family of Rayleigh fading channels, then θ reduces to the mean. Furthermore, the rate selection function is defined as $R_\epsilon(W^n)$ that maps from the set of n channel observations to a rate designed to satisfy the reliability requirement ϵ. In the simplest case, $R_\epsilon(W^n)$ selects the rate using the inverse distribution of the estimated channel

$$R_\epsilon(W^n) = \log_2\left(1 + F_W^{-1}(\epsilon\,;\,\hat{\theta})\right), \tag{2.16}$$

where $\hat{\theta} = g(W^n)$ is the estimated distribution parameter with estimator g. We clearly observe that the selected rate becomes a *random variable* itself. Naturally, as the number of observations n increases and the estimation error $\theta - \hat{\theta}$ decreases, we obtain the rate in (2.15).

As an example, under the assumption of Rayleigh fading (i.e., F is exponentially distributed with mean parameter θ), the maximum likelihood estimate is $\hat{\theta} = n^{-1}\sum_{i=1}^{n} W_i$ and the resulting rate selection function becomes [7]

$$R_\epsilon(W^n) = \log_2\left(1 - \frac{\ln(1-\epsilon)}{n}\sum_{i=1}^{n} W_i\right). \tag{2.17}$$

A rate selection function that makes use of the maximum likelihood estimator as in the example above has the significant downside that the estimate will only be correct *on average*. As a result, the selected rate may often be optimistic, resulting in a reliability potentially much lower than the target $1 - \epsilon$. In the following two sections, we will consider two ways to compute the actual reliability obtained using a rate selection function similar to the one above but based on statistical guarantees, namely *averaged reliability* and *probably correct reliability*.

2.3.2 Averaged Reliability

One way to account for the uncertainty in the sample set W^n is to consider the *averaged reliability*, denoted by $\overline{p}$, which is obtained by averaging the resulting reliability over the realizations of W^n such that [7]

$$\overline{p}_F = E_{W^n}\left[\Pr(R_{\epsilon_n}(W^n) > \log_2(1 + W) \,|\, W^n)\right], \tag{2.18}$$

where $W \sim F_W(\cdot\,; \theta)$ is the channel power under the true distribution. We use ϵ_n to emphasize that ϵ_n may be chosen differently from the reliability requirement ϵ to compensate for the uncertainty in the channel parameter estimate. Such compensation is generally required to ensure that the average reliability $\overline{p}_F$ is lower than ϵ. For asymmetric distributions, such as with Rayleigh fading, this occurs because one is more likely to observe a sample that supports a rate larger than the average than it is to observe a sample that supports a lower rate. One of the key elements of (2.18) is the random treatment to the outage probability. Hence, the averaged reliability arises as the outage probability that can be expected when using the rate selection function $R_{\epsilon_n}(W^n)$ for an average sample of channel observations.

While the expression in (2.18) allows us to define the reliability in a meaningful way, it is useful only if we know the true distribution F, which is usually not the case. However, we can turn it into a more useful reliability requirement by considering all distributions in $\mathcal{F}$, rendering

$$\sup_{F \in \mathcal{F}} \overline{p}_F \leq \epsilon. \tag{2.19}$$

For the case of Rayleigh fading, it can be shown that the averaged reliability can be written as [7]

$$\overline{p}_F = E_{W^n}\left[F_W\left(-\frac{\ln(1-\epsilon_n)}{n}\sum_{i=1}^{n} W_i\,;\,\theta\right)\right] \tag{2.20}$$

$$= 1 - \left(1 - \frac{\ln(1-\epsilon_n)}{n}\right)^{-n}. \tag{2.21}$$

Using this, the reliability requirement in (2.19) can be satisfied by using

$$\epsilon_n = 1 - e^{-n\left((1-\epsilon)^{-\frac{1}{n}}-1\right)} \tag{2.22}$$

in the maximum likelihood rate selection $R_{\epsilon_n}(W^n)$ from (2.17). Thus, a rate selected according to this function will, on average, result in a reliability of at least ϵ.

2.3.3 Probably Correct Reliability

The averaged reliability criterion presented in the previous section illustrates how the sampling uncertainty can be taken into account in the rate selection. However, in the context of URLLC, the fact that the reliability requirement is met only on average may not be satisfactory. Rather, we would like to guarantee that the reliability is satisfied with high probability. The *Probably Correct Reliability (PCR)* framework allows exactly for this. As also indicated by the name, PCR draws many parallels to the *Probably Approximately Correct (PAC)* learning framework that is studied in the field of statistical learning. Instead of taking the average over the reliability as in averaged reliability, in PCR we consider the *probability* that the reliability requirement will be satisfied, i.e.,

$$\widetilde{p}_F = \Pr\left(\Pr(R_{\epsilon_n}(W^n) > \log_2(1+W) \,|\, W^n) > \epsilon\right), \tag{2.23}$$

where the inner probability is with respect to the true channel W and outer probability is with respect to the sample W^n. As for the averaged reliability, we consider the PCR across all distributions in $\mathcal{F}$ and define the PCR reliability requirement

$$\sup_{F\in\mathcal{F}} \tilde{p}_F < \xi. \tag{2.24}$$

The confidence parameter ξ determines how confident we want to be in satisfying the reliability requirement. In more practical terms, it controls how much we want to back off the rate in order to compensate for unlikely bad channel samples W^n.

In general, meeting (2.24) is not straightforward. For the sake of illustration, and returning to the case of Rayleigh fading, the PCR $\widetilde{p}_F$ can be written as [7]

$$\widetilde{p}_F = \Pr\left(F_W\left(-\frac{\ln(1-\epsilon_n)}{n}\sum_{i=1}^{n} W_i\,;\,\theta\right) > \epsilon\right) \tag{2.25}$$

$$= 1 - \frac{\gamma\left(n, n\frac{\ln(1-\epsilon)}{\ln(1-\epsilon_n)}\right)}{(n-1)!}, \tag{2.26}$$

where $\gamma(\cdot, \cdot)$ is the lower incomplete gamma function. The PCR reliability requirement can then be satisfied by choosing the maximum ϵ_n satisfying

$$1 - \frac{\gamma\left(n, n\frac{\ln(1-\epsilon)}{\ln(1-\epsilon_n)}\right)}{(n-1)!} \leq \xi. \tag{2.27}$$

2.3.4 Non-parametric URLLC

So far, we have assumed that the family of the channel distribution is known. However, in practice the true distribution family is seldom known, and the distribution family is instead selected based on assumptions about the channel. While these assumptions impose structure to the channel distribution, which helps reduce the number of samples required to characterize it, there is a risk that the true channel cannot be accurately characterized by a distribution within the chosen family. This leads to a model mismatch, which can lead to a severe violation of the reliability guarantees [7].

To avoid specifying the channel distribution family, one can instead take a completely non-parametric approach to the rate selection problem. Instead of estimating the distribution using maximum likelihood, non-parametric estimation relies on the empirical CDF obtained from the channel samples W^n given as

$$\widehat{F}(W) = \frac{1}{n}\sum_{i=1}^{n} \mathbb{1}[W_i \leq W]. \tag{2.28}$$

We can define a corresponding rate selection function by estimating the ϵ-quantile of $\widehat{F}$ as the $\lfloor n\epsilon + 1 \rfloor$-th order statistic $W_{(\lfloor n\epsilon+1 \rfloor)}$ of W^n [7]

$$R_\epsilon(W^n) = \sup\{R > 0 : \widehat{F}(2^R - 1) \leq \epsilon\} \tag{2.29}$$

$$= \log_2\left(1 + W_{(\lfloor n\epsilon+1 \rfloor)}\right). \tag{2.30}$$

The averaged reliability of this estimator, for a given ϵ_n, can be shown to be [7]

$$\overline{p}_F = E_{W^n}\left[\hat{F}\left(W_{(\lfloor n\epsilon_n+1 \rfloor)}\right)\right] \tag{2.31}$$

$$= \frac{\lfloor n\epsilon_n + 1 \rfloor}{n+1}, \tag{2.32}$$

from which we can obtain the following rule for picking ϵ_n such that the reliability requirement ϵ is satisfied on average:

$$\lfloor n\epsilon_n + 1 \rfloor \leq \epsilon(n+1). \tag{2.33}$$

Note that this result implies that in order for $R_{\epsilon_n}(W^n)$ to be greater than zero, $\lfloor \epsilon(n+1) \rfloor > 0$ or equivalently $n \geq 1/\epsilon - 1$. This reflects the disadvantage of the non-parametric model over the parametric ones, since the number of required samples is significantly larger than when F is assumed to follow, e.g., an exponential distribution. In the context of URLLC, where ϵ might be in the order of 10^{-6}, obtaining sufficient samples for the non-parametric rate selection may be impractical.

Similarly, under the PCR framework it can be shown that the probability that a rate $R_{\epsilon_n}(W^n)$ will satisfy the reliability requirement ϵ can be expressed as [7]

$$\widetilde{p}_F = \Pr\left(\hat{F}\left(W_{(\lfloor n\epsilon_n+1\rfloor)}\right) > \epsilon\right) \tag{2.34}$$

$$= 1 - I_\epsilon\left(\lfloor n\epsilon_n + 1\rfloor, n + 1 - \lfloor n\epsilon_n + 1\rfloor\right), \tag{2.35}$$

where $I_x(a, b)$ is the regularized incomplete beta function — see [24, ch. 25] for further details. For a given confidence parameter ξ, we obtain the rule

$$\epsilon_n = \sup\{\varepsilon \in [0,1] : I_\epsilon\left(\lfloor n\varepsilon + 1\rfloor, n + 1 - \lfloor n\varepsilon + 1\rfloor\right) \geq 1 - \xi\}. \tag{2.36}$$

It can be shown by numerical evaluation that the PCR framework generally requires even more samples than under the averaged reliability criterion. Therefore, non-parametric channel estimation is practically infeasible for the reliability requirements that are typically assumed in the URLLC regime. In fact, even if one had the time to obtain the required number of samples, the channel might not remain stationary throughout the sampling window, thus violating one of the inherent assumptions in the non-parametric estimator. Nevertheless, the problem reflects the difficulty of providing universal reliability guarantees in wireless systems, and suggests that a proper reliability requirement should be specified not only in terms of the outage probability ϵ, but also in terms of the channel distribution under which the outage probability should be met.

We conclude this part of the section by illustrating the cost of the parametric and non-parametric estimators under the averaged reliability and PCR frameworks in terms of the normalized throughput under Rayleigh fading. We defined the normalized throughput achieved using a given rate selection rule $R_\epsilon(W^n)$ as

$$\omega_\epsilon(n) = \frac{E\left[R_\epsilon(W^n)\mathbb{1}_{R_\epsilon(W^n)\leq\log_2(1+W)}\right]}{R_\epsilon(1-\epsilon)}, \tag{2.37}$$

where R_ϵ is the ϵ-outage rate defined in Eq. (2.15) and the expectation is taken over both the samples W^n and the channel realizations W. The result is illustrated in Figure 2.2 for various numbers of samples n. As expected, the parameterized estimators, which are obtained by exploiting the knowledge that the fading is Rayleigh, require significantly fewer samples than the non-parametric ones in order to reach a throughput close to 1. Furthermore, the averaged reliability framework generally requires fewer samples than PCR, but its reliability guarantee is not as strong as the one provided by PCR. The oscillation in the non-parametric averaged reliability curve is due to the floor function in (2.33).

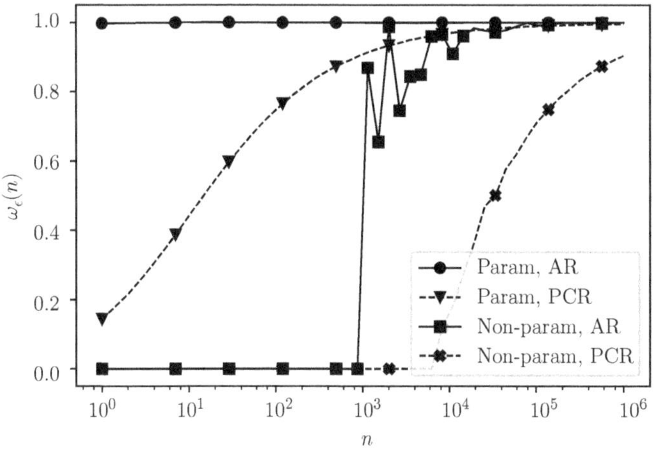

Figure 2.2 Normalized throughput with $\epsilon = 10^{-3}$ achieved by the parametric estimators under the averaged reliability and PCR frameworks for various number of samples n.

2.3.5 Tail Approximation

In the previous sections, we have considered two extreme cases of the rate selection problem, namely the parameterized one in which the family of the channel distribution is assumed to be known, and the non-parametric one where no assumptions about the channel distribution need to be specified. However, since URLLC targets very low probabilities of error, we are not interested in the whole channel distribution but in the lower tail[3]. This implies that, even though a model for the whole distribution is not available, a parametric approach can still be applied if we can characterize the lower tail.

Accounting for that, in this section we consider a different approach that is not based on knowledge about the specific distributions, but rather relies on an approximation of the lower tail for the channel distribution. One of the most used models is the so-called *power tail* approximation, which states that for a large class of CDFs, there exists $\kappa > 0$ such that for all $y > 0$

$$\lim_{t \to 0} \frac{F(ty)}{F(y)} = y^{1/\kappa}. \tag{2.38}$$

In other words, a large class of distributions exhibits an asymptotic power law behavior given by [48]

3 The lower tail (or sometimes *left tail*) refers to the lower quantiles of a random variable, i.e., values that the random variable is below with low probability (the order of 5% or even lower).

$$F(y) \approx \omega y^{1/\kappa} \tag{2.39}$$

for $\omega > 0$ and sufficiently small $y \geq 0$. Thus, by characterizing only the lower tail behavior, as opposed to the full distribution, we obtain a parameterized estimation problem that does not require us to assume a specific distribution family. The value of parameters ω and κ for a wide variety of fading models can be found in [14, 50]. Hence, using power tails approximations is a useful tool in statistical URLLC, since it allows a large number of models under the same parametric approach to be generalized.

From a practical point of view, and because the approximation is valid only in the lower tail, the estimation of ω and κ needs to be done based on samples from this region. This requires fixing a threshold $0 < \delta \leq 1$, such that only the smallest fraction δ of the samples are used to estimate ω and κ. The choice of δ has a high impact on the accuracy of the estimation. In particular, if δ is too high, then the power law approximation may not be accurate. Conversely, if δ is very small, the total number of required samples increases significantly. Thus, the selection of δ relies on *a priori* knowledge of the channel distributions, e.g., obtained through observations from similar scenarios.

Once δ has been selected, the parameters can be obtained from a set of samples W^n using the maximum likelihood estimates [7]

$$\widehat{\kappa} = \frac{1}{l} \sum_{i=1}^{l} \log\left(\frac{W_{(l)}}{W_{(i)}}\right) \tag{2.40}$$

$$\widehat{\omega} = \frac{l}{n} W_{(l)}^{-1/\widehat{\kappa}}, \tag{2.41}$$

where $l = \lceil \delta n \rceil$ is the number of samples that fall within the δ-quantile, and $W_{(i)}$ is the i-th order statistic of W^n. Using the estimated tail parameters, the rate selection function can be expressed as

$$R_{\epsilon}(W^n) = \log_2\left(1 + \left(\frac{\epsilon}{\widehat{\omega}}\right)^{\widehat{\kappa}}\right). \tag{2.42}$$

Because the rate selection function is obtained using a tail approximation that does not require knowledge of the distribution function of the samples, we cannot accurately characterize the uncertainty associated with the parameter estimates. As a consequence, the averaged reliability and the probably correct reliability frameworks presented earlier are not directly applicable. While it is possible to derive some approximations based on the asymptotic distribution of the parameters as the number of samples tends to infinity [7], the accuracy of the power law approximation, controlled by δ, often contributes to more uncertainty than the sampling itself. Unfortunately, characterizing the approximation error requires the family of the fading distribution to be specified, which would undermine the use of the power law approximation as a way to remove the assumption of the underlying distribution.

Another — more general — tail approximation is given by EVT [9], which is a solid framework used in statistics to characterize rare events. Within the EVT framework, for URLLC we are interested in the the Pickands–Balkeman–de Haan theorem, which characterizes the distribution of the samples above a large threshold, i.e., the higher tail of the distribution. However, for outage analysis the lower tail of the distribution is more useful, and thus the theorem can be reformulated to account for small values. Specifically, given a sequence of random variables W^n (samples) drawn from a CDF F_W, then for low enough threshold u the distribution of the values below u is approximately generalized Pareto [30]. Formally,

$$\Pr(u - W < y \,|\, W < u) = F_u(y) = 1 - \left(1 - \frac{\xi y}{\sigma_u}\right)^{-1/\xi}, \tag{2.43}$$

where the parameters ξ and σ_u need to be estimated. This result allows us to abstract from the true distribution of the channel, since it is valid regardless of F_W. However, choosing the parameters of the distribution and, more importantly, the threshold u is not trivial. The reader is referred to [30], where a premier algorithm to apply EVT to URLLC is presented.

2.4 Bayesian Statistics and URLLC

The previous section certainly highlights the difficulty of working in the ultra-reliable regime and the excessive number of training samples required to select a rate that meets given statistical assurances such as the average reliability or PCR. One interesting — albeit intuitive — observation is the fact that the throughput for parametric rate-selection methods is larger than for non-parametric methods, since the parametric approach has more information available. Naturally, a parametric approach may lead to *model mismatch*, introducing a bias in the estimation and rendering poor results (see [7] for an illustrative example).

Nevertheless, an important conclusion from the previous section is that the number of required samples to statistically guarantee certain performance is extremely high, which is in conflict with the latency aspect of URLLC. A potential solution to reduce the number of required samples is to include *prior information*, i.e., information that we already know about the system, into the learning process. Using prior information is an inherently human trait (we try to apply our knowledge from previous tasks to new ones), but it is also used heavily in statistics, engineering, and recently, in machine learning through meta-learning and transfer-learning principles — see, e.g., [25] for an application in wireless communication.

Since we are dealing with the outage event in a wireless communication link, our focus turns again to channel statistics. Prior information about the channel can be obtained, e.g., from previous measurements, from data obtained in similar scenarios or by partial knowledge of the propagation environment. In general, any information about the channel can be used as prior information and leveraged for inferring its statistics — even if the information is vague or uncertain. To make efficient use of prior information, we resort to *Bayesian inference* [19].

2.4.1 Bayesian Inference

Although not strictly necessary, Bayesian statistics often assumes parametric distributions for the involved random variables. In its most basic form, Bayesian inference takes advantage of both the prior information and the collected samples to provide a more refined and ideally accurate estimation than the frequentist approach, which only relies on the observed values. Specifically, given a sample data $X^n = (X_1, \dots, X_n)$ sampled from $f_{X^n}(X^n \mid \theta)$ with parameters $\theta \in \Theta$, parameter space Θ and a *prior distribution* $f_\theta(\theta \mid \phi)$, the objective is predicting the distribution of a new sample X, i.e., $f_X(X \mid \mathbf{x}, \phi)$. In our case, the observations may be samples of the received power W^n and the prior distribution is the quantification of the information we have about the channel, e.g., how the average power is distributed. Note, however, that other types of uninformative priors can be used, which are not obtained from previous experience (the reader is referred to [19] for a detailed description of Bayesian inference).

In the previous formulation, ϕ represents the set of *hyperparameters*, which characterizes the prior distribution. Given ϕ and the sample X^n, we state the Bayes' rule characterizing the distribution for θ. It is

$$f_\theta(\theta \mid X^n, \phi) = \frac{f_{X^n \mid \theta}(X^n \mid \theta) f_\theta(\theta \mid \phi)}{f_{X^n}(X^n \mid \phi)}, \tag{2.44}$$

where $f_\theta(\theta \mid X^n, \phi)$ is the *posterior* and $f_{X^n \mid \theta}(X^n \mid \theta)$ stands for the *likelihood* (i.e., the PDF of the sample conditioned on θ). Moreover, $f_\theta(\theta \mid \phi)$ is our prior as stated before, and

$$f_{X^n}(X^n \mid \phi) = \int_\Theta f_{X^n \mid \theta}(X^n \mid \theta) f_\theta(\theta \mid \phi)\, d\theta \tag{2.45}$$

is the *evidence*.

The posterior distribution in (2.44), although useful to characterize the impact of the uncertainty in the the estimation of θ, does not provide immediate insight into the distribution of the new sample X. This is straightforwardly done by averaging the likelihood over the posterior as

$$f_X(X\,|\,X^n, \phi) = \int_\Theta f_{X\,|\,\theta}(X\,|\,\theta) f_\theta(\theta\,|\,X^n, \phi)\, d\theta, \tag{2.46}$$

giving rise to the *posterior predictive.*

As we have seen, the main difference between Bayesian inference and a frequentist approach is that we work with distributions instead of parameters. This is useful not only to provide an estimator itself but also to account for the uncertainty in the estimation. Naturally, obtaining a closed-form expression for the posterior predictive is not simple in the majority of cases. Even in those situations, several approximations can be applied to the involved integrals so that, at the end, Bayesian inference can be applied in a numerical way. Popular methods in this regard includes the Laplace approximation and Markov Chain Monte Carlo (MCMC) methods — see [16] for further details.

2.4.2 Example: Rayleigh Fading with Shadowing Prior

Continuing our discussion on the outage analysis from an statistical point of view, we now exemplify how Bayesian inference can be used in this context. To that end, let us come back to our simple statistical communication model in (2.7), which is reproduced below for the sake of readability:

$$y = \sqrt{P_L}\, \beta\, \alpha\, x + n. \tag{2.47}$$

We have seen in Section 2.3 that a fading model is of principal importance to select a transmission rate and meet reliability requirements. In contrast to a frequentist approach, we illustrate here how Bayesian inference can be applied when some prior information is available.

Assume in this example that the fast fading, α, is characterized according to a Rayleigh model, i.e., $\alpha \sim \mathcal{CN}(0, \sigma_\alpha^2)$. Then, the squared modulus (instantaneous power) is exponentially distributed as [40]

$$f_{|\alpha|^2}(x; \Omega) = \frac{1}{\Omega} \mathrm{e}^{-x/\Omega}, \tag{2.48}$$

with $\Omega = E[|\alpha|^2] = \sigma_\alpha^2$ as its average power. According to (2.7), we agreed on normalizing the fast fading power, thus in principle $\Omega = 1$. However, considering we are interested in estimating the channel in the user's local area, and taking into account that the temporal variation of shadowing and pathloss occurs at a much larger scale, we can assume that both P_L and β are constant. Hence, we can define a new variable $W = P_L \beta^2 |\alpha|^2$ with PDF

$$f_W(w; \overline{W}) = \frac{1}{\overline{W}} \mathrm{e}^{-w/\overline{W}} = \lambda \mathrm{e}^{-\lambda w}, \tag{2.49}$$

where $\overline{W} = P_L \beta^2$ and, for convenience, $\lambda = 1/\overline{W}$.

In order to apply Bayesian inference, we need some prior information. In our example, note that the only parameter characterizing f_W is $\overline{W}$ (equivalently, λ), which has a clear physical interpretation: it accounts for both pathloss and shadowing. A simple prior is then assuming that $\overline{W}$, albeit constant, is sampled from one of the commonly-seen shadowing distributions. Specifically, let us stick to inverse gamma shadowing, which is supported by empirical evidences [36] and will lead to a simple mathematical analysis. Therefore, $\overline{W}$ is inverse gamma distributed with shape parameter m and scale parameter $\theta = P_L(m-1)$, i.e., $\overline{W} \sim \Gamma^{-1}(m, \theta)$. Hence, we have that λ is gamma distributed with shape m and scale $1/\theta$. In other words, $\lambda \sim \Gamma(m, 1/\theta)$, whose PDF is given by[4] [23]

$$f_\lambda(\lambda; m, \theta) = \frac{\theta^m}{\Gamma(m)} \lambda^{m-1} \mathrm{e}^{-\theta\lambda}. \tag{2.50}$$

As we know, $f_\lambda(\lambda; m, \theta)$ is the prior, since it is the additional information that we provide to the estimator, and m and θ are the corresponding hyperparameters, which in the most simple stage are assumed to be known. The objective is, hence, using this prior and a set of independent observations W^n, to predict the distribution of the received power W.

From (2.44), we have that the posterior distribution of λ is given by

$$f_\lambda(\lambda \,|\, W^n, m, \theta) = \frac{f_{W^n|\lambda}(W^n \,|\, \lambda) f_\lambda(\lambda \,|\, m, \theta)}{f_{W^n}(W^n \,|\, m, \theta)}, \tag{2.51}$$

which accounts for the uncertainty in the estimation, and the likelihood is simply

$$f_{W^n|\lambda}(W^n \,|\, \lambda) = \prod_{i=1}^{n} f_{W_i}(W_i \,|\, \lambda) = \lambda^n \exp\left(-\lambda \sum_{i=1}^{n} W_i\right), \tag{2.52}$$

since we assume independent observations and the fading is Rayleigh. Moreover, $f_{W^n}(W^n \,|\, m, \theta)$ is the evidence, computed as (2.45)

4 Note that we could also work directly with $\overline{W}$ and the inverse gamma distribution. However, using λ leads to simpler algebra without compromising the usefulness of the example here exposed.

$$f_{W^n}(W^n \mid m, \theta) = \int_0^\infty f_{W^n}(W^n \mid \lambda) f_\lambda(\lambda \mid m, \theta) d\lambda$$

$$= \frac{\theta^m \Gamma(n+m)}{\Gamma(m)\left(\sum_{i=1}^{n} W_i + \theta\right)^{n+m}}. \tag{2.53}$$

Introducing all the expressions in (2.51), we finally have

$$f_\lambda(\lambda \mid W^n, m, \theta) = \frac{\left(\sum_{i=1}^{n} W_i + \theta\right)^{n+m}}{\Gamma(n+m)} \lambda^{n+m-1} \exp\left(-\lambda\left[\sum_{i=1}^{n} W_i + \theta\right]\right), \tag{2.54}$$

which corresponds to a gamma distribution with shape parameter $m' = m + n$ and scale parameter $\theta' = 1/\left(\sum_{i=1}^{n} W_i + \theta\right)$. Interestingly, the posterior belongs to the same distribution as our prior (gamma), but with a different parameterization that, notably, depends on the sample data. We say that, in this case, the gamma distribution is a *conjugate prior* for the exponential likelihood in (2.52). In general, several distributions satisfy this property, and a detailed analysis of prior distributions can be found in [19, Sec. 2.4].

To better understand the meaning of the posterior, we can analyze the two extreme cases: *i)* when no samples are available ($n = 0$) and *ii)* when $n \to \infty$. In the former case, we observe that (2.54) reduces to the prior in (2.50), i.e., if no samples are available we just resort to the prior information. In the latter one, we have $\lim_{n\to\infty} E[\lambda \mid W^n, m, \theta] = \lim_{n\to\infty} m'\theta' = 1/\sum_{i=1}^{n} W_i$ and $\lim_{n\to\infty} \mathrm{Var}[\lambda \mid W^n, m, \theta] = \lim_{n\to\infty} m'(\theta')^2 = 0$, which is the maximum-likelihood estimator of λ.

Finally, the posterior predictive, which represents the distribution of a new observed sample W accounting for the uncertainty in the prior, is obtained as

$$f(W \mid W^n, m, \theta) = \int_0^\infty f_{W\mid\lambda}(W \mid \lambda) f_\lambda(\lambda \mid W^n, m, \theta)\, d\lambda \tag{2.55}$$

$$= \frac{m'}{(\theta')^{m'}(W + 1/\theta')^{m'+1}}, \tag{2.56}$$

corresponding to a Lomax distribution with shape m' and scale $1/\theta' = \sum_{i=1}^{n} W_i + \theta$ [23].

To illustrate the difference between Bayesian inference and the frequentist approach in Section 2.3.1, we depict in Figure 2.3 the distribution of the outage probability achieved with both a frequentist and a Bayesian approach. Specifically, and following the example presented here, we consider Rayleigh fading where

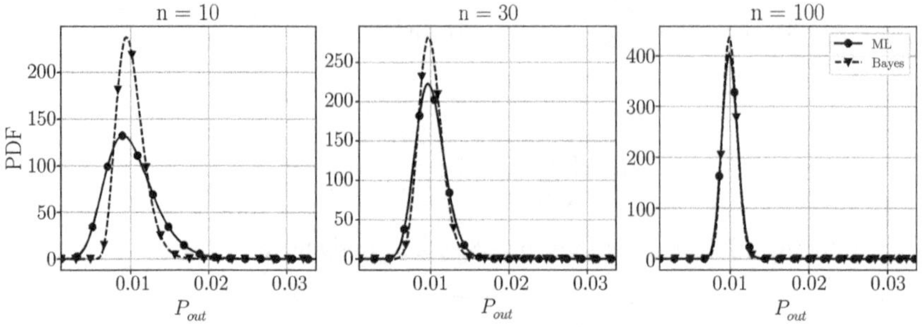

Figure 2.3 Probability density for the outage probabilities in (2.57) and (2.58) according to the joint distribution for (m, θ, W^n).

the average power is randomly sampled from an inverse gamma distribution with shape $m = 8$ and scale $\theta = 14$. Then, using a different number of samples n, the $\epsilon = 10^{-2}$ quantile is estimated according to a maximum likelihood estimator as in (2.17), i.e.,

$$P_{\text{out}}^{\text{ML}}(\epsilon) = 1 - \exp\left(\lambda \log(1-\epsilon)\frac{1}{n}\sum_{i=1}^{n} W_i\right), \tag{2.57}$$

and the predictive posterior in (2.56). In this latter case, we explicitly have that

$$P_{\text{out}}^{\text{Bay}}(\epsilon) = 1 - \exp\left(-\frac{\lambda}{\theta'}\left[\frac{1}{(1-\epsilon)^{1/m'}} - 1\right]\right), \tag{2.58}$$

which is obtained by inverting (2.56) and using the exponential CDF corresponding to Rayleigh fading. Note that these two outage probabilities are random variables on the sample vector W^n. Hence, their PDFs are shown in Figure 2.3, where they have been averaged over multiple realizations of the average power $\overline{W}$ (equivalently, λ). Interestingly, we observe that for large sample size n, both estimators render the same distribution — the sample size is large enough to trust the data and therefore the impact of the prior is negligible. In turn, when we have scarce samples, the use of prior information gives a narrow distribution and, thus, a more predictable system.

2.4.3 Reliability for Bayesian URLLC

Revising the theme from Section 2.3.1, we can also define reliability in the context of Bayesian inference through statistical guarantees. The key difference from the frequentist approach is the utilization of prior information, which necessitates that a definition for reliability, in addition to accounting for the uncertainty of learning channel statistics, must also account for the scenario's variability with respect to the prior distribution. For example, in Section 2.4.2, a statistical guarantee for reliability must include both the uncertainty of the observed values W^n and the variability in the average received power $\overline{W}$.

In a general setup, n observations W^n are drawn from the parametric distribution with CDF $F_W(W \mid \boldsymbol{\theta})$. The parameter $\boldsymbol{\theta}$ is assumed to follow some prior distribution with known hyperparameters $\boldsymbol{\phi}$. A simple approach to define reliability in this context is the outage probability evaluated over the posterior predictive distribution, i.e.,

$$\begin{aligned} P_{\text{out}}(R) &= \Pr(R \geq \log_2(1+W) \mid W^n, \boldsymbol{\phi}) \\ &= \Pr(2^R - 1 \geq W \mid W^n, \boldsymbol{\phi}) \\ &= F_{W \mid W^n, \boldsymbol{\phi}}(2^R - 1 \mid W^n, \boldsymbol{\phi}), \end{aligned} \tag{2.59}$$

where R is the selected rate and $F_{W \mid W^n, \boldsymbol{\phi}}$ is the CDF for the posterior predictive distribution (e.g., integrating (2.56) from 0 to $2^R - 1$). Setting

$$R_\epsilon(W^n, \boldsymbol{\phi}) = \log_2(1 + F^{-1}_{W \mid W^n, \boldsymbol{\phi}}(\epsilon \mid W^n, \boldsymbol{\phi})) \tag{2.60}$$

then yields a conditional outage probability of ϵ where the interval $[R_\epsilon, \infty)$ is refereed to as the $1-\epsilon$ *credible interval* for the outage capacity. The intuition behind the definition of reliability in (2.59), is the probability of exceeding the outage capacity given the information at hand, namely W^n and $\boldsymbol{\phi}$, which is arguably not as strong as the guarantees based on statistical learning we saw in Section 2.3.1. To that end, the outage probability is now defined as the conditional outage probability with respect to $\boldsymbol{\theta}$ rather than $\boldsymbol{\phi}$ such that

$$\begin{aligned} P_{\text{out}}(W^n, \boldsymbol{\theta}) &= \Pr(R_\epsilon(W^n, \boldsymbol{\phi}) > \log_2(1+W) \mid W^n, \boldsymbol{\theta}) \\ &= F_W(2^{R(W^n, \boldsymbol{\phi})} - 1 \mid \boldsymbol{\theta}), \end{aligned} \tag{2.61}$$

which is a generalization of (2.58) where $R_\epsilon(W^n, \boldsymbol{\phi})$ is some rate selection function. The average reliability is defined similarly to (2.18), but with respect to the joint distribution of $(W^n, \boldsymbol{\theta})$

$$\begin{aligned} \overline{p}_{F_{W,\boldsymbol{\theta}}} &= E_{W^n, \boldsymbol{\theta}}\left[\Pr(R_\epsilon(W^n, \boldsymbol{\phi}) > \log_2(1+W) \mid W^n, \boldsymbol{\theta})\right] \\ &= E_{W^n, \boldsymbol{\theta}}\left[P_{\text{out}}(W^n, \boldsymbol{\theta})\right] \\ &= E_{\boldsymbol{\theta}}\left[E_{W^n \mid \boldsymbol{\theta}}\left[P_{\text{out}}(W^n, \boldsymbol{\theta}) \mid \boldsymbol{\theta}\right]\right], \end{aligned} \tag{2.62}$$

where the last equality uses the law of total expectation. We see that the inner expectation in (2.62) is identical to the definition of average reliability in the frequentist context, but the variability in the scenario is also accounted for by averaging over the fading parameters $\boldsymbol{\theta}$. The PCR is also defined by considering the joint distribution of $(W^n, \boldsymbol{\theta})$ such that

$$\begin{aligned} \widetilde{p}_{F_{W,\theta}} &= \Pr\left(\Pr\left(R_\epsilon(W^n, \boldsymbol{\phi}) > \log_2(1+W) \,|\, W^n, \boldsymbol{\theta}\right) > \epsilon\right) \\ &= \Pr(P_{\text{out}}(W^n, \boldsymbol{\theta}) > \epsilon) \\ &= E_{\boldsymbol{\theta}}\left[\Pr(P_{\text{out}}(W^n, \boldsymbol{\theta}) > \epsilon \,|\, \boldsymbol{\theta})\right], \end{aligned} \tag{2.63}$$

where the last equality uses the law of total probability, which gives the expectation over $\boldsymbol{\theta}$. Again, the inner probability in (2.63) (with respect to W^n given $\boldsymbol{\theta}$) aligns with the frequentist definition. By designing the rate selection function $R_\epsilon(W^n, \boldsymbol{\phi})$ such that $\overline{p}_{F_{W,\theta}} \le \epsilon$ or $\widetilde{p}_{F_{W,\theta}} \le \delta$, a statistical guarantee for reliability is obtained. We omit any examples here, and in practice, closed-form solutions for the selected rate are unavailable in most cases due to the added complexity from the outer expectations. The issue with modeling mismatch is also further exacerbated due to relying on the prior distribution, which may lead to unrealistic reliability guarantees if the prior is wrong.

As a final note relating to statistical guarantees, we mention another branch of statistical learning known as PAC-Bayesian learning. PAC-Bayesian learning features two interesting innovations, namely the use of *generalized posteriors* as well as theoretical results for PAC inequalities (such as the PCR in (2.23) and (2.63)). With a generalized posterior, the likelihood term has been replaced by a generic loss term, which has the advantages of being model-free and tuneable to put more or less weight on the likelihood versus the prior. PAC-Bayesian learning is then about providing bounds for the expected loss for some choice of prior. The framework can also be used when no prior information is available, in which case the prior is often chosen for its analytical or computational properties. We refer the reader to [22] for a primer on PAC-Bayesian learning.

2.5 Prior Information via Statistical Radio Maps

Following the demonstration of the advantages that prior information can give when inferring channel statistics, one may ask how do we obtain such prior information? This section will explore how *spatial correlation* of channel statistics can be used in this regard. We start by considering the following motivating examples.

User-generated radio maps

In some areas, mobile users, e.g., autonomous vehicles, require agile and efficient resource allocation for URLLC (a similar scenario is considered in [8]). Data from uplink transmissions are collected at the base station, including CSI and location information. This allows the base station to construct a *radio map*, where CSI can be provided to the mobile users requiring only a few samples to estimate their locations. In some cases, the location could even be predicted without any measurements, e.g., for predictable routes in public transport or simply based on the last known location and velocity of a vehicle. For the narrowband channels analyzed in this chapter, CSI could be the average SNR, but other statistics such as SNR variance would also be informative for resource allocation. Assuming the environment is stationary in time, the radio map at observed locations is obtained directly. However, predictions are needed for unobserved locations. This is where spatial correlation of channel statistics can be leveraged under the assumption that channel statistics are similar (i.e., correlated) for locations that are close in space. The (average) received signal strength (RSS) is an example of spatially correlated channel statistics. Numerous use cases and prediction methods relating to the spatial behavior of RSS are seen in the literature — see for example [10, 15, 47]. Thus, using location estimates and leveraging spatial correlations, the generated radio map can provide prior information about CSI to mobile users.

Drone surveying

Imagine a drone surveying an area, e.g., for a *drone mapping* task, where a map is stitched together from many photographs. The drone communicates with an operator on the ground sending its status and receiving commands. The channel between drone and operator experiences outages at some locations. The drone can use its previous CSI measurements combined with GPS location to predict these outages — again leveraging spatial correlations of channel statistics.

Using radio maps for various tasks in wireless communication is by no means a novel idea, but most efforts focus on average statistics, which are not useful for URLLC. Going back to Section 2.2, the pathloss P_L and shadowing β can be considered average statistics in the sense that they contribute to $\overline{\gamma}$, the constant SNR, whereas the small-scale fading coefficient α introduces the instantaneous SNR through $\gamma = |\alpha|^2\overline{\gamma}$. Shadowing is known to have spatial correlation [21] which makes it predictable but the correlation for the small-scale fading coefficient quickly vanishes over short distances [10]. This makes spatial prediction of small-scale fading infeasible in practice, and as a result, the literature tends to focus on shadowing [8, 10]. However, the observation that small-scale fading decorrelates over short distances does not account for the underlying fading distribution — i.e., the statistics for the fading distribution could have a significant spatial correlation. For example, the variance $\mathrm{Var}[|\alpha|^2] = \sigma_\alpha^2$ is unlikely to be completely random, and in fact, the variance should change somewhat smoothly in space depending on the propagation environment. This line of thought opens up the idea of a *statistical radio map* wherein the spatial behavior of various statistics are modeled. For URLLC, the statistical radio map could model statistics relating to the tail of the fading distribution, such as parameters for the power-law approximation introduced in Section 2.3.5. In general, a statistical radio map could model any relevant fading statistic, e.g., a single quantile or the entire distribution. This section will introduce the concept of statistical radio maps by studying a generic framework that allows us to explore spatial correlation of fading statistics. After a primer on Gaussian processes (GPs), we show how these can be used to model and estimate the statistical radio map. Finally, we use the map to obtain a predictive fading distribution using Bayesian statistics and how this can be used to guarantee reliability.

The setup has d transmitters at locations $\mathbf{s}_1, \dots, \mathbf{s}_d \in \mathbb{R}^D$ with a receiver at location $\mathbf{s}_{\text{rx}} \in \mathbb{R}^D$ for dimension D. The transmitter locations correspond to user locations and previous drone locations in the examples above. We assume narrowband channels between transmitters and receiver and model then received power W. Each channel between $\mathbf{s}_i$ and $\mathbf{s}_{\text{rx}}$ experiences fading, modeled statistically through some fading distribution $F_{\boldsymbol{\theta}}$ parameterized by $\boldsymbol{\theta}$, e.g., Rayleigh or Rician, such that $W_i \sim F_{\boldsymbol{\theta}_i}$ for the transmitter at $\mathbf{s}_i$[5]. The fading distribution has been fitted at each location $\mathbf{s}_i$ for $i = 1, \dots, d$ through previous measurements. The goal is to infer channel statistics at an unobserved location $\mathbf{s}_0$ (corresponding to a new user and new drone location in the examples above).

The problem formulation is: given $\{\mathbf{s}_i, \boldsymbol{\theta}_i\}_{i=1}^d$, how can we estimate $\boldsymbol{\theta}_0$ only given $\mathbf{s}_0$? Furthermore, how can this be used to achieve reliability by selecting an appropriate communication rate R for the transmitter at $\mathbf{s}_0$? Figure 2.4 illustrates the setup. This general framework poses several challenges, including, to mention

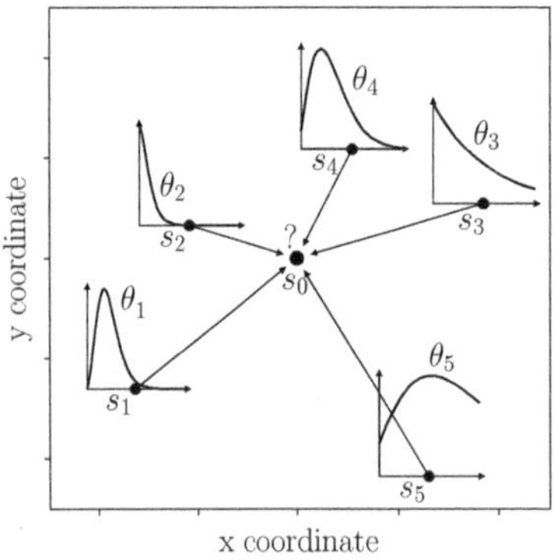

Figure 2.4 Map with 5 observed locations $\mathbf{s}_1, \dots, \mathbf{s}_5$ and an unobserved location $\mathbf{s}_0$. The black dots are the locations in 2-D.The fading distribution $F_{\boldsymbol{\theta}_i}$ for the received power W_i has been fitted at the observed locations – the figure shows the corresponding PDFs. This framework aims to predict $\boldsymbol{\theta}_0$ at the unobserved location $\mathbf{s}_0$ and then use this to select a communication rate.

5 Note the strong assumption that fading can be modeled with the same parametric distribution for all locations. This is convenient for statistical inference but not necessarily realistic. However, by using flexible fading distributions such as the $\alpha - \kappa - \mu$ fading distribution [17], the assumption may be reasonable.

just a few, selecting an appropriate fading distribution, fitting the distribution to observed data, accounting for temporal changes in the propagation environment, accounting for interference, predicting the fading parameters at the unobserved location, and using this in a Bayesian setting to select the communication rate. This section will only cover the two latter, that is, predicting fading parameters and selecting communication rate from this. To predict the fading parameters at $\mathbf{s}_0$, we will make use of GPs. Before applying GP to the problem, we will give an overview of the theory required to solve it.

2.5.1 A Primer on Gaussian Processes

Gaussian processes are, in essence, an extension of the multivariate Gaussian distribution from a discrete set to a continuum of random variables. We start by defining the multivariate Gaussian distribution.

Consider the d-dimensional random variable $\mathbf{X} \in \mathbb{R}^d$ with mean $\boldsymbol{\mu} \in \mathbb{R}^d$ and positive-definite covariance matrix $\boldsymbol{\Sigma} \in \mathbb{R}^{d\times d}$. When the PDF for $\mathbf{X}$ is given by [28]

$$f(\mathbf{x}; \boldsymbol{\mu}, \boldsymbol{\Sigma}) = \frac{1}{(2\pi)^{d/2} |\boldsymbol{\Sigma}|^{1/2}} \exp\left(-\frac{1}{2}(\mathbf{x}-\boldsymbol{\mu})^{\mathsf{T}} \boldsymbol{\Sigma}^{-1} (\mathbf{x}-\boldsymbol{\mu})\right) \tag{2.64}$$

we say that $\mathbf{X}$ follows a (non-degenerate) multivariate Gaussian distribution, denoted $\mathbf{X} \sim \mathcal{N}(\boldsymbol{\mu}, \boldsymbol{\Sigma})$. The multivariate Gaussian distribution is a flexible distribution celebrated in the statistical community for its analytical tractability and prevalence in experimental data due to the central limit theorem [29]. However, since the random variable in the multivariate distribution has a finite size, it does not model systems with values for every continuous point in some space. Such systems include time-continuous stochastic signals, the concentrations of different minerals in the earth, and the RSS in a wireless communication network for different locations on a map. We introduce a special class of stochastic processes to model these systems, namely the Gaussian processes over the real numbers.

The formal definition for a *Gaussian process* in $\mathbb{R}^D$ is a stochastic process $\mathcal{X} = \{X(\mathbf{s});\ \mathbf{s} \in \mathbb{R}^D\}$ where, for any finite subset $\boldsymbol{S} = \{\mathbf{s}_1, \dots, \mathbf{s}_d\} \subset \mathbb{R}^D$, the random variable

$$\mathbf{X}(\mathbf{S}) = [X(\mathbf{s}_1) \quad \dots \quad X(\mathbf{s}_d)]^{\mathsf{T}}$$

follows a multivariate Gaussian distribution. GPs are characterized by their *mean* and *kernel* functions. The mean function $m : \mathbb{R}^D \to \mathbb{R}$ maps points in $\mathbb{R}^D$ to the mean value while the kernel function $k : \mathbb{R}^D \times \mathbb{R}^D \to \mathbb{R}_+$ maps pairs of points to the associated covariance. Given m, k and any finite subset $\mathbf{S} \subset \mathbb{R}^D$, we can construct the mean vector as

$$\boldsymbol{\mu}(\mathbf{S}) = E[\mathbf{X}(\mathbf{S})] \quad \text{where} \quad \boldsymbol{\mu}_i(\mathbf{S}) = m(\mathbf{s}_i) \tag{2.65}$$

and covariance matrix as

$$\boldsymbol{\Sigma}(\mathbf{S}) = \text{Cov}\,(\mathbf{X}(\mathbf{S}), \mathbf{X}(\mathbf{S})) \quad \text{where} \quad \boldsymbol{\Sigma}_{i,j}(\mathbf{S}) = k(\mathbf{s}_i, \mathbf{s}_j) \tag{2.66}$$

such that $\mathbf{X}(\mathbf{S}) \sim \mathcal{N}(\boldsymbol{\mu}(\mathbf{S}), \boldsymbol{\Sigma}(\mathbf{S}))$. The notation $\mathcal{X} = \mathcal{GP}(m(\mathbf{s}), k(\mathbf{s}, \mathbf{s}'))$ is introduced as a convenient way to characterize GPs.

Example 1. A simple example of a GP in $\mathbb{R}$ has zero mean $m(s) = 0$ and *exponential* kernel function $k(s, s') = \sigma^2 \exp(-(s - s')^2/l)$ such that

$$\mathcal{X} = \mathcal{GP}\left(0, \sigma^2 \exp\left(-\frac{(s - s')^2}{l}\right)\right) \tag{2.67}$$

The intuition behind the exponential kernel function is a variance of σ^2 and a smooth decrease in the correlation with the distance of two points. l controls the rate that the correlation decreases and is sometimes referred to as the correlation distance [10]. Figure 2.5 shows realizations of $\mathcal{X}$ for different values of l.

Selecting the kernel function is a crucial step when modeling GPs to make sure that correlation in the underlying process is well modeled. Another aspect is the mathematical properties for different kernel functions, such as stationarity of the process and non-degenerate covariance matrices. See [37, ch. 4] for more on kernel functions for GPs.

2.5.1.1 Regression in Gaussian processes

A common problem in GPs is regression where, given a set of observed values $\mathbf{X} = \mathbf{x} \in \mathbb{R}^d$ and their d associated points $\mathbf{S}$ from process $\mathcal{X}$, we want to predict

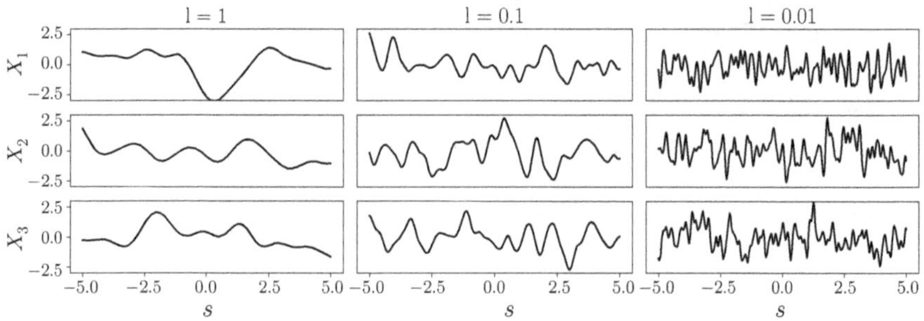

Figure 2.5 Realization of GPs in $\mathbb{R}$ with zero mean and exponential kernel. Here, $\sigma^2 = 1$ and realizations for $l = 1$, $l = 0.1$ and $l = 0.01$ are shown.

the values of the process $\mathbf{X}^* \in \mathbb{R}^m$ at m unobserved points $\mathbf{S}^*$[6]. The prediction is on the conditional distribution $\mathbf{X}\,|\,\mathbf{X},\mathbf{S},\mathbf{S}^*$, which is available on closed form due to the aforementioned analytical tractability of the multivariate Gaussian distribution. We start by considering the joint distribution for $\left[\mathbf{X}^\top \quad (\mathbf{X}^*)^\top\right]^\top$, which is multivariate Gaussian distribution with mean

$$\begin{bmatrix} \mu_{\mathbf{X}} \\ \mu_{\mathbf{X}^*} \end{bmatrix} \quad \text{and covariance} \quad \begin{bmatrix} \Sigma_{\mathbf{XX}} & \Sigma_{\mathbf{XX}^*} \\ \Sigma_{\mathbf{X}^*\mathbf{X}} & \Sigma_{\mathbf{X}^*\mathbf{X}^*} \end{bmatrix}. \tag{2.68}$$

Then, the distribution for $\mathbf{X}^*\,|\,\mathbf{X},\mathbf{S},\mathbf{S}^*$ is also a multivariate Gaussian distribution with mean [27]

$$E[\mathbf{X}^*\,|\,\mathbf{X},\mathbf{S},\mathbf{S}^*] = \mu_{\mathbf{X}^*} + \Sigma_{\mathbf{X}^*\mathbf{X}}\Sigma_{\mathbf{XX}}^{-1}\left(\mathbf{X} - \mu_{\mathbf{X}}\right) \tag{2.69}$$

and covariance [27]

$$\mathrm{Cov}[\mathbf{X}^*\,|\,\mathbf{X},\mathbf{S},\mathbf{S}^*] = \Sigma_{\mathbf{X}^*\mathbf{X}^*} - \Sigma_{\mathbf{X}^*\mathbf{X}}\Sigma_{\mathbf{XX}}^{-1}\Sigma_{\mathbf{XX}^*} \tag{2.70}$$

Here, $E[\mathbf{X}\,|\,\mathbf{X},\mathbf{S},\mathbf{S}^*]$ and $\mathrm{Cov}[\mathbf{X}\,|\,\mathbf{X},\mathbf{S},\mathbf{S}^*]$ are referred to as the *predictive mean* and *covariance*, respectively, which combined completely characterizes the distribution for $\mathbf{X}^*\,|\,\mathbf{X},\mathbf{S},\mathbf{S}^*$ known as the *predictive distribution*. Regression relying on the predictive distribution often uses the mean, which is also the minimum mean square error (MMSE) estimate [27]. Predictions can also be based on other criteria since the entire distribution is known, e.g., a conservative prediction lower than $\mathbf{X}(\mathbf{S}^*)$ with high confidence. Note that the conditioning on the points $\mathbf{S},\mathbf{S}^*$ refers to the fact that the mean vectors and covariance matrices in (2.69) and (2.70) can be computed using these when the mean and kernel functions are known. Of course, the functions m, k and their associated parameters (such as σ^2 and l in example 1) are not always known which is covered in section 2.5.1.2. In figure 2.6 (left), the predictive distribution for the GP introduced in example 1 is computed using (2.69)-(2.70) based on 5 observations.

A common extension of the regression problem is the case where $\mathbf{X}$ is observed with additive Gaussian noise, i.e., $\mathbf{Y} = \mathbf{X} + \mathbf{n}$ is observed where $\mathbf{X}$ and $\mathbf{n}$ are independent and $\mathbf{n} \sim \mathcal{N}(\mathbf{0}, \sigma_n^2\mathbf{I})$. In this case, the joint distribution becomes

$$\begin{bmatrix} \mathbf{Y} \\ \mathbf{X}^* \end{bmatrix} \sim \mathcal{N}\left(\begin{bmatrix} \mu_{\mathbf{Y}} \\ \mu_{\mathbf{X}^*} \end{bmatrix}, \begin{bmatrix} \Sigma_{\mathbf{XX}} + \sigma_n^2\mathbf{I} & \Sigma_{\mathbf{XX}^*} \\ \Sigma_{\mathbf{X}^*\mathbf{X}} & \Sigma_{\mathbf{X}^*\mathbf{X}^*} \end{bmatrix}\right), \tag{2.71}$$

and the predictive distribution for $\mathbf{X}^*\,|\,\mathbf{Y},\mathbf{S},\mathbf{S}^*$ is found by substituting $\Sigma_{\mathbf{XX}} + \sigma_n^2\mathbf{I}$ as $\Sigma_{\mathbf{XX}}$ in (2.69)-(2.70). Figure 2.6 extends the previous example in the setting of

6 As seen in figure 2.5, a GP looks different for each realization. In regression and most other inference problems, we assume that the process for the observed points $\mathbf{X}(\mathbf{S}) = \mathbf{x}$ and the unobserved points of interest $\mathbf{X}(\mathbf{S}^*)$ belongs to the same realization - i.e., $\mathbf{X}(\mathbf{S})$, $\mathbf{X}(\mathbf{S}^*)$ were drawn dependently from a multivariate Gaussian distribution according to the kernel function.

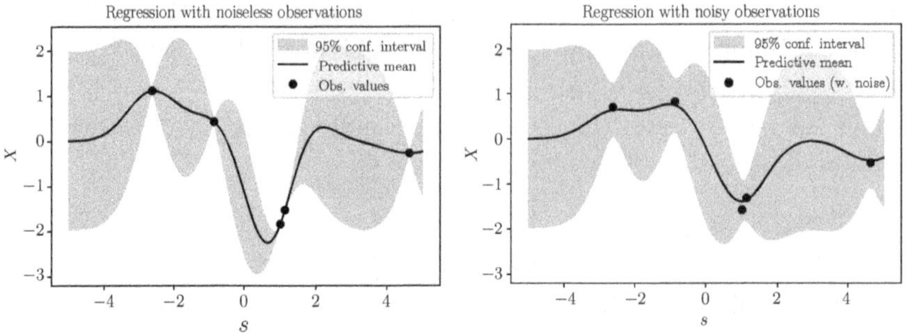

Figure 2.6 Regression for the GP introduced in example 1 with $\sigma^2 = 1$ and $l = 1$. In the left plot, 5 values of **X** are observed directly and in the right plot, the observations **Y** are noisy with noise variance $\sigma_n^2 = 0.1$. In both examples, the predictive distribution is computed for a range of points $s \in [-5,5]$ where the line shows the predictive mean according to (2.69). The gray area is the 95% confidence interval for the predictions based on the variance along the main diagonal of (2.70).

noisy observations. Notice that, due to the observation noise, the predictive mean does not align with the observations and the confidence interval has non-zero width at the observation points.

2.5.1.2 Maximum likelihood estimation of hyper parameters

Given points $\mathbf{S}$, the associated mean $\boldsymbol{\mu}(\mathbf{S})$ and covariance $\boldsymbol{\Sigma}(\mathbf{S})$ for $\mathbf{X}(\mathbf{S})$ is considered the parameters for the GP $\mathcal{X}$. On the other hand, the hyperparameters are the parameters for the mean and kernel functions, such as the variance σ^2 and correlation distance l in example 1. In general, we denote the hyperparameters $\boldsymbol{\phi}_m$ and $\boldsymbol{\phi}_k$ for the mean and kernel function, respectively, and the complete set of hyperparameters is $\boldsymbol{\phi} = \begin{bmatrix} \boldsymbol{\phi}_m^{\mathsf{T}} & \boldsymbol{\phi}_k^{\mathsf{T}} \end{bmatrix}^{\mathsf{T}}$ and possibly noise variance σ_n^2. This section will show how the hyperparameters can be estimated when the models for mean and kernel functions are known, e.g., it is known that the GP has s constant mean and exponential kernel, but the hyperparameters are unknown. The estimation method shown here is maximum likelihood estimation (MLE) which only relies on observed values for optimization — see [37, ch. 5] for the case when prior information about the hyperparameters is known.

Consider d the noisy observations $\mathbf{Y} = \mathbf{y}$ at points $\mathbf{S}$ from the GP $\mathcal{X}$ with

$$\boldsymbol{\mu}(\mathbf{S}; \boldsymbol{\phi}_m) = E[\mathbf{X}(\mathbf{S})] \qquad \text{where} \quad \mu_i(\mathbf{S}; \boldsymbol{\phi}_m) = m(\mathbf{s}_i; \boldsymbol{\phi}_m), \tag{2.72}$$

$$\boldsymbol{\Sigma}(\mathbf{S}; \boldsymbol{\phi}_k) = \mathrm{Cov}\,(\mathbf{X}(\mathbf{S}), \mathbf{X}(\mathbf{S})) \qquad \text{where} \quad \Sigma_{i,j}(\mathbf{S}; \boldsymbol{\phi}_k) = k(\mathbf{s}_i, \mathbf{s}_j; \boldsymbol{\phi}_k). \tag{2.73}$$

Since $\mathbf{Y} = \mathbf{X} + \mathbf{n}$ and $\mathbf{n} \sim \mathcal{N}(\mathbf{0}, \sigma_n^2\mathbf{I})$, it follows that

$$\mathbf{Y} \sim \mathcal{N}(\underbrace{\boldsymbol{\mu}(\mathbf{S};\boldsymbol{\phi}_m)}_{\boldsymbol{\mu}_\mathbf{Y}}, \underbrace{\boldsymbol{\Sigma}(\mathbf{S};\boldsymbol{\phi}_k) + \sigma_n^2\mathbf{I}}_{\boldsymbol{\Sigma}_\mathbf{Y}}). \tag{2.74}$$

with likelihood denoted as $f(\mathbf{y};\mathbf{S},\boldsymbol{\phi}_m,\boldsymbol{\phi}_k,\sigma_n^2)$ given by (2.64). The unknown parameters $\boldsymbol{\phi},\sigma_n^2$ are estimated by maximizing the likelihood, or equivalently the log-likelihood, with respect to these. The log-likelihood is

$$\begin{aligned}\ell(\mathbf{y};\mathbf{S},\boldsymbol{\phi}_m,\boldsymbol{\phi}_k,\sigma_n^2) &= \log(f(\mathbf{y};\mathbf{S},\boldsymbol{\phi}_m,\boldsymbol{\phi}_k,\sigma_n^2)) \\ &= -\frac{d}{2}\log(2\pi) - \frac{1}{2}\log|\boldsymbol{\Sigma}_\mathbf{Y}| - \frac{1}{2}\left(\mathbf{y}-\boldsymbol{\mu}_\mathbf{Y}\right)^\mathsf{T}\boldsymbol{\Sigma}_\mathbf{Y}^{-1}\left(\mathbf{y}-\boldsymbol{\mu}_\mathbf{Y}\right).\end{aligned} \tag{2.75}$$

Maximizing the log-likelihood in (2.75) with respect to the parameters is generally not possible in closed form due to the non-linear dependence on the hyperparameters. Numerical methods such as gradient descent are therefore used for parameter estimation. We state the derivative of the log-likelihood for the various hyperparameters, which is useful as input to numerical methods alongside the log-likelihood itself. For the mean parameters $(\phi_m)_i \in \boldsymbol{\phi}_m$ it is [31]

$$\frac{\partial \ell}{\partial(\phi_m)_i} = (\mathbf{y}-\boldsymbol{\mu}_\mathbf{Y})^\mathsf{T}\boldsymbol{\Sigma}_\mathbf{Y}^{-1}\frac{\partial \boldsymbol{\mu}_\mathbf{Y}}{\partial(\phi_m)_i}, \tag{2.76}$$

and for the kernel parameters $(\phi_k)_i \in \boldsymbol{\phi}_k$ it is [37]

$$\frac{\partial \ell}{\partial(\phi_k)_i} = \frac{1}{2}\operatorname{tr}\left(\left(\mathbf{c}\mathbf{c}^\mathsf{T} - \boldsymbol{\Sigma}_\mathbf{Y}^{-1}\right)\frac{\partial \boldsymbol{\Sigma}_\mathbf{Y}}{\partial(\phi_k)_i}\right), \quad \mathbf{c} = \boldsymbol{\Sigma}_\mathbf{Y}^{-1}\mathbf{y} \tag{2.77}$$

where $\operatorname{tr}(\cdot)$ is the trace operator. Finally, for the noise variance σ_n^2, the derivative is

$$\frac{\partial \ell}{\partial \sigma_n^2} = \frac{1}{2}\operatorname{tr}\left(\mathbf{c}\mathbf{c}^\mathsf{T} - \boldsymbol{\Sigma}_\mathbf{Y}^{-1}\right). \tag{2.78}$$

Equation (2.75)-(2.78) allows for first-order numerical optimization with respect to the hyperparameters. One notable exception, where the hyperparameters are available on closed form, is the case when only $\boldsymbol{\phi}_m$ is unknown and the mean function is linear in the parameters — i.e., m is a linear function of $\boldsymbol{\phi}_m$. This allows us to write $\boldsymbol{\mu}(\boldsymbol{\phi}_m) = \mathbf{M}\boldsymbol{\phi}_m$ where $\mathbf{M} = \mathbf{M}(\boldsymbol{S})$, known as the *design matrix*, only depends on $\mathbf{S}$. In this case, the maximum likelihood estimate for $\boldsymbol{\phi}_m$ is uniquely determined by [27]

$$\hat{\boldsymbol{\phi}}_m = \left(\mathbf{M}^\mathsf{T}\boldsymbol{\Sigma}_\mathbf{Y}^{-1}\mathbf{M}\right)^{-1}\mathbf{M}^\mathsf{T}\boldsymbol{\Sigma}_\mathbf{Y}^{-1}\mathbf{y}, \tag{2.79}$$

which the reader may recognize as the solution to the *normal equations* with correlated noise. When the mean function is linear, but the covariance parameters are unknown, we can still use (2.79) by expressing $\hat{\boldsymbol{\phi}}_m$ as a function of $\mathbf{y}, \mathbf{S}, \boldsymbol{\phi}_k$ and σ_n^2, which effectively reduces the problem to an estimation of $\boldsymbol{\phi}_k$ and σ_n^2.

Example 2. Consider the GP in $\mathbb{R}$

$$\mathcal{X} = \mathcal{GP}\left(c_0 + c_1 s, \sigma^2 \exp\left(-\frac{(s-s')^2}{l}\right)\right), \tag{2.80}$$

with parameters $\boldsymbol{\phi} = \begin{bmatrix} c_0 & c_1 & \sigma^2 & l \end{bmatrix}^{\mathsf{T}}$. d noisy values $\mathbf{Y} = \mathbf{y}$ are observed at points $\mathbf{s}$ in the interval $[s_{\min}, s_{\max}]$ with noise variance σ_n^2. To estimate $\boldsymbol{\phi}$ and σ_n^2, we notice that the mean function is linear in c_0 and c_1. The design matrix is

$$\mathbf{M}(\mathbf{s}) = \begin{bmatrix} 1 & s_1 \\ \vdots & \vdots \\ 1 & s_d \end{bmatrix}, \quad \text{with} \quad \boldsymbol{\mu}_{\mathbf{Y}} = \begin{bmatrix} 1 & s_1 \\ \vdots & \vdots \\ 1 & s_d \end{bmatrix} \begin{bmatrix} c_0 \\ c_1 \end{bmatrix} = \mathbf{M}(\mathbf{s})\boldsymbol{\phi}_m, \tag{2.81}$$

and the maximum likelihood estimation for $\boldsymbol{\phi}_m$ given $\mathbf{s}, \sigma^2, l$ and σ_n^2 is obtained from (2.79). The remaining parameters are obtained numerically using the log-likelihood (2.75), the gradient (2.77) with respect to the kernel parameters $\boldsymbol{\phi}_k = \begin{bmatrix} \sigma^2 & l \end{bmatrix}^{\mathsf{T}}$ and the derivative (2.78) with respect to the noise parameter σ_n^2. Here, the derivative for the kernel parameters is

$$\nabla_{\boldsymbol{\phi}_k} k(s, s') = \begin{bmatrix} \dfrac{\partial k(s,s')}{\partial \sigma^2} \\ \dfrac{\partial k(s,s')}{\partial l} \end{bmatrix} = \begin{bmatrix} \exp\left(-\dfrac{(s-s')^2}{l}\right) \\ \dfrac{\sigma^2}{l^2}(s-s')^2 \exp\left(-\dfrac{(s-s')^2}{l}\right) \end{bmatrix} \tag{2.82}$$

such that $\frac{\partial}{\partial(\boldsymbol{\phi}_k)_i}(\boldsymbol{\Sigma}_{\mathbf{Y}})_{jp} = (\nabla_{\boldsymbol{\phi}_k} k(s_j, s_p))_i$. All parameters should be positive, so the optimization is performed over the search space

$$\begin{bmatrix} \sigma^2 & l & \sigma_n^2 \end{bmatrix} \in [\epsilon_{\sigma^2}, \infty) \times [\epsilon_l, \infty) \times [\epsilon_{\sigma_n^2}, \infty), \tag{2.83}$$

where $\epsilon_{\sigma^2}, \epsilon_l, \epsilon_l > 0$ ensures that $\boldsymbol{\Sigma}_{\mathbf{Y}}$ is invertible. Sequential least square programming (SLSQP), which allows for bounded optimization, is used for optimization [11]. The settings are $s_{\min} = s_{\max} = 200$, $c_0 = 1$, $c_1 = 0.1$, $\sigma^2 = l = 1$ and $\sigma_n^2 = 0.1$. Figure 2.7 shows an example with $d = 200$ observations, and figure 2.8 shows the estimation error as the number of observations d increases.

This concludes the primer on Gaussian processes. See [37] for more information including details for numerical implementation.

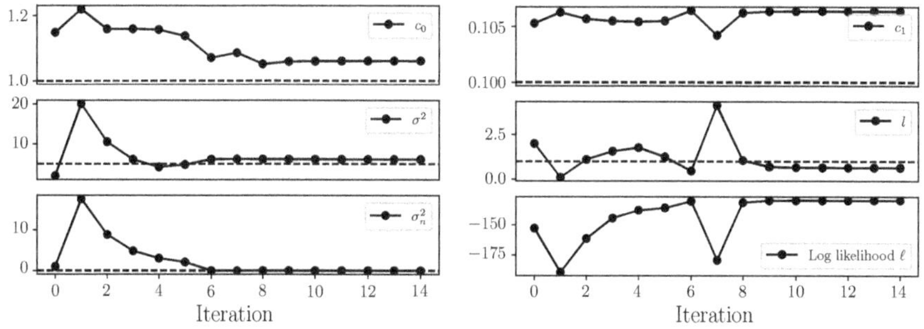

Figure 2.7 Estimated parameters and log-likelihood for different iterations in the SLSQP algorithm, which converges after 14 iterations. The dashed lines show the true parameters.

Figure 2.8 Estimation error for the hyperparameters with different number of observations d. The statistic shown is the mean relative error defined as $\text{MRE} = \frac{1}{d}\sum_{i=1}^{d} |(\hat{\phi} - \phi)/\phi|$ for hyperparameter ϕ and estimate $\hat{\phi}$.

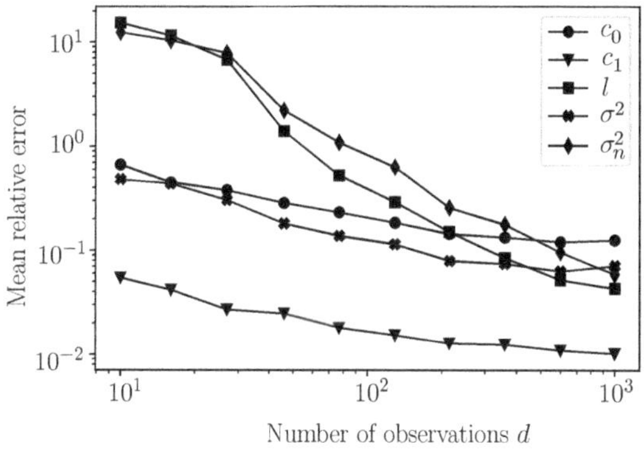

2.5.2 Statistical Radio Maps with Rician Fading

Continuing with the setup introduced in Section 2.5 (illustrated in Figure 2.4), we now assume that the fading distribution for the channel between transmitter at different locations $\mathbf{s}$ and the receiver at location $\mathbf{s}_{\text{rx}}$ is Rician. The PDF for the channel coefficient power $W = |h|^2$ is given by (2.6) denoted $p_W(w; \overline{W}, K)$ with parameters $\boldsymbol{\theta} = \begin{bmatrix} \overline{W} & K \end{bmatrix}^{\text{T}}$. We repeat the problem formulation: how can $\boldsymbol{\theta}_0$ be predicted given $\{\mathbf{s}_i, \boldsymbol{\theta}_i\}_{i=1}^{d}$ and how can this be used to select the communication rate R?

The first step is to model the spatial correlation of $\boldsymbol{\theta}$. For simplicity, we assume that $\overline{W}$ and K are independent and proceed to model them as separate Gaussian processes. The baseband model in (2.7) combining pathloss, shadowing, and small-scale fading is used to model the parameters. With

$$W(\mathbf{s}) = |h(\mathbf{s})|^2 = P_L(\mathbf{s})|\beta(\mathbf{s})|^2|\alpha(\mathbf{s})|^2, \tag{2.84}$$

the following observations are noted:

- The pathloss $P_L(\mathbf{s})$ and shadowing $\beta(\mathbf{s})$ are constant in time, assuming the propagation environment is static, whereas the small-scale fading $\alpha(\mathbf{s})$ varies over time.
- The pathloss is can be modeled with the simplified model $P_L(\mathbf{s}) = G||\mathbf{s} - \mathbf{s}_{\text{rx}}||^{-\eta}$ where G is some constant and η is the pathloss exponent.
- Shadowing and small-scale fading can be modeled to have unit power, i.e., $E[|\beta(\mathbf{s})|^2] = E[|\alpha(\mathbf{s})|^2] = 1$, by absorbing any factors into G in the pathloss term.
- The temporal variation of small-scale fading is modeled as $\alpha \sim \mathcal{CN}(\mu, \sigma^2)$, where $E[|\alpha|^2] = 1$ is fulfilled by setting $\mu^2 = \frac{K}{K+1}, \sigma^2 = \frac{1}{K+1}$ for some $K > 0$.

The temporal variation for the channel coefficient is now given by

$$h = \sqrt{P_L}\beta\alpha \sim \mathcal{CN}\left(\sqrt{P_L}\beta\sqrt{\frac{K}{1+K}}, P_L|\beta|^2\frac{1}{1+K}\right), \tag{2.85}$$

from which it follows that W is Rician with total power $\overline{W} = E[W] + \text{Var}[W] = P_l|\beta|^2$ and ratio $K = E[W]/\text{Var}[W]$. This motivates a model of $\overline{W}$ in dB since then

$$Z(\mathbf{s}) \triangleq 10\log_{10}(\overline{W}(\mathbf{s})) = 10\log_{10}(G) - 10\eta\log_{10}||\mathbf{s} - \mathbf{s}_{\text{rx}}|| + 20\log_{10}(|\beta(\mathbf{s})|) \tag{2.86}$$

where the first two terms are deterministic and the last term containing the shadowing coefficient experiences spatial correlation. Since $E[|\beta(\mathbf{s})|^2] = 1$, it does not affect the mean of Z, and we get the model

$$Z(\mathbf{s}) = \mathcal{GP}\left(c - 10\eta\log_{10}||\mathbf{s} - \mathbf{s}_{\text{rx}}||, k(\mathbf{s}, \mathbf{s}')\right), \tag{2.87}$$

where $c = 10\log_{10}(G)$ is a constant and $k(\mathbf{s}, \mathbf{s}')$ is the kernel function describing the correlation for shadowing in dB. The model (2.87) is the exact model seen in many radio map papers, e.g., [10], [38] and [45]. Here, the *Gudmundson* model is often used for the kernel function with [21]

$$k(\mathbf{s}, \mathbf{s}') = \sigma^2\exp\left(-\frac{||\mathbf{s} - \mathbf{s}'||}{l}\right). \tag{2.88}$$

Other kernel functions could be applicable depending on what fits the data well. For the K parameter, which models the small-scale fading, no immediate information could be incorporated into its Gaussian process. We chose the generic model

$$K(\mathbf{s}) = \mathcal{GP}\left(c, k(\mathbf{s}, \mathbf{s}')\right), \tag{2.89}$$

where c is a constant and $k(\mathbf{s}, \mathbf{s}')$ is any kernel chosen to fit the data. Note that K is modeled in the linear domain.

After selecting models for the GPs, the next step is to find the hyperparameters through maximum likelihood estimation as described in Section 2.5.1.2. Then, the predictive distribution for $\boldsymbol{\Theta}_0$ at $\mathbf{s}_0$ is computed using regression as described in Section 2.5.1.1. Denoting the locations and known distribution parameters

$$\mathbf{S} = \begin{bmatrix} \mathbf{s}_1 & \dots & \mathbf{s}_d \end{bmatrix}, \mathbf{Z} = \begin{bmatrix} Z_1 & \dots & Z_d \end{bmatrix}^\mathsf{T}, \text{ and } \mathbf{K} = \begin{bmatrix} K_1 & \dots & K_d \end{bmatrix}^\mathsf{T} \tag{2.90}$$

the predictive distributions are

$$Z_0 \,|\, \mathbf{Z}, \mathbf{S}, \mathbf{s}_0 \sim \mathcal{N}\left(\mu_{Z_0}, \sigma^2_{Z_0}\right), \quad \text{and} \tag{2.91}$$

$$K_0 \,|\, \mathbf{K},\ \mathbf{S},\ \mathbf{s}_0 \sim \mathcal{N}\left(\mu_{K_0}, \sigma^2_{K_0}\right) \tag{2.92}$$

The Gaussian distributions in (2.91) and (2.92) are not immediately applicable. Firstly, Z_0 needs conversion from dB to the linear domain through $\overline{W}_0 = 10^{Z_0/10}$. Applying the conversion results in a *log-normal* distribution for $\overline{W}_0$ with PDF [23]

$$f_{\overline{W}_0}(\overline{W}) = \frac{1}{\overline{W}\sqrt{2\pi\sigma^2_{\overline{W}_0}}} \exp\left(-\frac{(\log(\overline{W}) - \mu_{\overline{W}_0})^2}{2\sigma^2_{\overline{W}_0}}\right), \quad W \geq 0 \tag{2.93}$$

where $\mu_{\overline{W}_0} = \frac{\log(10)}{10}\mu_{Z_0}$ and $\sigma^2_{\overline{W}_0} = \frac{(\log(10))^2}{100}\sigma^2_{Z_0}$. Note that the $\overline{W}$ parameter in the Rician distribution must be positive, and this is conveniently fulfilled for $\overline{W}_0$ due to the exponential conversion. K must also be positive, which is not fulfilled by (2.92). To solve this, we approximate the distribution for K_0 with a *truncated normal distribution* which is the distribution for K_0 conditioned on $K_0 \geq 0$. The PDF for the truncated normal is denoted $f_{K_0}(K)$ for $K \in \mathbb{R}$ — see [23, p. 156] for further details. With these adaptations and the assumed independence between $\overline{W}$ and K, the predictive distribution for $\boldsymbol{\Theta}_0$ becomes $f(\boldsymbol{\Theta}_0) = f_{\overline{W}_0, K_0}(\overline{W}, K) = f_{\overline{W}_0}(\overline{W}) f_{K_0}(K)$.

The final step is rate selection. The usual Bayesian framework of obtaining the posterior predictive distribution for W_0 is not applicable since this requires observations of W_0. Instead, we compute the *prior* predictive distribution, which is the predictive distribution for W_0 given only the prior. $f(\Theta_0 \mid \phi_0)$ acts as the parameter prior in this context with hyperparameters

$$\phi_0 = \begin{bmatrix} \phi_{\overline{W}_0}^{\mathsf{T}} & \phi_{K_0}^{\mathsf{T}} \end{bmatrix}^{\mathsf{T}} = \begin{bmatrix} \mu_{\overline{W}_0} & \sigma_{\overline{W}_0}^2 & \mu_{K_0} & \sigma_{K_0}^2 \end{bmatrix}^{\mathsf{T}}. \tag{2.94}$$

Dropping the subscript 0, the posterior predictive PDF is the marginal

$$\begin{aligned} f(W \mid \phi) &= \int f(W \mid \Theta) f(\Theta \mid \phi) d\Theta \\ &= \int_0^\infty \int_0^\infty f_W(W \mid \overline{W}, K) f_{\overline{W}}(\overline{W} \mid \phi_{\overline{W}}) f_K(K \mid \phi_K)\, d\overline{W}\, dK. \end{aligned} \tag{2.95}$$

The integrand in (2.95) is the product of a Rician, a log-normal, and a truncated normal distribution making the integral infeasible to compute. Fortunately, each factor is a valid distribution that is easily drawn from, which allows us to simulate $W \mid \phi$ in the two-step procedure:

1. Draw parameters $\overline{W} \sim f_{\overline{W}}(\cdot \mid \phi_{\overline{W}})$ and $K \sim f_K(\cdot \mid \phi_K)$
2. Draw channel power $W \sim f_W(\cdot \mid \overline{W}, K)$

The procedure is repeated n times and relevant statistics can be estimated from W^n. For target reliability ϵ, the rate is now selected according to the distribution for $W \mid \phi$ through

$$R_\epsilon = \log_2(1 + F_{W \mid \phi}^{-1}(\epsilon)) \approx \log_2(1 + W_{(\lfloor \epsilon n + 1 \rfloor)}) \tag{2.96}$$

where $F_{W \mid \phi}^{-1}(\epsilon)$ is the inverse CDF for $W \mid \phi$ which is approximated by the order statistic $W_{(\lfloor \epsilon n + 1 \rfloor)}$ from W^n. Note that n can be selected to be any number such that the error introduced by the approximation in (2.96) is arbitrarily low. The selected rate R_ϵ ideally fulfills that the outage probability

$$P_{\text{out}} = \Pr(R_\epsilon > \log_2(1 + W)) \approx \epsilon \tag{2.97}$$

with high probability in some sense — this is studied further in the following section. The numerous steps to select the communication rate from a statistical radio map are summarized in algorithm 1.

Algorithm 1 Rate selection using statistical radio maps with Rician fading.

Require: Locations and fading distribution parameters $\left\{\mathbf{s}_i, \overline{W}_i, K_i\right\}_{i=1}^{d}$, kernel function models $k_{\overline{W}}$ and k_K, unobserved location $\mathbf{s}_0$, target reliability ϵ and number of simulations n.

Estimate hyperparameters for GPs $Z(\mathbf{s})$ and $K(\mathbf{s})$ with models (2.87) and (2.89), respectively (see Section 2.5.1.2).

Calculate the predictive distribution for Z_0, K_0 (see Section 2.5.1.1) and apply transformations to obtain $\boldsymbol{\phi}_0 = \begin{bmatrix}\mu_{\overline{W}_0} & \sigma^2_{\overline{W}_0} & \mu_{K_0} & \sigma^2_{K_0}\end{bmatrix}^{\mathsf{T}}$.

for $i = 1 \dots, n$ **do**

  Draw $\overline{W}_i \sim f_{\overline{W}}\left(\cdot \mid \mu_{\overline{W}_0}, \sigma^2_{\overline{W}_0}\right)$ (log-normal)

  Draw $K_i \sim f_K\left(\cdot \mid \mu_{K_0}, \sigma^2_{K_0}\right)$ (truncated normal)

  Draw $W_i \sim f_W\left(\cdot \mid \overline{W}_i, K_i\right)$ (Rician)

end for

Select communication rate $R_\epsilon = \log_2(1 + W_{(\lfloor \epsilon n + 1 \rfloor)})$

2.5.3 Example: Rate Selection Using Statistical Radio Map with Simulated Data

This chapter concludes with a numerical example with the method described in algorithm 1. To show that the method has some practical merit, we will use a simulation library to simulate the spatial behavior of the channel coefficients. The ray-tracing based channel simulator known as *QuaDRiGa* [18] will be used to generate data as described in the following.

Simulating a statistical radio map

The receiver is located at $\mathbf{s}_{\text{rx}} = \begin{bmatrix}0 & 0 & 10\end{bmatrix}^{\mathsf{T}}$ m with non-interfering users transmitting from locations $\mathbf{s}_i$. The channel coefficients are simulated according to the 3GPP NR Urban Micro-Cell scenario with line of sight (see [18, p. 81] for further details). Transmitting at 2.6 GHz, the channel simulator gives 12 channel coefficients for different delays that are summed to get the narrowband channel coefficient. The model in (2.4) is applied to simulate small-scale fading by multiplying each channel coefficient with a random phase before summation. This method allows us to simulate an arbitrary number of narrowband channel coefficients. 2601 channels are simulated at locations from a uniform grid with x and y coordinate in $[-50, -48, \dots, 48, 50]$ m and z coordinate 1.5 m. For each channel $n = 10^5$, narrowband channel coefficients are simulated. The output is $\{\mathbf{s}_i, W_i^n\}_{i=1}^{2601}$, where $W^n \in \mathbb{R}^n$ are the generated powers for the channel at $\mathbf{s}_i$.

Rician distributions are now fitted at each location using a maximum likelihood estimator $(\overline{W}_i, K_i) = g(W^n)$. Of course, the Rician distribution does not necessarily model the fading well, but for the purpose of illustrating the usefulness of statistical radio maps, we will consider the fitted distributions as the ground truth. Figure 2.9 shows the fitted parameters. Visual inspection of the maps shows that both parameters indeed have spatial correlation, i.e., smooth variation in space for $\overline{W}$ in particular and K to a lesser extent. $\overline{W}$ and K seem to have some positive correlation (the estimated correlation coefficient is 0.28), but they are not strongly correlated, making both parameters informative as prior information.

500 samples are now drawn from the radio map to set up the problem. We sample locations using an inhomogeneous sample process to model, e.g., an urban environment where some locations are more likely to have transmitters than others. The transmitter locations $\mathbf{s}_i$ for $i = 1, \dots, 500$ are shown in Figure 2.10.

The exponential kernel introduced in example 1 is used to model the GPs for both parameters[7]. To avoid over-fitting the GPs to the data, we also assume that the parameters are observed with noise as described in Section 2.5.1.2. Finally, the data is normalized by subtracting the mean and dividing by the standard deviation, which helps stabilize the numerical methods (the predictions are later transformed back into the original scale). The hyperparameters are estimated from the 500 observations and summarized in Table 2.1. Next, the predictive distributions for $\overline{W}$ and K are obtained at every location in the original grid, i.e.,

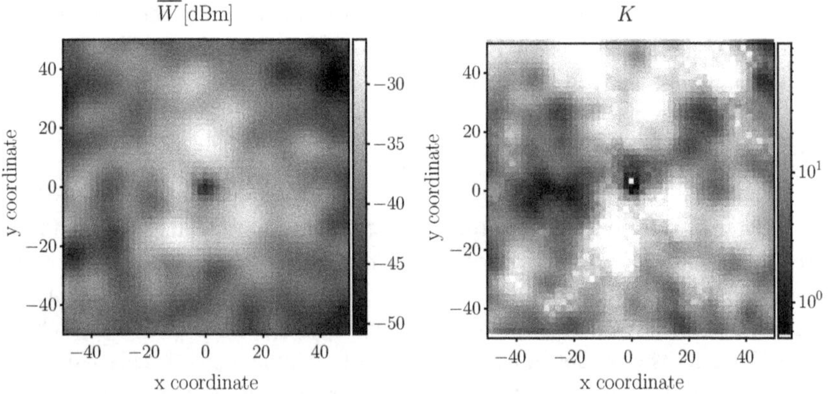

Figure 2.9 Statistical radio map shown as heat maps with average power $\overline{W}$ on the left and K factor on the right.

---

7 The exponential kernel is good at modeling smooth processes due to it being infinitely differentiable [37]. K seems to have some sharp transitions and is possibly modeled better using another kernel.

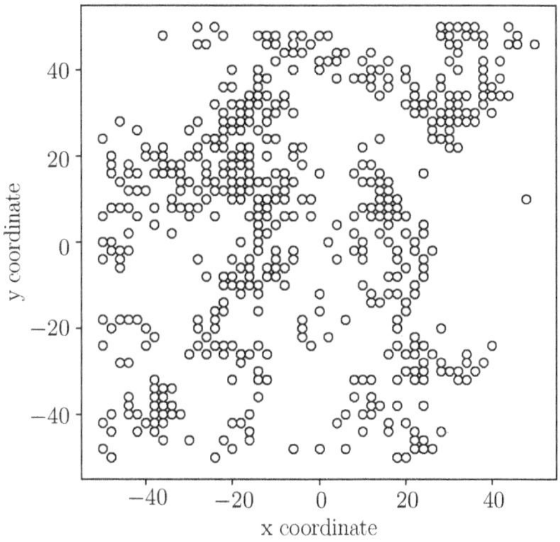

Figure 2.10 500 transmitter locations $\mathbf{s}_i$ drawn from an inhomogeneous sample process.

Table 2.1 Estimated hyperparameters.

GP	c	η	σ^2	l	σ_n^2
Z[1]	−62.67	0.37	8.44	77.82	0.14
K	25.47		1080.57	46.61	123.02

[1] $Z = 10\log_{10}(\overline{W})$
c is the constant mean, η is the pathloss exponent, σ^2 is the variance for the processes, l is the correlation distance and σ_n^2 is the noise variance.

2601 locations in total. Figure 2.11 compares the original maps and the predictive means according to (2.69). Notice that the predicted values are less accurate in the right part of the map where very few observations are available — in particular for the K parameter. Of course, the predictive variance will also reflect this by being larger in areas with fewer observations, similar to Figure 2.6. The communication rate R_ϵ is now selected for every point on the map based on the predictive distribution. $n = 10^7$ simulations are used to estimate the ϵ-quantile at each location for $\epsilon = 0.01$. The resulting outage is probability computed as:

$$P_{\text{out}} = P(R_\epsilon > \log_2(1 + W)) = P\left(2^{R_\epsilon} - 1 > W\right) = F_W\left(2^{R_\epsilon} - 1; \overline{W}, K\right) \tag{2.98}$$

where F_W is the Rician CDF and $\overline{W}$ and K are the fitted values used as ground truth. The data set contains $2601 - 500 = 2101$ unobserved locations, where the outage probability is analyzed. Figure 2.12 shows a histogram for P_{out} where it

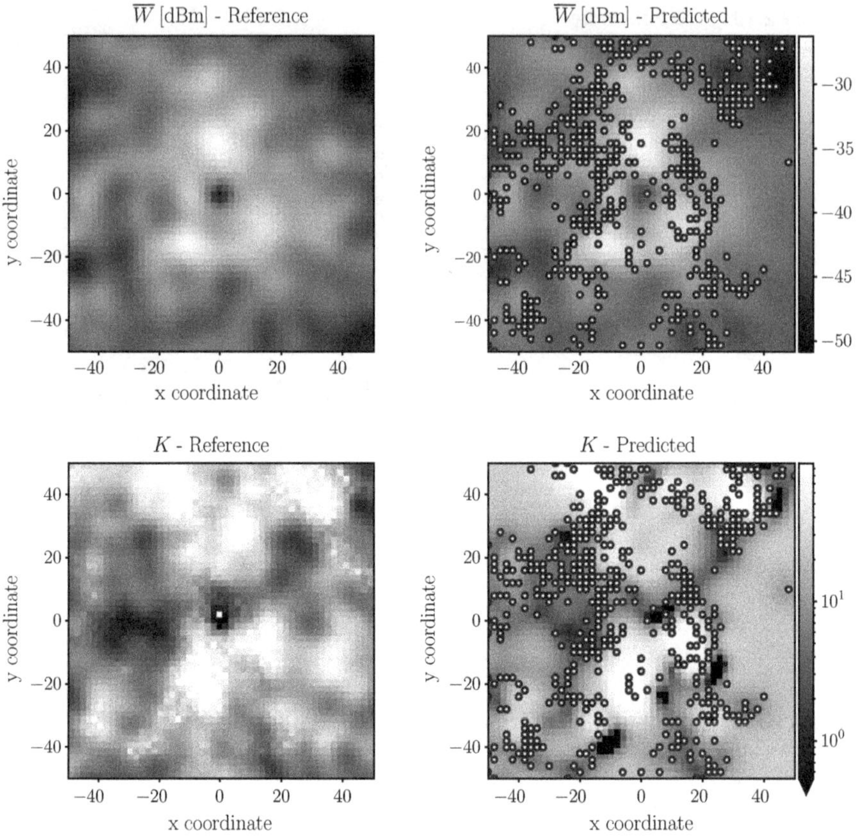

Figure 2.11 Predictive means for $\overline{W}$ and K obtained from the GPs (right) versus the reference values from the fitted fading distributions (left). The dots show the observed locations from which the predictions are made.

is seen that the average for the unobserved locations $\overline{P}_{\text{out}} = 1.36 \cdot 10^{-2}$ (corresponding to the average reliability) aligns well with the target reliability. A statistic similar to the PCR reliability requirement is the probability that the outage requirement is violated for different locations. This is estimated by[8]

$$\tilde{p}_{\epsilon} \approx \frac{\sum_i \mathbb{1}(P_{\text{out},i} > \epsilon)}{2101} = 0.31. \tag{2.99}$$

The fact that the estimated PCR is less than 0.5 shows that the rate selection scheme is conservative. Rate selection is conservative since the GPs accounts for

8 Statistical dependencies for the outage probabilities are not accounted for in (2.99).

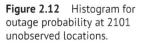

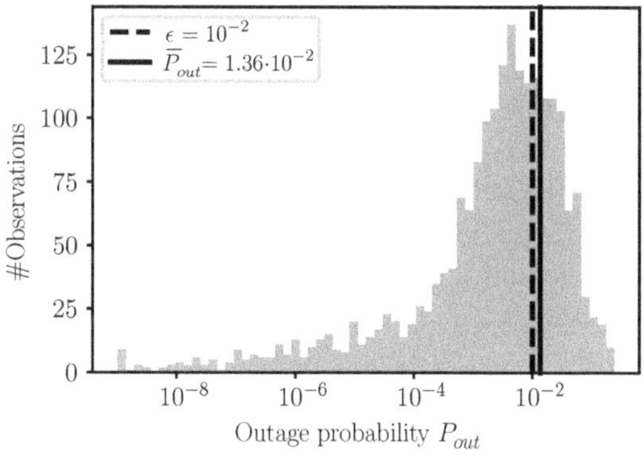

Figure 2.12 Histogram for outage probability at 2101 unobserved locations.

estimation error through the predictive variance. In fact, as the number of observations increases, the PCR will converge to 0.5 while the average outage probability goes to ϵ, assuming the models are correct.

2.6 Summary

Considering the critical role of URLLC for wireless communication in general and in 3GPP in particular, this chapter has given an overview of the technology from a statistical point of view. The chapter investigated why the requirements imposed by URLLC are notoriously difficult to fulfill, where one key observation is the excessive number of observations required to characterize ultra-rare events, which necessitates careful statistical analysis and specialized methods. URLLC has different definitions for reliability, availability, latency, etc., depending on which communication layers are included. This chapter paid attention to the physical layer since it is fundamental to understanding higher layer performances. A narrowband channel consisting of pathloss, shadowing, small-scale fading, and noise was introduced as the primary example, followed by a characterization of the outage probability as a measure of reliability at the physical layer. The outage probability was analyzed as a random variable, and reliability was defined by putting different statistical guarantees on the outage event — here, specifically the average reliability and probability correct reliability. Adhering to these guarantees, it was shown that rate selection based on parametric models requires many fewer samples than non-parametric to reach the same throughput (assuming that the parametric models are known and correct). However, the models cannot be known in practice, and a few methods to model the lower tail of fading distributions were introduced to model rare events directly. Inspired by the observation that prior information (e.g., knowing the distribution) reduces the number of samples to achieve some target reliability, Bayesian statistics were introduced to incorporate prior information formally in the rate-selection problem. It was shown that in the ideal case where the parameter prior is perfectly known, the communication rate could be selected with higher precision compared to the case with no prior information — in particular for few observations. Finally, statistical radio maps were introduced to obtain prior information about the fading distribution for new transmissions. Combining Bayesian statistics and Gaussian processes, which leverage spatial correlation of fading statistics, a method to select communication rate only relying on observations from nearby locations was introduced. This was tested in a synthetic scenario with ray-traced simulated data, where the resulting outage probability aligned well with the target reliability even for locations farther away from observations, although the selected rate is consequently more conservative here.

There are still many open problems to meet the strict requirements posed by URLLC. Challenges omitted here include analyzing broadband communication, multiple-input multiple-output (MIMO) systems, time-varying channel statistics and, importantly, the performance at higher communication layers. Using prior information and Bayesian statistics in the context of URLLC have only been explored sparsely in the literature so far, but we expect that it will have a pivotal role in the coming years for 5G and beyond.

Bibliography

[1] 3rd Generation Partnership Project. Study on scenarios and requirements for next generation access technologies. Technical Report 138 913, V 15.0.0, 3GPP, 2018.

[2] 3rd Generation Partnership Project. Service requirements for cyber-physical control applications in vertical domains. Technical Report 22.104, V 17.4.0, 3GPP, 2020.

[3] 3rd Generation Partnership Project. Service requirements for the 5g system. Technical Report 22.261, V 18.0.0, 3GPP, 2020.

[4] A. Abdi and M. Kaveh. On the utility of gamma PDF in modeling shadow fading (slow fading). In *1999 IEEE 49th Vehicular Technology Conference (Cat. No.99CH36363)*, volume 3, pages 2308–2312 vol.3, 1999.

[5] Ali Abdi and Mostafa Kaveh. A comparative study of two shadow fading models in ultrawideband and other wireless systems. *IEEE Transactions on Wireless Communications*, 10(5):1428–1434, 2011.

[6] Madyan Alsenwi, Nguyen H. Tran, Mehdi Bennis, Anupam Kumar Bairagi, and Choong Seon Hong. embb-urllc resource slicing: A risk-sensitive approach. *IEEE Communications Letters*, 23(4):740–743, 2019.

[7] Marko Angjelichinoski, Kasper Fløe Trillingsgaard, and Petar Popovski. A statistical learning approach to ultra-reliable low latency communication. *IEEE Transactions on Communications*, 67(7):5153–5166, 2019.

[8] Amin Azari, Mustafa Ozger, and Cicek Cavdar. Risk-aware resource allocation for URLLC: Challenges and strategies with machine learning. *IEEE Communications Magazine*, 57(3):42–48, 2019. Conference Name: IEEE Communications Magazine.

[9] Mehdi Bennis, Mérouane Debbah, and H. Vincent Poor. Ultrareliable and low-latency wireless communication: Tail, risk, and scale. *Proceedings of the IEEE*, 106(10):1834–1853, 2018.

[10] Vinay-Prasad Chowdappa, Carmen Botella, Juan Javier Samper-Zapater, and Rafael J. Martinez. Distributed radio map reconstruction for 5g automotive. *IEEE Intelligent Transportation Systems Magazine*, 10(2):36–49, 2018. Conference Name: IEEE Intelligent Transportation Systems Magazine.

[11] SciPy Documentation. https://docs.scipy.org/doc/scipy/reference/generated/scipy.optimize.minimize.html. Accessed: 2021-12-12.

[12] Gregory D. Durgin, Theodore S. Rappaport, and David A. de Wolf. New analytical models and probability density functions for fading in wireless communications. *IEEE Transactions on Communications*, 50(6):1005–1015, Jun. 2002.

[13] Gregory David Durgin. *Theory of stochastic local area channel modeling for wireless communications*. PhD thesis, Virginia Tech, 2000.

[14] Patrick C. F. Eggers, Marko Angjelichinoski, and Petar Popovski. Wireless channel modeling perspectives for ultra-reliable communications. *IEEE Transactions on Wireless Communications*, 18(4):2229–2243, 2019.

[15] Xiaochen Fan, Xiangjian He, Chaocan Xiang, Deepak Puthal, Liangyi Gong, Priyadarsi Nanda, and Gengfa Fang. Towards system implementation and data analysis for crowdsensing based outdoor rss maps. *IEEE Access*, 6:47535–47545, 2018.

[16] Gaussian Processes for Machine Learning. *Pattern Recognition and Machine Learning*. Springer, 2006.

[17] Gustavo Fraidenraich and Michel Daoud Yacoub. The α-η-μ and α-κ-μ fading distributions. In *2006 IEEE Ninth International Symposium on Spread Spectrum Techniques and Applications*, pages 16–20, 2006.

[18] Fraunhofer Heinrich Hertz Institute. *Quasi Deterministic Radio Channel Generator User Manual and Documentation*, v2.4.0 edition, October 2020.

[19] A. Gelman, J.B. Carlin, H.S. Stern, D.B. Dunson, A. Vehtari, and D.B. Rubin. *Bayesian Data Analysis, Third Edition*. Chapman & Hall/CRC Texts in Statistical Science. Taylor & Francis, 2013.

[20] Andrea Goldsmith. *Wireless Communications*. Cambridge University Press, 2005.

[21] M. Gudmundson. Correlation model for shadow fading in mobile radio systems. *Electronics Letters*, 27(03), 1991.

[22] Benjamin Guedj. A primer on pac-bayesian learning, 2019.

[23] Normal L. Johnson, Samuel Kotz, and N. Balakrishnan. *Continuous univariate distributions*, volume 1. John Wiley, 2 edition, 1994.

[24] Normal L. Johnson, Samuel Kotz, and N. Balakrishnan. *Continuous univariate distributions*, volume 2. John Wiley, 2 edition, 1995.

[25] Anders E. Kalør, Osvaldo Simeone, and Petar Popovski. Prediction of mmwave/thz link blockages through meta-learning and recurrent neural networks. *IEEE Wireless Communications Letters*, 10(12):2815–2819, 2021.

[26] Roy Karasik, Osvaldo Simeone, Hyeryung Jang, and Shlomo Shamai. Learning to broadcast for ultra-reliable communication with differential quality of service via the conditional value at risk, 2021.

[27] Steven M. Kay. *Fundementals of Statistical Signal Processing*, volume II. Pearson, 1993.

[28] Steven M. Kay. *Fundementals of Statistical Signal Processing*, volume II. Pearson, 1998.

[29] Steven M. Kay. *Intuitive Probability and Random Processes using Matlab ®*. Springer, 2006.

[30] Niloofar Mehrnia and Sinem Coleri. Wireless channel modeling based on extreme value theory for ultra-reliable communications. *IEEE Transactions on Wireless Communications (Early access)*, pages 1–1, 2021.

[31] K. B. Petersen and M. S. Pedersen. The matrix cookbook, nov 2012. Version 20121115.

[32] Yury Polyanskiy, H. Vincent Poor, and Sergio Verdu. Channel coding rate in the finite blocklength regime. *IEEE Transactions on Information Theory*, 56(5):2307–2359, 2010.

[33] Petar Popovski. *Wireless Connectivity: An Intuitive and Fundamental Guide.* Wiley, 1 edition, 2020.

[34] Petar Popovski, Cedomir Stefanovic, J. Jimmy Nielsen, Elisabeth de Carvalo, Marko Angjelichinoski, and Alexandru-Sabin Trillingsgaard, Kasper F. Bana. Wireless access in ultra-reliable low-latency communication (urllc). *IEEE Transactions on Communications*, 67(8), 08 2018.

[35] John G. Proakis and Masoud Salehi. *Digital Communications.* McGraw Hill, 5 edition, 2007.

[36] Pablo Ramírez-Espinosa and F. Javier López-Martínez. Composite fading models based on inverse gamma shadowing: Theory and validation. *IEEE Transactions on Wireless Communications*, 20(8):5034–5045, 2021.

[37] C. E. Rasmussen Rasmussen and C. K. I. Williams. *Gaussian Processes for Machine Learning.* The MIT Press, 2006.

[38] Berna Sayrac, Ana Galindo-Serrano, Sana Ben Jemaa, Janne Riihijärvi, and Petri Mähönen. Bayesian spatial interpolation as an emerging cognitive radio application for coverage analysis in cellular networks. *Transactions on Emerging Telecommunications Technologies*, 24, 11 2013.

[39] C. E. Shannon. A mathematical theory of communication. *The Bell System Technical Journal*, 27(3):379–423, 1948.

[40] Marvin K. Simon and Mohamed-Slim Alouini. *Digital communication over fading channels*. Wiley-IEEE Press, 2005.

[41] Bernard Sklar. *Digital communications: Fundamentals and applications*. Prentice-Hall PTR, Upper Saddle River, N.J, 2 edition, 2001.

[42] B. Soret, P. Mogensen, K. I. Pedersen, and M. C. Aguayo-Torres. Fundamental tradeoffs among reliability, latency and throughput in cellular networks. In *2014 IEEE Globecom Workshops (GC Wkshps)*, pages 1391–1396, 2014.

[43] Gordon L. Stüber. *Principles of Mobile Communication*. Springer Nature, 4 edition, 2017.

[44] V. N. Swamy, P. Rigge, G. Ranade, B. Nikolić, and A. Sahai. Wireless channel dynamics and robustness for ultra-reliable low-latency communications. *IEEE Journal on Selected Areas in Communications*, 37(4):705–720, 2019.

[45] Rocco Di Taranto, Srikar Muppirisetty, Ronald Raulefs, Dirk T. M. Slock, Tommy Svensson, and Henk Wymeersch. Location-aware communications for 5g networks. *IEEE Signal Processing*, 2014.

[46] David Tse and Pramod Viswanath. *Fundamentals of Wireless Communication*. Cambridge University Press, 2005.

[47] Xiangyu Wang, Xuyu Wang, Shiwen Mao, Jian Zhang, Senthilkumar C. G. Periaswamy, and Justin Patton. Indoor radio map construction and localization with deep gaussian processes. *IEEE Internet of Things Journal*, 7(11): 11238–11249, 2020. Conference Name: IEEE Internet of Things Journal.

[48] Zhengdao Wang and G.B. Giannakis. A simple and general parameterization quantifying performance in fading channels. *IEEE Transactions on Communications*, 51(8):1389–1398, 2003.

[49] Michel Daoud Yacoub. The κ-μ distribution and the η-μ distribution. *IEEE Antennas and Propagation Magazine*, 49(1):68–81, 2007.

[50] Bingcheng Zhu. Asymptotic performance of composite lognormal-x fading channels. *IEEE Transactions on Communications*, 66(12):6570–6585, 2018.

3

Characterizing and Taming the Tail in URLLC

Chen-Feng Liu[1,], Yung-Lin Hsu[2], Mehdi Bennis[3], and Hung-Yu Wei[4]*

[1] *Technology Innovation Institute, Masdar City, Abu Dhabi, United Arab Emirates*
[2] *Graduate Institute of Communication Engineering, National Taiwan University, Taipei, Taiwan*
[3] *Centre for Wireless Communications, University of Oulu, Oulu, Finland*
[4] *Department of Electrical Engineering, National Taiwan University, Taipei, Taiwan*
* *Corresponding Author*

3.1 Beyond Average

In order to support mission-critical and latency-sensitive applications by 5G networks, the 3rd Generation Partnership Project (3GPP) Release 15 defines the goal of Ultra-Reliable Low Latency Communication (URLLC) as information delivery within 1 ms End-to-End (E2E) latency (measured at the ingress and egress points between the data link and network layers) with an outage probability less than 10^{-5} for a small payload size, e.g. 32 bytes [1]. To meet this target, the majority of URLLC research has been focused on the investigation of the average latency and delay outage probability performance.[1] However, for the URLLC system design, dealing with latency through the lens of average-based metrics is ill-suited [7] while analyzing reliability with threshold violation is limited. To gain insights into URLLC systems, a further examination of the statistical information and/or metrics, rooted in the tail behavior of probability distributions, is needed. These include:

- tail distribution of the delay, channel fading, or packet inter-arrival time;
- variance and higher-order statistics;
- threshold deviation with a very low occurrence probability;
- worst-case metrics;
- Age of Information (AoI).

1 This chapter has been presented in part in [2–6].

Ultra-Reliable and Low-Latency Communications (URLLC) Theory and Practice: Advances in 5G and Beyond, First Edition. Edited by Trung Q. Duong, Saeed R. Khosravirad, Changyang She, Petar Popovski, Mehdi Bennis and Tony Q.S. Quek.
© 2023 John Wiley & Sons Ltd. Published 2023 by John Wiley & Sons Ltd.

The rationale and ideas are elaborated in the remainder of Chapter 3.1. Moreover, some helpful methodologies for analyzing the aforementioned statistics and metrics will be introduced in this chapter.

3.1.1 Tail Distribution

The definition of the URLLC target in 3GPP can be expressed as a constraint on the delay outage probability, i.e.

$$\Pr(X > x_{th}) \leq \epsilon, \tag{3.1}$$

where X, x_{th}, and ϵ denote the delay, delay threshold (e.g. 1 ms), and tolerable probability (e.g. 10^{-5}), respectively. While satisfying this constraint, we only need the Complementary Cumulative Distribution Function (CCDF) value at the specific point x_{th}. However, to design an agile system in which we can target various values of x_{th}, having the CCDF profile is preferable instead of knowing one specific value. Since the target region of ϵ is very low, we can solely focus on the (right) tail distribution, i.e. $\bar{F}_X(x), \forall\, x \geq \bar{F}_X^{-1}(\delta)$ with $\delta \ll 1$. In current communication system design, the tail distributions are acquired by extrapolation methods [8]. In this regard, a large amount of data are collected and fit to the practical probability distributions which are effective in an average manner. Then the tail of the obtained distribution (by data fitting) is directly applied to the URLLC regime. Since the amount of extremity data is limited, rare events with a very low occurrence probability may not be well captured by the obtained distribution so that the acquired tail distribution is not valid for URLLC. Hence, a better modeling of the tail distributions is needed.

3.1.2 Variance and Higher-order Statistics

Variance is mathematically defined as

$$\mathrm{Var}(X) = \mathbb{E}\left[(X - \mathbb{E}[X])^2\right] \geq 0 \tag{3.2}$$

which measures how far the realizations of a random variable are spread out from its mean value $\mathbb{E}[X]$. In other words, the variance reflects the dispersion of a probability distribution. The higher the variance is, the more dispersive the distribution.

While the variance represents the second central moment, higher-order statistics include the skewness, kurtosis, fifth moment, sixth moment, and so forth. Among them, skewness is mathematically defined as

$$\mathrm{Skew}[X] = \mathbb{E}\left[\left(\frac{X - \mathbb{E}[X]}{\sqrt{\mathrm{Var}(X)}}\right)^3\right] \in \mathbb{R} \tag{3.3}$$

which is also called the third standardized moment. Skewness measures the asymmetry of a probability distribution with respect to the mean value. In this regard, generally, the skewness is negative when the left tail of the distribution is longer than the right tail of the distribution. If the right tail is longer than the left tail, the skewness is positive. Additionally, a symmetric distribution has zero skewness, but the latter does not imply the existence of the former.

Kurtosis, the fourth standardized moment, is mathematically defined as

$$\mathrm{Kurt}[X] = \mathbb{E}\left[\left(\frac{X - \mathbb{E}[X]}{\sqrt{\mathrm{Var}(X)}}\right)^4\right] \geq 0 \tag{3.4}$$

in which a higher kurtosis is incurred by the stronger extremity in the outliers of random variable realizations. The kurtosis can reflect the shape of a probability distribution. In this regard, the larger/smaller kurtosis has the fatter/thinner tail of the probability distribution.

Note that the larger variance, skewness, and kurtosis give the higher occurrence probabilities of threshold violation events, i.e. outage probabilities in the right tail, resulting in the worse reliability performance.

3.1.3 Threshold Deviation

In many wireless systems, it is implicitly assumed that when the delay outage occurs, the threshold-violating packets will be dropped. However, in some mission-critical or time-sensitive applications, dropping packets deteriorates the user's quality of experience severely. For example, in contrast with keeping the delayed image frames in virtual reality, throwing away those frames intermittently interrupts the displayed video and badly discomforts the human user. Therefore, instead of dropping the threshold-violating packets, keeping them and mitigating their effects can provide a better quality of service for some applications. To this goal, the threshold deviation

$$Y|_{X>x_{th}} = X - x_{th} > 0 \tag{3.5}$$

and its details such as the mean, variance, higher-order statistics, probability distribution function, and Cumulative Distribution Function (CDF) need to be further taken into consideration in addition to the outage probability.

3.1.4 Worst-case Metrics

The primary URLLC design considers the delay experienced in a single transmission link. Since traffic accidents are caused not only by ourselves but also others, traffic safety issues concern all vehicles, cyclists, and pedestrians on the road. Therefore, in vehicular communication, which is one enabler for the Intelligent Transportation System (ITS), the highest delay in the network is a critical concern. Depending on the considered scenarios, some other worst-case metrics such as the strongest interference, the lowest received power, and the severest fading channel among the network or over the communication timeline require dedicated attention.

The largest and smallest values among a set of numbers are, respectively, denoted by

$$Z = \max\{X_1, \cdots, X_n\}, \tag{3.6}$$

$$W = \min\{X_1, \cdots, X_n\}. \tag{3.7}$$

If $X_1, \cdots, X_n$ are independent and identically distributed (*i.i.d.*), the CDFs of Z and W are expressed as

$$F_Z(z) = F_{X_1}(z) \times \cdots \times F_{X_n}(z) = [F_X(z)]^n, \tag{3.8}$$

$$F_W(w) = 1 - \bar{F}_{X_1}(w) \times \cdots \times \bar{F}_{X_n}(w) = 1 - [\bar{F}_X(w)]^n. \tag{3.9}$$

In this case, the statistics of the worst-case metric can be found given the availability of $F_X(\cdot)$. Nevertheless, since wireless networks are persistently becoming more heterogeneous and complex, assuming the full information of $F_X(\cdot)$ becomes impractical. Due to the heterogeneity, the network entities are entangled together but not independent. Hence, as the network density is ever-increasing, analyzing the performance of the worst-case metric with a very large n is difficult.

3.1.5 Age of Information

In the Internet of Things (IoT), the performance of the provided services is affected by the freshness of the available information of the IoT devices. For example, in industrial automation, wireless sensors monitor the environmental status. After processing the received status data from sensors, the controller in control systems issues commands to the actuators. Since the factory environment varies with time, the bigger mismatch between the real status value and the controller's available one happens in the staler information. Therefore, the effectiveness of the control commands relies on the freshness of the controller's available information. In ITSs, vehicles share their driving information and safety messages via vehicular communication in order to maintain traffic safety. Since the road traffic status changes dynamically and significantly, the validity of the shared information, analogously, depends on its freshness. Briefly speaking, the fresher the available information is, the more effective the provided services in IoT systems.

The freshness of the available information can be measured by its age [9]. Specifically, the AoI is defined as the elapsed time since the data was generated at the source until the current time instant. After the new data arrives at the source, it is stored in the data buffer if the source is not granted the transmission, or the previous data has not been completely delivered to the destination. In this situation, a queuing delay is incurred for the new data. After the source has access to the wireless channel, the new data will experience a transmission delay before it is successfully delivered to the destination. Here, the transmission delay may incorporate the delays caused by the retransmissions due to data collision, received signal-to-interference-plus-noise ratio outage, and/or decoding errors in finite blocklength transmission [10] (i.e. reliability issues). Besides the queuing and transmission delays, the E2E delay includes the computation delay, processing delay, propagation delay, and other delays, depending on the considered system.

Note that the AoI, which is a function of time, is measured with respect to the latest information at the destination. Let the destination's sequentially updated data be indexed by $n \in \mathbb{Z}^+$ and denote the time instant of the nth successful data reception at the destination as $t_n > 0$. At time instant t_n, the age of the destination's newly received information is the E2E delay of the nth data, i.e. D_n. Afterwards, the information's age increases linearly with time. By denoting the AoI as a function $a(t), \forall t \geq 0$, the general form of the AoI function is given by

$$a(t) = D_n + t - t_n, \ \forall t \in [t_n, t_{n+1}), n \in \mathbb{Z}^+. \tag{3.10}$$

The details of the E2E delay depend on the transmission and scheduling schemes. Furthermore, when the new data is completely delivered to the destination, the peak AoI [11], i.e. lifetime, of the original one is expressed as

$$A_n = \lim_{\tau \to 0^+} a(t_{n+1} - \tau) = D_n + t_{n+1} - t_n. \tag{3.11}$$

In summary, having up-to-date information is important in IoT systems, and information freshness is coupled with the E2E delay. While we design URLLC for IoT systems, the impacts of reliability and latency on the AoI or other URLLC-related metrics should be considered.

3.2 Methodologies for Taming the Tail

3.2.1 Entropic Risk Measure

As pointed out in Chapter 3.1, the variance, skewness, kurtosis, and other higher-order statistics need to be investigated. Besides solely considering a single statistic, some of them can be jointly taken into account, e.g. the mean-variance trade-off [12]

$$\mathbb{E}[X] + \alpha \mathrm{Var}(X) \tag{3.12}$$

with the trade-off parameter $\alpha \in \mathbb{R}$. Furthermore, the mean, variance, and all higher-order statistics are incorporated together in the entropic risk measure [13]

$$\frac{1}{\rho} \ln\left(\mathbb{E}\left[e^{\rho X}\right]\right) \tag{3.13}$$

with the risk-sensitive parameter $\rho \in \mathbb{R}$. To manifest this advantage, we express $\ln(\mathbb{E}[e^{\rho X}])$ as its Maclaurin series expansion and have

$$\begin{aligned}(3.13) &= \mathbb{E}[X] + \frac{\rho}{2!}\mathrm{Var}(X) + \frac{\rho^2}{3!}\mathbb{E}\left[(X - \mathbb{E}[X])^3\right] \\ &\quad + \frac{\rho^3}{4!}\left\{\mathbb{E}\left[(X - \mathbb{E}[X])^4\right] - 3[Var(X)]^2\right\} + \cdots \\ &= \mathbb{E}[X] + \frac{\rho}{2!}\mathrm{Var}(X) + \frac{\rho^2}{3!}\mathrm{Skew}[X][Var(M)]^{3/2} \\ &\quad + \frac{\rho^3}{4!}(\mathrm{Kurt}[X] - 3)[Var(X)]^2 + \cdots.\end{aligned} \tag{3.14}$$

3.2.2 Extreme Value Theory

Providing a powerful and robust framework to investigate the asymptotic statistics and study the tail behavior of a distribution, extreme value theory [14, 15] is helpful to analyze the tail distribution, threshold deviation, and worst-case metrics. Next, some of the fundamental principles in extreme value theory are introduced.

3.2.2.1 Generalized Extreme Value Distributions

Theorem 3.1 (Fisher–Tippett–Gnedenko theorem): Consider n *i.i.d.* random variables $X_1, \cdots, X_n$ with the CDF $F_X(x)$ and define $Z_n = \max\{X_1, \cdots, X_n\}$. If $F_{Z_n}(z)$ converge to a non-degenerate distribution function as $n \to \infty$, Z_n can be asymptotically characterized by a Generalized Extreme Value (GEV) distribution, i.e.

$$\lim_{n\to\infty} F_{Z_n}(z) = \begin{cases} e^{-(1+\xi(z-\mu)/\sigma)^{-1/\xi}}, & \text{if } \xi \neq 0, \\ e^{-e^{-(z-\mu)/\sigma}}, & \text{if } \xi = 0. \end{cases} \tag{3.15}$$

The GEV distribution, with the support $\{z : 1+\xi(z-\mu)/\sigma \geq 0\}$, is characterized by a location parameter $\mu \in \mathbb{R}$, a scale parameter $\sigma > 0$, and a shape parameter $\xi \in \mathbb{R}$.

Among the three characteristic parameters, the shape parameter governs the GEV distribution's right tail, i.e. the tail/decay behavior of the CCDF $\bar{F}_{Z_n}(z)$. In this regard, the GEV distributions are categorized into three types according to the value of ξ.

- **Short-tailed:** when $\xi < 0$, the GEV distribution is short-tailed, with a finite right endpoint at $\bar{F}_Z^{-1}(0) = \mu - \sigma/\xi < \infty$.
- **Light-tailed:** when $\xi = 0$, the CCDF decays to zero exponentially or faster. In this case, the GEV distribution is light-tailed.
- **Heavy-tailed:** when $\xi > 0$, the GEV distribution, whose tail is more weighted than an exponential function, is heavy-tailed.

In addition, the right endpoints of the light-tailed and heavy-tailed CCDFs approach infinity, i.e. $\bar{F}_Z^{-1}(0) \to \infty$.

Theorem 3.2: Given a stationary process $\{X_1, X_2, \cdots\}$ with the same marginal CDF $F_X(x)$, we define $M_n = \max\{X_1, \cdots, X_n\}$ and assume that if $X_1, \cdots, X_n$ are independent, M_n satisfies Theorem 3.1 with the characteristic parameters μ, σ, and ξ. Thus, as $n \to \infty$, M_n can be asymptotically characterized by the GEV distribution with the location parameter $\mu' = \mu - \sigma(1-\theta^{\xi})/\xi \in \mathbb{R}$, the scale parameter $\sigma' = \sigma\theta^{\xi} > 0$, and the shape parameter $\xi \in \mathbb{R}$. Here, $\theta \in (0, 1]$ is called the extremal index which is equal to 1 if $X_1, \cdots, X_n$ are independent.

3.2.2.2 Generalized Pareto Distributions

Theorem 3.3 (Pickands–Balkema–de Haan theorem)**:** Consider a random variable X with the CDF $F_X(x)$ and suppose that $F_X(x)$ satisfies Theorem 3.1. Given a threshold value d, as $d \to F_X^{-1}(1)$, the conditional CDF of the excess value $Y|_{X>d} = X - d > 0$, i.e. $F_{Y|X>d}(y) = \Pr(X - d \leq y | X > d)$, is asymptotically converged to a generalized Pareto distribution (GPD), i.e.

$$\lim_{d \to F_X^{-1}(1)} F_{Y|X>d}(y) = \begin{cases} 1 - (1 + \xi y/\tilde{\sigma})^{-1/\xi}, & \text{if } \xi \neq 0, \\ 1 - e^{-y/\tilde{\sigma}}, & \text{if } \xi = 0. \end{cases} \tag{3.16}$$

The GPD, with the support $\{y : 1 + \xi y/\tilde{\sigma} \geq 0\}$, is characterized by a scale parameter $\tilde{\sigma} > 0$ and a shape parameter $\xi \in \mathbb{R}$.

Analogously, the GPD is

- **short-tailed** when $\xi < 0$ with the finite right endpoint $\bar{F}_Z^{-1}(0) = -\tilde{\sigma}/\xi < \infty$;

- **light-tailed** when $\xi = 0$;
- **heavy-tailed** when $\xi > 0$.

While the shape parameters of the GEV distribution (in Theorem 3.1) and GPD (in Theorem 3.3) are identical, the remaining parameters are related via

$$\tilde{\sigma} = \sigma + \xi(d - \mu). \tag{3.17}$$

Note that the worst-case metrics can be analyzed by leveraging Theorems 3.1 and 3.2 while Theorem 3.3 characterizes the statistics of threshold deviation. Moreover, given a fixed value a with $\bar{F}_X(a) = \delta \ll 1$ and by applying (3.16), the tail distribution $\bar{F}_X(x), \forall\, x \geq a$, can be approximated as follows:

$$\begin{aligned}\bar{F}_X(x) &= \Pr(X > a) \cdot \frac{\Pr(X > x)}{\Pr(X > a)}\\ &= \Pr(X > a) \cdot \frac{\Pr(X > x, X > a)}{\Pr(X > a)}\\ &= \bar{F}_X(a) \cdot \bar{F}_{X|X>a}(x)\\ &\approx \delta(1 + \xi(x - a)/\tilde{\sigma})^{-1/\xi}.\end{aligned} \tag{3.18}$$

Instead of the extrapolation methods mentioned in Chapter 3.1, fitting the extremity data to GPDs provides a more precise mathematical model for URLLC design.

3.3 Distribution Model Characterization for the Tail

The GEV distribution in Theorem 3.1 can be fully characterized by resorting to the von Mises conditions [15].

Theorem 3.4 (von Mises conditions): Given the differentiable probability density function $f_X(x)$ of X, the characteristic parameters of the GEV distribution in Theorem 3.1 can be found as per

$$\mu = \lim_{n\to\infty} F_X^{-1}\left(1 - \frac{1}{n}\right), \tag{3.19a}$$

$$\sigma = \lim_{n\to\infty} \frac{1}{n \cdot f_X\left(F_X^{-1}\left(1 - \frac{1}{n}\right)\right)}, \tag{3.19b}$$

$$\xi = -1 - \lim_{x\to F_X^{-1}(1)} \frac{(1 - F_X(x))f_X'(x)}{\left(f_X(x)\right)^2}. \tag{3.19c}$$

Once all parameters of the GEV distribution are available, the parameters of the GPD can be found as per (3.17). However, as we have commented on (3.8) and

(3.9), the closed-form expression of $f_X(x)$ may not always be available. Although the applicability of Theorem 3.4 is restricted, the characteristic parameters can be learned from statistical methods with the empirical data.

3.3.1 Statistical Learning Methods

3.3.1.1 Generalized Extreme Value Distributions

Given a set of empirical data $\{z_1, \cdots, z_M\}$ which follows the distribution $f_Z(\cdot)$, we aim to find the parameters (μ, σ, ξ) of the GEV distribution, which is the closest to the empirical distribution with respect to the Kullback–Leibler (KL) divergence. To this end, provided a sufficiently large M and $\triangle_z \to 0$, we solve the minimization problem

$$\begin{aligned}\underset{\mu,\sigma,\xi}{\arg\text{minimize}} \sum_{i=1}^{M} \frac{1}{M} \ln\left(\frac{1/M}{\int_{z_i}^{z_i+\Delta_z} f_Z(\mu,\sigma,\xi|z)\mathrm{d}z}\right) \\ = \underset{\mu,\sigma,\xi}{\arg\text{minimize}} \quad -\sum_{i=1}^{M} \frac{1}{M} \ln\left(f_Z(\mu,\sigma,\xi|z_i)\right) \quad (3.20)\end{aligned}$$

in which

$$f_Z(\mu,\sigma,\xi|z_i) = \frac{1}{\sigma} \cdot e^{-(1+\xi(z_i-\mu)/\sigma)^{-1/\xi}} \cdot (1+\xi(z_i-\mu)/\sigma)^{-1/\xi-1} \quad (3.21)$$

is the likelihood function of the GEV distribution. Since tractably deriving the closed-form solution to the KL divergence minimization problem (3.20) is not feasible, we invoke the (stochastic) gradient descent approach in which (μ, σ, ξ) are iteratively updated as per

$$\mu_{j+1} = \mu_j + \frac{\nu}{M} \sum_{i=1}^{M} \frac{\partial}{\partial \mu} \ln\left(f_Z(\mu_j,\sigma_j,\xi_j|z_i)\right), \quad (3.22a)$$

$$\sigma_{j+1} = \sigma_j + \frac{\nu}{M} \sum_{i=1}^{M} \frac{\partial}{\partial \sigma} \ln\left(f_Z(\mu_j,\sigma_j,\xi_j|z_i)\right), \quad (3.22b)$$

$$\xi_{j+1} = \xi_j + \frac{\nu}{M} \sum_{i=1}^{M} \frac{\partial}{\partial \xi} \ln\left(f_Z(\mu_j,\sigma_j,\xi_j|z_i)\right), \quad (3.22c)$$

with the learning rate ν. Here, j denotes the iteration index, and the gradients are

$$\frac{\partial}{\partial\mu}\ln\left(f_Z(\mu_j,\sigma_j,\xi_j|z_i)\right) = \frac{\xi_j+1}{\sigma_j+\xi_j(z_i-\mu_j)} - \frac{1}{\sigma_j}(1+\xi_j(z_i-\mu_j)/\sigma_j)^{-1/\xi_j-1}, \tag{3.23a}$$

$$\frac{\partial}{\partial\sigma}\ln\left(f_Z(\mu_j,\sigma_j,\xi_j|z_i)\right) = \frac{(\xi_j+1)(z_i-\mu_j)}{\sigma_j^2+\xi_j(z_i-\mu_j)\sigma_j} - \frac{(z_i-\mu_j)}{\sigma_j^2}(1+\xi_j(z_i-\mu_j)/\sigma_j)^{-1/\xi_j-1} - \frac{1}{\sigma_j}, \tag{3.23b}$$

$$\frac{\partial}{\partial\xi}\ln\left(f_Z(\mu_j,\sigma_j,\xi_j|z_i)\right) = \frac{1}{\xi_j^2}\ln(1+\xi_j(z_i-\mu_j)/\sigma_j) - \frac{(1/\xi_j+1)(z_i-\mu_j)}{\sigma_j+\xi_j(z_i-\mu_j)} - \frac{(1+\xi_j(z_i-\tilde{\mu}_j)/\sigma_j)^{-1/\xi_j}}{\xi_j^2}\left[\ln(1+\xi_j(z_i-\mu_j)/\sigma_j) - \frac{\xi_j(z_i-\mu_j)}{\sigma_j+\xi_j(z_i-\mu_j)}\right]. \tag{3.23c}$$

3.3.1.2 Generalized Pareto Distributions

Given a set of empirical exceedance data $\{y_1, \cdots, y_M\}$ which follows the distribution $f_Y(\cdot)$, we find the GPD's characteristic parameters $(\tilde{\sigma}, \xi)$ by minimizing the KL divergence between the empirical distribution and a GPD. The KL divergence minimization problem for the GPD is given by

$$\underset{\tilde{\sigma},\xi}{\arg\text{minimize}} \sum_{i=1}^{M}\frac{1}{M}\ln\left(\frac{1/M}{\int_{y_i}^{y_i+\Delta_y} f_Y(\tilde{\sigma},\xi|y)\mathrm{d}y}\right) = \underset{\tilde{\sigma},\xi}{\arg\text{minimize}} \; -\sum_{i=1}^{M}\frac{1}{M}\ln\left(f_Y(\tilde{\sigma},\xi|y_i)\right) \tag{3.24}$$

with a sufficiently large M and $\triangle_y \to 0$, in which

$$f_Y(\tilde{\sigma},\xi|y_i) = \frac{1}{\tilde{\sigma}}\left(1+\frac{\xi y_i}{\tilde{\sigma}}\right)^{-(1+1/\xi)} \tag{3.25}$$

is the likelihood function of the GPD. Analogously, due to the unavailability of the closed-form solution to problem (3.24), we utilize the (stochastic) gradient descent method and iteratively update $\tilde{\sigma}$ and ξ as per

$$\tilde{\sigma}_{j+1} = \tilde{\sigma}_j + \frac{\nu}{M}\sum_{i=1}^{M}\frac{\partial}{\partial\tilde{\sigma}}\ln\left(f_Y(\tilde{\sigma}_j,\xi_j|y_i)\right), \tag{3.26a}$$

$$\xi_{j+1} = \xi_j + \frac{\nu}{M}\sum_{i=1}^{M}\frac{\partial}{\partial\xi}\ln\left(f_Y(\tilde{\sigma}_j,\xi_j|y_i)\right). \tag{3.26b}$$

The gradients are shown as follows:

$$\frac{\partial}{\partial \tilde{\sigma}} \ln\left(f_Y(\tilde{\sigma}_j, \xi_j | y_i)\right) = \frac{\xi_j + 1}{\frac{\tilde{\sigma}_j^{\,2}}{y_i} + \tilde{\sigma}_j \xi_j} - \frac{1}{\tilde{\sigma}_j}, \tag{3.27a}$$

$$\frac{\partial}{\partial \xi} \ln\left(f_Y(\tilde{\sigma}_j, \xi_j | y_i)\right) = \frac{1}{\xi_j^2} \ln\left(1 + \frac{\xi_j y_i}{\tilde{\sigma}_j}\right) - \frac{1 + \frac{1}{\xi_j}}{\frac{\tilde{\sigma}_j}{y_i} + \xi_j}. \tag{3.27b}$$

3.3.2 Federated Learning

Note that a sufficient amount of empirical/historical/training data is required for statistical learning. However, before learning the characteristic parameters (i.e. model) of the GEV distributions and GPDs, collecting the training data may be relatively time-consuming in some situations. Let us explain as follows. Consider that the GEV distribution is utilized to characterize the maximum of a large number (e.g. n) of time-varying values. In this case, only one training data z_i is obtained after n time units, i.e. $z_i = \{x_{1i}, \cdots, x_{ni}\}$. When we leverage the GPD to approximate a tail distribution, only the exceedances among the empirical data are applicable to learning the GPD model. To be more specific, if we set $d = \bar{F}_X^{-1}(10^{-2})$ as the threshold for exceedances, there is one excess value out of, on average, 100 data points. Briefly speaking, only a small portion of the collected data is usable for model training. If the data collection and model training of the GEV distribution/GPD need to be executed in an online manner, having the precise model information to guarantee URLLC will be a critical issue.

Assume that there are multiple devices in the network which have identical objectives and learn the same model of the GEV distribution or GPD. One solution to alleviate the time overheads of data collection is that the devices upload their local training data to a remote server or data center. The server/-data center learns the model of the GEV distribution/GPD. However, uploading the local data consumes extra transmit power, causing energy concerns to battery-limited IoT devices. In addition, users might have data privacy concerns and not be willing to upload their local data. To address these concerns, we resort to Federated Learning (FL), a collaborative learning paradigm [16]. Instead of centrally training the model of the GEV distribution/GPD by collecting data from all devices, the device trains a model using the individual data experienced at its location. Then the local models are aggregated at the remote server and calculated as a global model, which will be subsequently fed back to the local devices and used as the initial point of the following model training procedure.

Let us briefly explain the principles of FL as follows. The objective function (3.20) is first rewritten as

$$\frac{1}{M}\sum_{i=1}^{M}\ln(f_Z(\mu,\sigma,\xi|z_i)) = \sum_{k=1}^{K}\frac{|\mathcal{M}_k|}{M}\sum_{i\in\mathcal{M}_k}\frac{1}{|\mathcal{M}_k|}\ln(f_Z(\mu,\sigma,\xi|z_i)) \tag{3.28}$$

in which $\cap_{k=1}^{K}\mathcal{M}_k = \emptyset$, $\cup_{k=1}^{K}\mathcal{M}_k = \{z_1,\cdots,z_M\}$, and $M = \sum_{k=1}^{K}|\mathcal{M}_k|$. The right-hand side of (3.28) can be treated as the scenario in which there are K devices in the network (i.e. the first summation), and the kth device has the set $\mathcal{M}_k$ of its local training data (i.e. the second summation). Then based on the second summation, each device uses its local data in $\mathcal{M}_k$ to find a local (GEV distribution) model, following the iterative gradient descent (3.22) in which M is replaced by $|\mathcal{M}_k|$. After a certain number of iterations (i.e. one epoch), each device k has its own parameters $(\mu_J^k, \sigma_J^k, \xi_J^k)$ which are sent to the remote server. By referring to the first summation on the right-hand side of (3.28), the remote server calculates the weighted average

$$\mu^* = \sum_{k=1}^{K}\frac{|\mathcal{M}_k|\cdot\mu_J^k}{M}, \tag{3.29a}$$

$$\sigma^* = \sum_{k=1}^{K}\frac{|\mathcal{M}_k|\cdot\sigma_J^k}{M}, \tag{3.29b}$$

$$\xi^* = \sum_{k=1}^{K}\frac{|\mathcal{M}_k|\cdot\xi_J^k}{M}, \tag{3.29c}$$

as the global model (μ^*, σ^*, ξ^*) which are subsequently fed back to all devices for their local model training in the next epoch. For the next epoch, the global model is used as the initial point in the first iteration, i.e.

$$\mu_1 = \mu^* + \frac{\nu}{|\mathcal{M}_k|}\sum_{i\in\mathcal{M}_k}\frac{\partial}{\partial\mu}\ln(f_Z(\mu^*,\sigma^*,\xi^*|z_i)), \tag{3.30a}$$

$$\sigma_1 = \sigma^* + \frac{\nu}{|\mathcal{M}_k|}\sum_{i\in\mathcal{M}_k}\frac{\partial}{\partial\sigma}\ln(f_Z(\mu^*,\sigma^*,\xi^*|z_i)), \tag{3.30b}$$

$$\xi_1 = \xi^* + \frac{\nu}{|\mathcal{M}_k|}\sum_{i\in\mathcal{M}_k}\frac{\partial}{\partial\xi}\ln(f_Z(\mu^*,\sigma^*,\xi^*|z_i)), \tag{3.30c}$$

and the steps in (3.22) are used in the remaining iterations. The interactive processes between the devices and remote server repeat until convergence is achieved. Similarly, for the model training of the GPD, the local GPD model

$(\tilde{\sigma}_J^k, \xi_J^k)$ is uploaded from each device to the remote server, and the global GPD model $(\tilde{\sigma}^*, \xi^*)$ is fed back to all devices after the weighted average.

The pros and cons of the FL approach [16–18] are, briefly, as follows. The data privacy is preserved since only the local model but not the original data is shared. The global model is obtained by incorporating the local models. Even FL can ease the bias of the local training results and improve the system performance, the accuracy of the local models for weighted average should be considered carefully. Regarding energy consumption, when the size of the local data is greater than the model size, FL can save communication energy by training at the device and uploading only the local models. However, if the model training at the device dominates in the system-wide energy consumption, FL might not be a good option for saving energy. On the contrary, if the model size is massive in comparison with the size of the local data, FL may consume more energy for model exchanging compared to the centralized training approach. Moreover, the computation capabilities of the local devices are, in general, heterogeneous and weaker than the capability of the remote server. The synchronization between devices for FL as well as the training time at the devices and server are also crucial points.

3.4 Performance Evaluation

3.4.1 Entropic Risk Measure Analysis

The entropic risk measure is analyzed in a vehicular edge computing network [2]. Therein, the Vehicular User Equipment (VUEs) distributedly decide whether their computation tasks are offloaded to the Multi-access Edge Computing (MEC) server or computed locally. The objective is to minimize the entropic risk measure of the E2E delay which consists of the communication delay and computation delay. Given the positive delay and non-negative risk-sensitive parameter $\rho \geq 0$, the focus on decreasing the mean of the delay is highlighted when ρ approaches zero as per (3.14). Oppositely, as ρ increases, the minimization focus is gradually shifted to the variance and higher-order statistics. These impacts of varying ρ on the E2E delay are shown in Figure 3.1. As introduced in Chapter 3.1, with the smaller variance, skewness, and kurtosis, the distribution is less dispersive, the right tail is shorter, and the tail is thinner, respectively. Hence, in Figure 3.1(a), the CCDF with a larger ρ has a sharper tail behavior. Further in Figure 3.1(b), the mean of the E2E delay increases with ρ, whereas the delay variance decreases.

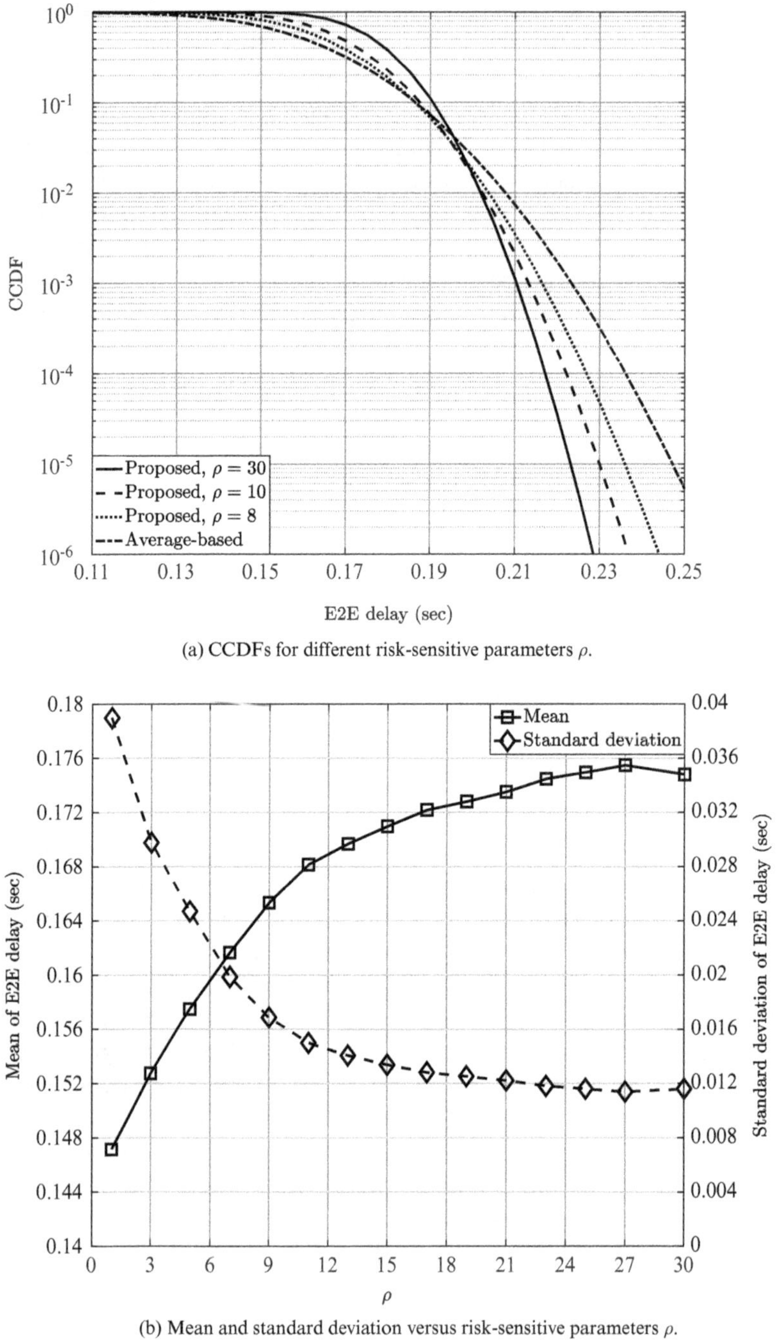

Figure 3.1 Analysis of the entropic risk measure. ©IEEE 2020. Reprinted with permission from [2].

3.4.2 Effectiveness of Applying GEV Distributions

A distributed and dynamic power allocation mechanism is proposed for Vehicular-to-Vehicular (V2V) networks [3]. While allocating its transmit power, the VUE needs to characterize the statistics of the network-wide maximal queue length among all data buffers in an online manner. To this end, the VUE asymptotically characterizes the statistics of the maximal queue length by a GEV distribution, referring to Theorem 3.1. As shown in Figure 3.2, the effectiveness of resorting to the GEV distribution is verified when there are more than 40 VUEs in the studied V2V scenario. However, the number, which is sufficiently large for applying the GEV distribution, is not deterministic but depends on the considered scenario.

3.4.3 Effectiveness of Applying GPDs

A dynamic task offloading problem is studied in a multi-user and multi-server MEC framework [4], where the threshold deviation event with respect to the queue length of the computation task buffer is taken into account. The statistics of threshold deviation are approximately characterized by a GPD. Afterwards, the constraints on the mean and variance of the approximate GPD are imposed in

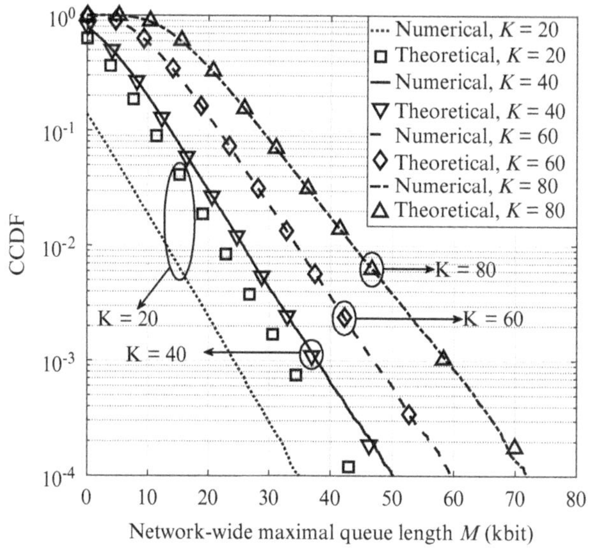

Figure 3.2 Effectiveness of characterizing the worst-case metrics by GEV distributions. ©IEEE 2018. Reprinted with permission from [3].

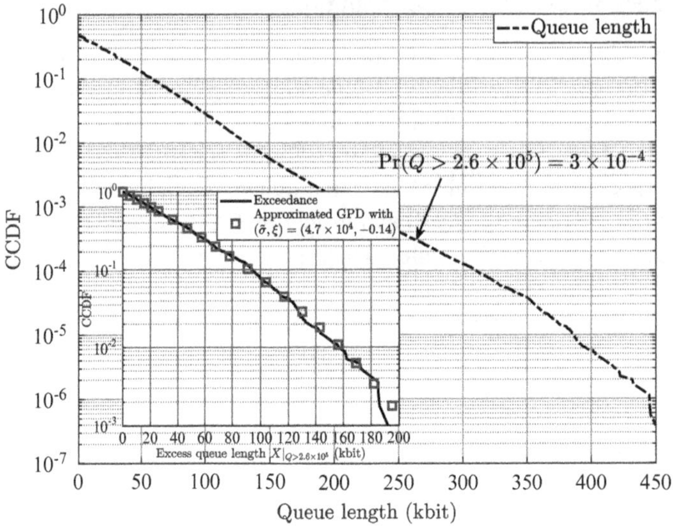

Figure 3.3 Effectiveness of characterizing threshold deviation by GPD. ©IEEE 2017. Reprinted with permission from [4].

the formulated optimization problem. In the numerical results of this MEC work [4], the effectiveness of approximating the distribution of threshold deviation as a GPD is verified and shown in Figure 3.3, where the threshold deviation event is defined as $Q > 2.6 \times 10^5$, and its occurrence probability is 3×10^{-4}. Additionally, the scale and shape parameters of the approximate GPD are 4.7×10^4 and -0.14, respectively.

3.4.4 Statistical Learning for GPD

By considering a GPD with the characteristic parameters $(\tilde{\sigma}, \xi) = (1, 0)$, the convergence speed of the iterative gradient descent approach is investigated in Figure 3.4. Therein, a higher learning rate gives a faster convergence. However, the convergence may not be achieved with a very high learning rate. The convergence speed is also affected by the initial values. The closer the initial values to the real ones, the faster the convergence is achieved.

The impacts of the time overhead and the number of training data on the estimation accuracy of the GPD model are investigated in the extension work [5] of the above MEC studies [4]. In Figure 3.5, each value of the parameter curves is estimated with the available training data obtained at that time index. Note that

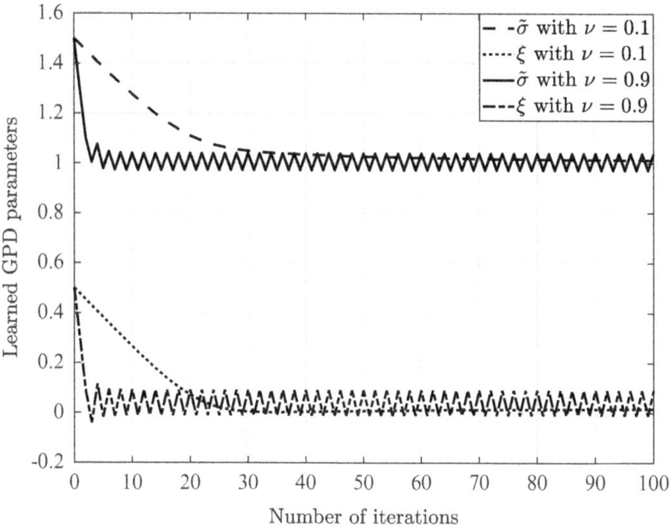

(a) Different learning rates ν with the initial GPD parameters $(\tilde{\sigma}_0, \xi_0) = (1.5, 0.5)$.

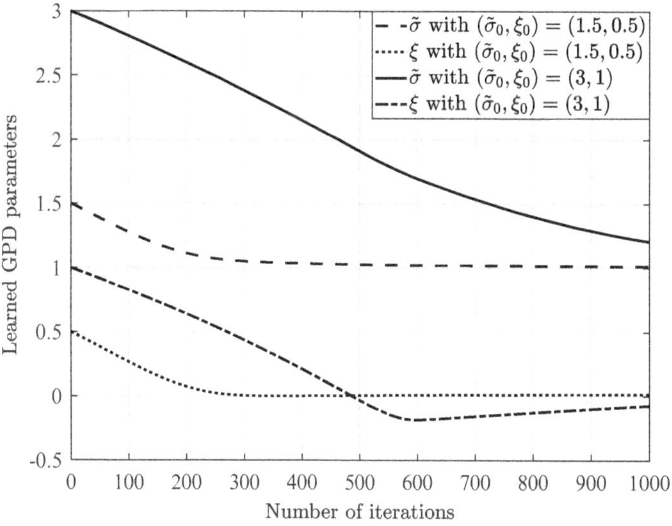

(b) Different initial GPD parameters $(\tilde{\sigma}_0, \xi_0)$ with the learning rate $\nu = 0.01$.

Figure 3.4 Learning accuracy versus number of iterations.

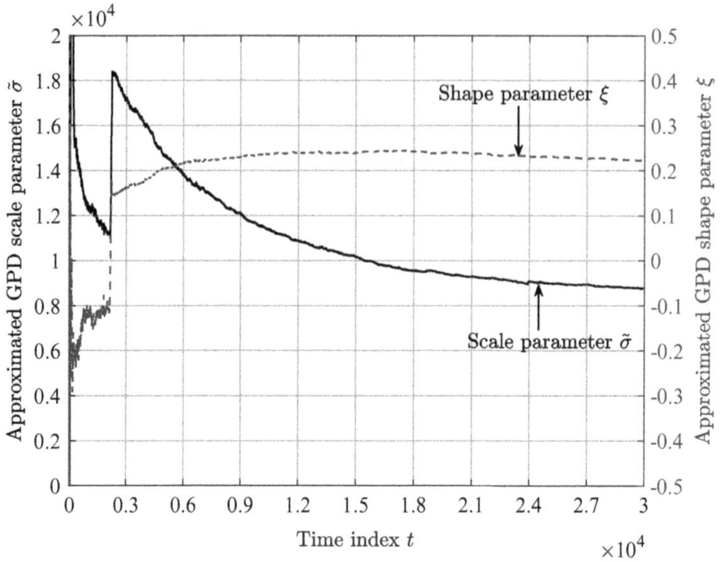

Figure 3.5 Time overheads and the number of training data for the GPD model characterization. ©IEEE 2019. Reprinted with permission from [5].

since the threshold for exceedances is set as $d = \bar{F}_X^{-1}(10^{-2})$, the number of available training data is equal to the time index divided by 100. When there is little training data, the estimation is not stable and vulnerable to the data values. The instability is shown by the ripples of the curves when $t < 0.22 \times 10^4$. The vulnerability can be seen from the big change around $t = 0.22 \times 10^4$ due to an extreme value among the training data. The convergence to the real parameters is achieved when the data number increases, but the time overhead, i.e. 3×10^4 time units, is tremendous.

3.4.5 Federated Learning for GPD

The benefit of applying FL is shown in Figure 3.6 by considering the estimation of the GPD model [6]. In this work, the GPD is used to characterize the tail distribution of a function of the packet inter-arrival time. Among the CCDFs in Figure 3.6, the GPD model $\theta_k^{corr} = (\tilde{\sigma}_k^{corr}, \xi_k^{corr})$ is obtained by using only the local data. With the aid of FL, the estimated global GPD model $\theta_{\mathrm{GL}}^{\mathrm{corr}} = (\tilde{\sigma}_{\mathrm{GL}}^{\mathrm{corr}}, \xi_{\mathrm{GL}}^{\mathrm{corr}})$ is closer to the ideal one $\theta = (\tilde{\sigma}, \xi)$. Thus, the time overhead can be reduced by incorporating the FL method. This advantage is influential for guaranteeing URLLC in an online manner.

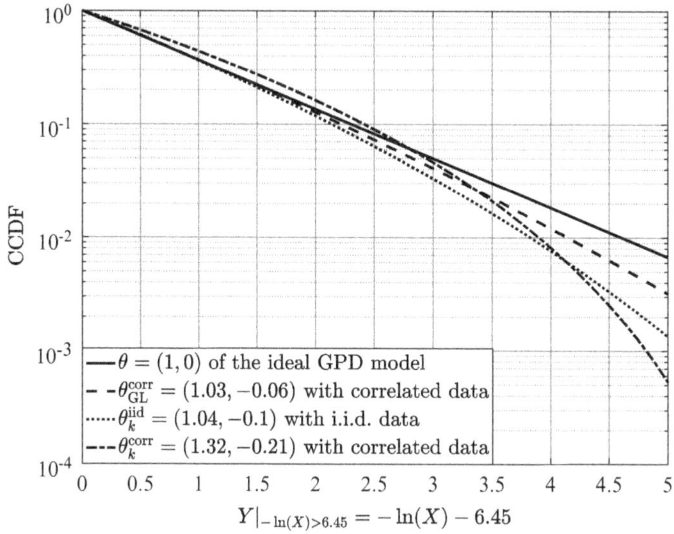

Figure 3.6 Improvement of the estimations of the GPD model by FL. ©IEEE 2021. Reprinted with permission from [6].

3.4.6 Trade-offs Between the Average AoI and Extreme AoI

The trade-offs between the average AoI and the tail behavior of the extreme AoI are jointly investigated in an industrial IoT network [19]. The objective therein is to optimize the status-updating frequency which affects the average AoI performance. A tail behavior constraint, e.g. heavy tail, light tail, or short tail, is imposed on the maximal AoI over a time period in the formulated problem. In the numerical results as shown in Figure 3.7, a short-tailed distribution of the maximal AoI is guaranteed at the expense of the higher average AoI. The heavy-tailed maximal AoI has the lower average AoI performance. In contrast with the heavy-tailed distribution, the short-tailed one has a smaller occurrence probability of extreme AoI values.

3.4.7 Impacts of Reliability on the AoI

The impacts of reliability on the AoI are investigated in a V2V network [20], where reliability is measured in terms of the decoding error probability in finite blocklength transmission [21]. As shown in Figure 3.8, when the error probability is very high, the high AoI is incurred due to multiple times of data retransmission. Regarding finite blocklength transmission, a low decoding error probability is achieved at the expense of a low transmission rate, which makes a longer delay

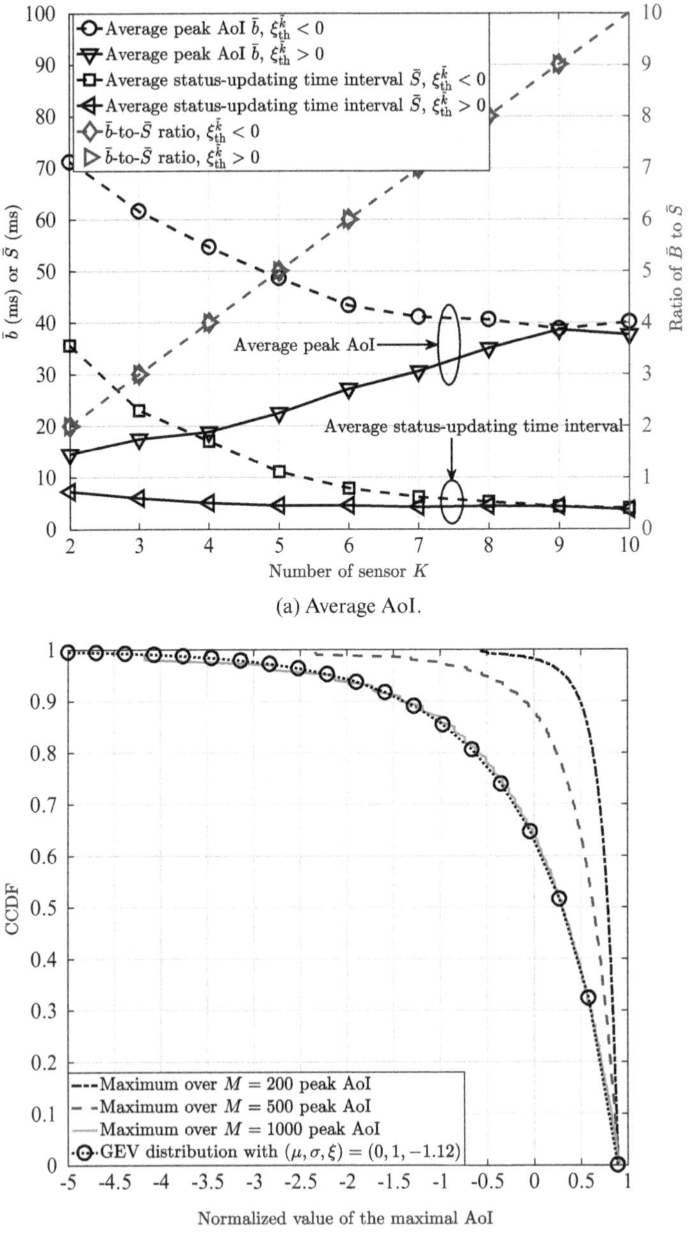

(a) Average AoI.

(b) CCDFs of the short-tailed maximal AoI.

Figure 3.7 Trade-offs between the average AoI and the tail behavior of the maximal AoI ©IEEE 2019. Reprinted with permission from [19].

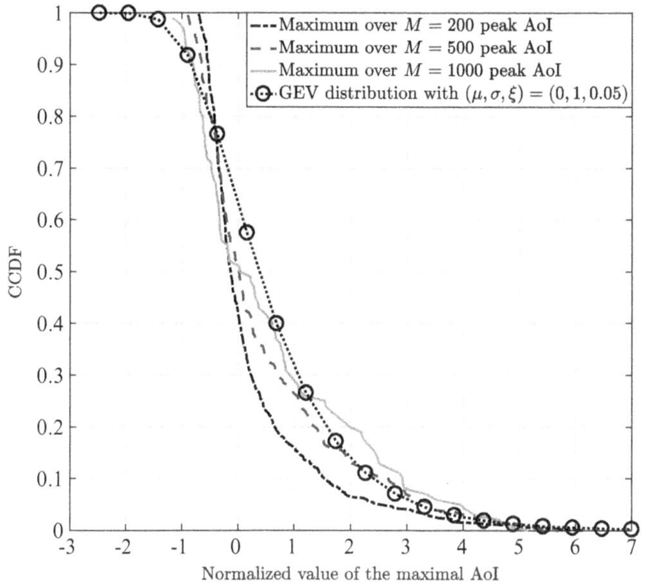

(c) CCDFs of the heavy-tailed maximal AoI.

Figure 3.7 *(Con't)*

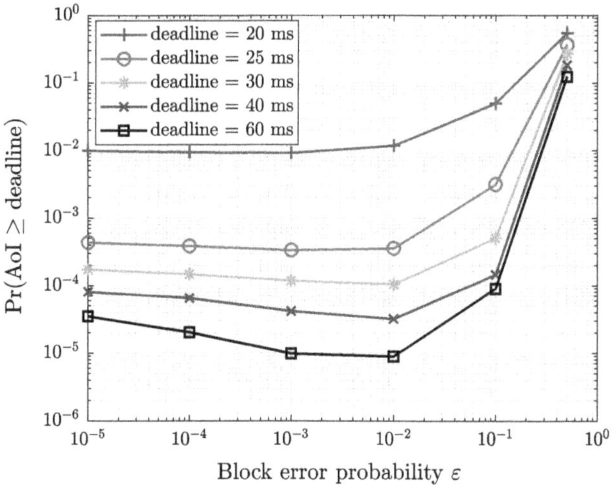

Figure 3.8 AoI versus decoding error rate. ©IEEE 2020. Reprinted with permission from [20].

for information delivery and higher AoI. Therefore, when the error probability decreases from 10^{-2} to 10^{-5}, the AoI increases with the improved reliability. In summary, the AoI is dominated by data retransmission and the lower transmission rate when the error probability is high and low, respectively.

Bibliography

[1] Z. Li, M. A. Uusitalo, H. Shariatmadari, and B. Singh, "5G URLLC: Design challenges and system concepts," in *Proc. 15th Int. Symp. Wireless Commun. Syst.*, Aug. 2018, pp. 1–6.

[2] S. Batewela, C.-F. Liu, M. Bennis, H. A. Suraweera, and C. S. Hong, "Risk-sensitive task fetching and offloading for vehicular edge computing," *IEEE Commun. Lett.*, vol. 24, no. 3, pp. 617–621, Mar. 2020.

[3] C.-F. Liu and M. Bennis, "Ultra-reliable and low-latency vehicular transmission: An extreme value theory approach," *IEEE Commun. Lett.*, vol. 22, no. 6, pp. 1292–1295, Jun. 2018.

[4] C.-F. Liu, M. Bennis, and H. V. Poor, "Latency and reliability-aware task offloading and resource allocation for mobile edge computing," in *Proc. IEEE Global Commun. Conf. Workshops*, Dec. 2017, pp. 1–7.

[5] C.-F. Liu, M. Bennis, M. Debbah, and H. V. Poor, "Dynamic task offloading and resource allocation for ultra-reliable low-latency edge computing," *IEEE Trans. Commun.*, vol. 67, no. 6, pp. 4132–4150, Jun. 2019.

[6] C.-F. Liu and M. Bennis, "Federated learning with correlated data: Taming the tail for age-optimal industrial IoT," in *Proc. 19th Int. Symp. Modeling Optim. Mobile, Ad Hoc, Wireless Netw.*, Oct. 2021, pp. 1–6.

[7] M. Bennis, M. Debbah, and H. V. Poor, "Ultrareliable and low-latency wireless communication: Tail, risk, and scale," *Proc. IEEE*, vol. 106, no. 10, pp. 1834–1853, Oct. 2018.

[8] N. Mehrnia and S. Coleri, "Wireless channel modeling based on extreme value theory for ultra-reliable communications," *IEEE Trans. Wireless Commun.*, vol. 21, no. 2, pp. 1064–1076, Feb. 2022.

[9] S. Kaul, M. Gruteser, V. Rai, and J. Kenney, "Minimizing age of information in vehicular networks," in *Proc. 8th Annu. IEEE Commun. Soc. Conf. Sensor, Mesh Ad Hoc Commun. Netw.*, Jun. 2011, pp. 350–358.

[10] G. Durisi, T. Koch, and P. Popovski, "Toward massive, ultrareliable, and low-latency wireless communication with short packets," *Proc. IEEE*, vol. 104, no. 9, pp. 1711–1726, Sep. 2016.

[11] M. Costa, M. Codreanu, and A. Ephremides, "On the age of information in status update systems with packet management," *IEEE Trans. Inf. Theory*, vol. 62, no. 4, pp. 1897–1910, Apr. 2016.

[12] A. J. Nagengast, D. A. Braun, and D. M. Wolpert, “Risk-sensitivity and the mean-variance trade-off: Decision making in sensorimotor control,” *Proc. Royal Soc. B.*, vol. 278, no. 1716, pp. 2325–2332, Aug. 2011.

[13] H. Föllmer and A. Schied, *Stochastic Finance: An Introduction in Discrete Time*, 4th ed. Berlin, Germany: Walter de Gruyter, 2016.

[14] S. Coles, *An Introduction to Statistical Modeling of Extreme Values.* London, U.K.: Springer, 2001.

[15] L. de Haan and A. F. Ferreira, *Extreme Value Theory: An Introduction.* New York, NY, USA: Springer, 2006.

[16] B. McMahan, E. Moore, D. Ramage, S. Hampson, and B. A. y Arcas, “Communication-efficient learning of deep networks from decentralized data,” in *Proc. 20th Int. Conf. Artificial Intell. Statistics*, vol. 54, Apr. 2017, pp. 1273–1282.

[17] X. Qiu, T. Parcollet, D. J. Beutel, T. Topal, A. Mathur, and N. D. Lane, “Can federated learning save the planet?” in *Proc. NeurIPS - Tackling Climate Change with Machine Learning*, Dec. 2020.

[18] P. Kairouz *et al.*, “Advances and open problems in federated learning,” *Foundations and Trends® in Machine Learning*, vol. 14, no. 1–2, pp. 1–210, 2021.

[19] C.-F. Liu and M. Bennis, “Taming the tail of maximal information age in wireless industrial networks,” *IEEE Commun. Lett.*, vol. 23, no. 12, pp. 2442–2446, Dec. 2019.

[20] M. K. Abdel-Aziz, S. Samarakoon, C.-F. Liu, M. Bennis, and W. Saad, “Optimized age of information tail for ultra-reliable low-latency communications in vehicular networks,” *IEEE Trans. Commun.*, vol. 68, no. 3, pp. 1911–1924, Mar. 2020.

[21] Y. Polyanskiy, H. V. Poor, and S. Verdú, “Channel coding rate in the finite blocklength regime,” *IEEE Trans. Inf. Theory*, vol. 56, no. 5, pp. 2307–2359, May 2010.

4

Unsupervised Deep Learning for Optimizing Wireless Systems with Instantaneous and Statistic Constraints

Chengjian Sun[1], Changyang She[2,], and Chenyang Yang[3]*

[1] *School of Electronics and Information Engineering, Beihang University, 100191, Beijing, Beijing, Xueyuan Road, China*
[2] *School of Electrical and Information Engineering, University of Sydney, 2006, NSW, Sydney, Street Name, Australia*
[*] *Corresponding Author*

4.1 Introduction

Beyond 5th generation (B5G) cellular systems are expected to support diverse Quality-of-Service (QoS) requirements of various applications, say video streaming and Ultra-Reliable and Low-Latency Communications (URLLC) [3]. To efficiently use network resources to satisfy the QoS requirements in a dynamic environment, a Base Station (BS) needs to optimize its transmission policy according to the environment parameters before they change. A typical wireless policy adapts to small-scale channels, e.g. power allocation, beamforming, and user scheduling, where the BS needs to find the optimal policy every few milliseconds, depending on the channel coherence time. If the policy cannot be obtained in closed-form, which is the case for many problems in wireless communications, numerical algorithms (e.g. interior-point method) have to be used to find the solution. This incurs high computational overheads. If the computing time used for searching for the optimal solution is longer than the channel coherence time, the obtained solution cannot guarantee the QoS with the current channel realization. This issue becomes more critical for URLLC with 1 ms End-to-End (E2E) latency [29].

To avoid executing traditional numerical algorithms repeatedly whenever the environment status changes, the novel idea of "learning to optimize" was proposed in [33], which finds a mapping from the environmental parameters to the

Ultra-Reliable and Low-Latency Communications (URLLC) Theory and Practice: Advances in 5G and Beyond, First Edition. Edited by Trung Q. Duong, Saeed R. Khosravirad, Changyang She, Petar Popovski, Mehdi Bennis and Tony Q.S. Quek.
© 2023 John Wiley & Sons Ltd. Published 2023 by John Wiley & Sons Ltd.

optimal decision by approximating the mapping with a Deep Neural Network (DNN). While promising, the method proposed therein needs a large number of labels to train the DNN, which are obtained by finding the solutions of the original optimization problem for given realizations of random environment parameters. This is possible if the original problem is a variable optimization problem, which aims to find a scalar (e.g. transmit power) or a vector (e.g. beamforming vector), but is very hard if not impossible when the original problem should be formulated as functional optimization problem [14].

Most of existing problems in wireless communications are formulated as constrained variable optimization problems, e.g. finding beamforming vectors to maximize the sum-rate or minimize the total power subjects to the maximal transmit power constraint and the QoS constraint such as the data rate or the Signal-to-Noise Ratio (SNR) exceeding a threshold. In these variable optimization problems, the objective function, constraints, and the variables to be optimized change in the same timescale. If they change in different timescales, the problems turn out to be functional optimization problems [19].

An optimization problem that finds a function to maximize or minimize an objective belongs to functional optimization problems [14], which are quite common for optimal control problems but are less familiar to the wireless community. When the timescale of the performance metric concerned by a wireless policy is much longer than the timescale of the environment parameters that the optimization depends on or the policy itself is in multi-timescale, which is often the case in cross-layer design, the policy should be found from a functional optimization problem. In general, the solutions of functional optimization problems can hardly be obtained in closed-form, which are usually obtained numerically. One of the widely applied numerical methods for solving functional optimization problems is the Finite Element Method (FEM) [37]. As a mesh-based method, FEM suffers from the curse of dimensionality, especially in the multi-user scenarios in wireless networks, where the dimension increases with the number of users. To overcome the difficulty in generating labels for training, an unsupervised learning approach has recently been proposed to solve functional optimization problems with statistical constraints [10].

4.1.1 Related Works

Two branches of deep learning techniques have been proposed to solve the wireless optimization problems: supervised deep learning [9, 15, 21, 33] and unsupervised deep learning [10, 17, 19].

The idea of "learning to optimize" was first proposed for variable optimization problems by the authors in [33], where it was proved that approximated solutions were able to be obtained from fully-connected DNNs. A deep learning framework was proposed in [21] to find the relationship between flow information and link usage by learning from past computation experience. To learn the optimal predictive resource allocation under the QoS constraint of video streaming, a DNN was designed and active learning was used to decrease the required labels in [15]. To improve the approximation accuracy, a cascaded neural network was introduced to approximate optimal resource allocation policies and deep transfer learning was applied to fine-tune the DNN in non-stationary wireless networks [9]. By training the DNNs offline, an approximated decision can be obtained with low complexity online [9, 15, 21, 33], say about 1% of the original numerical optimization [15]. Such an idea can be regarded as a kind of computing offloading over time, which shifts the computations from online to offline. However, two issues remain unresolved in this approach of supervised learning: (1) the labels may be obtained in unaffordable complexity, which is especially true for functional optimization problems, and (2) the QoS violations caused by the approximation errors are not controlled, which makes the approach inapplicable for the wireless systems requiring stringent QoS such as URLLC.

To find the optimal policy without labeled training samples, unsupervised deep learning was introduced in [8] to solve a variable optimization problem with maximal power constraint and proposed in [10, 17] to solve functional optimization problems subject to statistic constraints. In [8], the previous two issues are circumvented by using the empirical average of the objective function of the optimization problem as the cost function for training a DNN and by selecting a proper activation function in the output layer of the DNN. However, whether or not using the empirically averaged objective function as the cost function can give rise to the optimal solution was not explained, and using activation function can only satisfy simple constraints such as maximum or non-negative resource constraints. In [10], the primal-dual method was applied to maximize and minimize the Lagrangian function of the constrained optimization problem in the primal domain and the dual domain, respectively. In the primal domain, the optimal policy is approximated by a DNN. The parameters of the DNN and the Lagrangian multipliers, i.e. the optimization variable in the dual domain, are updated iteratively. The same method was introduced to solve distributed optimization in [17]. By considering the original problems in its dual domain, complex constraints can be satisfied. However, the proposed method in [10, 17] is only applicable to the functional optimization problems with statistical constraints.

When optimizing transmission policies in wireless communications, there exist both variable optimizations and functional optimizations, and both instantaneous constraints and statistic constraints exist. A theoretic interpretation of why variable optimization problems can be learned without supervision remains unclear.

While some resource constraints can be satisfied by choosing proper activation functions [10], many (especially QoS) constraints are complex and hence cannot be satisfied by activation functions. How to guarantee instantaneous constraints with unsupervised deep learning remains an open problem.

4.1.2 Motivation and Contributions

In this chapter, we investigate how to establish a unified framework for learning to optimize both variable and functional optimizations subject to both instantaneous and statistic constraints, and for solving functional optimizations subject to both types of constraints with unsupervised deep learning.

Since the QoS requirement in URLLC is complex and stringent, we take a Downlink (DL) URLLC system as an example to show how to apply the proposed framework. In particular, we formulate two resource allocation problems with delay and reliability constraints. One is variable optimization, where a BS allocates bandwidth according to large-scale channel gains. Another is a hybrid variable and functional optimization with both instantaneous and statistic constraints, where a BS jointly allocates bandwidth according to large-scale channel gains and transmit power according to small-scale channel gains. The main contributions are summarized as follows.

- We prove that the mapping from environment parameters to the solution of a constrained variable optimization problem can be formulated as a proper functional optimization problem with instantaneous constraints. Then, we develop a unified framework for using unsupervised deep learning to find the approximated optimal policy from both variable and functional optimization problems. Different from the method in [10, 17] that only considers statistic constraints, both instantaneous and statistic constraints are considered in our framework.
- We illustrate how to solve functional optimization problems with the bandwidth and power allocation problem in URLLC. We derive global optimal solution of the problem from its first-order necessary conditions in a symmetric scenario, where the QoS requirements, packet arrival rates, and large-scale channel gains of all users are identical. Simulation and numerical results show that performance achieved by the unsupervised learning is very close to that of the optimal policy, and is superior to supervised deep learning in terms of QoS guarantee and the policy approximation accuracy.

The remainder of the chapter is organized as follows. In Section 2, we show how to convert a variable optimization problem into a functional optimization problem and how to solve functional optimization problems subject to both instantaneous and statistic constraints with unsupervised deep learning. In Section 3, we consider two resource allocation problems in URLLC systems to illustrate how to use the proposed framework. Simulation and numerical results are provided in Section 4. We conclude the Chapter in Section 5.

4.2 Unsupervised Deep Learning for Variable and Functional Optimizations

In this section, we first introduce the definitions of functional and functional optimization. Then, we prove that a constrained continuous variable optimization problem can be equivalently converted into a functional optimization problem with instantaneous constraints. Next, we introduce a functional optimization problem with statistic constraint in wireless networks with an example, the classical water-filling power control. Finally, we present a framework to solve functional optimization problems with both instantaneous and statistic constraints using unsupervised deep learning.

4.2.1 Functional and Functional Optimization

According to the definition in [18], a *functional* is a function of a function, which maps a function into a scalar. Functional is a kind of function where the "variable" itself is a function. A general type of functional can be expressed as an integral of functions, say

$$\mathcal{F}[\boldsymbol{x}(\theta)] = \int_{\theta \in \mathcal{D}_\theta} F_0\left[\boldsymbol{x}(\theta); \theta\right] \mathrm{d}\theta,$$

where $\mathcal{F}[\boldsymbol{x}(\theta)]$ is a functional since its "variable" $\boldsymbol{x}(\theta)$ is a function of θ, and $F_0\left[\boldsymbol{x}(\theta); \theta\right]$ is a function of two group of variables, a specific value of θ and the corresponding value of $\boldsymbol{x}(\theta)$.

An optimization problem is a *functional optimization problem* if either the objective function or the constraint is a functional.

4.2.2 Functional Optimization Problem with Instantaneous Constraints

Consider a continuous variable optimization problem that finds a vector $\boldsymbol{x} \in \mathcal{D}_x \subseteq \mathbb{R}^{N_x}$ consisting of N_x variables to minimize objective $f(\boldsymbol{x}; \theta)$ under constraints $C_i(\boldsymbol{x}; \theta)$,

$$\min_{\boldsymbol{x}} \quad f(\boldsymbol{x}; \theta) \tag{4.1}$$

$$\text{s.t.} \quad C_i(\boldsymbol{x}; \theta) \leq 0, i = 1, \ldots, I, \tag{4.1a}$$

where $\boldsymbol{\theta} \in \mathcal{D}_{\theta} \subseteq \mathbb{R}^{N_{\theta}}$ is a vector of N_{θ} environmental parameters, which is a realization of continuous random variables and is assumed known for optimization, $\mathcal{D}_{\theta}$ is a compact set, and $f(\boldsymbol{x}; \boldsymbol{\theta})$ and $C_i(\boldsymbol{x}; \boldsymbol{\theta})$ are differentiable with respect to (w.r.t.) $\boldsymbol{x}$ and $\boldsymbol{\theta}$. Since the constraint $\boldsymbol{x} \in \mathcal{D}_x$ can be considered as a special case of (4.1a), it is not listed explicitly.

For example, $\boldsymbol{x}$ is a beamforming vector, and $\boldsymbol{\theta}$ is a channel vector that is known by estimation at the BS before optimizing beamforming. Another example is the predictive resource allocation problem in [15], where $\boldsymbol{x}$ is a matrix composed of the fractions of bandwidth assigned to several mobile users in the frames of a prediction window, and $\boldsymbol{\theta}$ is a matrix consisting of future average data rates in the frames of these users that are known by prediction before the optimization. In most of the cases, the closed-form optimal solution of problem (4.1) can hardly be obtained from the Karush–Kuhn–Tucker (KKT) conditions. As a result, one needs to search for the optimal solution numerically again whenever the value of $\boldsymbol{\theta}$ changes and hence needs to be updated by estimation or prediction. For the example of beamforming, the update duration is the channel coherence time. For the example in [15], the update duration is the duration of the prediction window, within which the large scale channel gains (and hence the average data rates) may change among frames. To facilitate practical use for wireless applications with fast changing environmental parameters, a promising approach is to find the mapping from $\boldsymbol{\theta}$ to the optimal solution, i.e. find the function $\boldsymbol{x}^*(\boldsymbol{\theta})$. This can be obtained by supervised learning, where a DNN is used to approximate $\boldsymbol{x}^*(\boldsymbol{\theta})$ and is trained with the labels generated by solving problem (4.1) for a large number of realizations of $\boldsymbol{\theta}$ [33].

To avoid generating labels by solving a variable optimization problem, one can resort to unsupervised deep learning by using the objective function of the problem as the loss function for training the DNN. Yet this is not straightforward since the objective function in (4.1) is a function of $\boldsymbol{x}$ and $\boldsymbol{\theta}$, rather than a function of *the function to be optimized, i.e.* $\boldsymbol{x}(\boldsymbol{\theta})$.

In fact, the mapping from the environmental parameters to the optimal solution of problem (4.1) can be found from a functional optimization problem. The issue then becomes: how to formulate such a functional optimization problem?

In order to find the function $\boldsymbol{x}^*(\boldsymbol{\theta})$, we construct the following functional optimization problem,

$$\min_{\boldsymbol{x}(\boldsymbol{\theta})} \quad \mathbb{E}_{\theta}\{f[\boldsymbol{x}(\boldsymbol{\theta}); \boldsymbol{\theta}]\} = \int_{\boldsymbol{\theta} \in \mathcal{D}_{\theta}} f[\boldsymbol{x}(\boldsymbol{\theta}); \boldsymbol{\theta}]\, p(\boldsymbol{\theta}) \mathrm{d}\boldsymbol{\theta} \tag{4.2}$$

$$\text{s.t.} \quad C_i[\boldsymbol{x}(\boldsymbol{\theta}); \boldsymbol{\theta}] \le 0, \ \forall \boldsymbol{\theta} \in \mathcal{D}_{\theta}, i = 1, \dots, I, \tag{4.2a}$$

where $\boldsymbol{x}(\boldsymbol{\theta})$ is optimized to minimize the expectation of the objective function in problem (4.1) over $\boldsymbol{\theta}$, and $p(\boldsymbol{\theta})$ is the Probability Density Function (PDF) of $\boldsymbol{\theta}$.

This is a functional optimization problem since the objective function in (4.2) is a function of the function $\boldsymbol{x}(\boldsymbol{\theta})$.

The constraints in problems (4.1) and (4.2) are not functionals, because the left-hand sides of them only depend on *the realizations of the random environment parameters* $\boldsymbol{\theta}$. We refer to these kinds of constraint as *instantaneous constraints.* For example, when the beamforming vector is optimized according to the channel vector known at a BS, the instantaneous data rate constraint or the transmit power constraint belongs to the instantaneous constraints.

It is worth noting that the constraints in the two problems are different. The constraints in (4.1a) needs to be ensured for a specific realization of $\boldsymbol{\theta}$. As a result, the solution of problem (4.1) is optimal only for the given realization of $\boldsymbol{\theta}$. Once the value of $\boldsymbol{\theta}$ varies, the problem needs to be solved again. However, the constraints in (4.2a) should be satisfied for all the possible realizations of $\boldsymbol{\theta} \in \mathcal{D}_{\boldsymbol{\theta}}$. Therefore, the solution of problem (4.2), denoted by $\boldsymbol{x}_{opt}(\boldsymbol{\theta})$, is optimal for arbitrary realization of $\boldsymbol{\theta}$. When the environment status changes, the optimal solution can be immediately obtained from $\boldsymbol{x}_{opt}(\boldsymbol{\theta})$, and there is no need to solve the problem again.

Proposition 1: $\boldsymbol{x}^*(\boldsymbol{\theta})$ is optimal for problem (4.2), and the value of $\boldsymbol{x}_{opt}(\boldsymbol{\theta})$ given arbitrary realization of $\boldsymbol{\theta}$ is optimal for problem (4.1) with probability one.

This proposition is proved in Appendix A.1. It indicates that a constrained continuous variable optimization problem can be equivalently converted into a functional optimization problem with instantaneous constraints in the sense of almost surely finding the same mapping.

4.2.3 Functional Optimization Problem with Statistic Constraints

If the timescale in a wireless application for measuring the system performance or the QoS is much longer than the update duration of the environment parameters for the optimization, or the timescales of the "variables" to be optimized differ, then the objective function or the constraint will be a functional. To help understand, we re-visit the classic power control problem [13], which adjusts transmit power $P(g)$ according to small-scale channel gain g. The goal is to maximize the ergodic capacity subject to the average transmit power constraint,

$$\max_{P(g)} \quad \mathbb{E}_g\left\{W\log_2\left[1+\frac{\alpha g P(g)}{N_0 W}\right]\right\} = \int_0^{\infty} W\log_2\left[1+\frac{\alpha g P(g)}{N_0 W}\right] p(g)\,\mathrm{d}g, \quad (4.3)$$

$$\text{s.t.} \quad \int_0^{\infty} P(g) p(g)\,\mathrm{d}g \le P_{\text{ave}}, \quad (4.3a)$$

where W is the bandwidth, P_{ave} is the maximal average transmit power, α is the large-scale channel gain, $p(g)$ is the PDF of the small-scale channel gain, and N_0 is the single-side noise spectral density. This is a functional optimization problem, since both the objective and the constraint are functional, which are measured in a timescale much longer than the update duration of the environment parameter for the optimization, i.e. channel coherence time.

Unlike the instantaneous constraints in (4.1a) and (4.2a), the left-hand side of constraint in (4.3a) relies on the *distribution rather than a specific realization of the environmental parameter* g. We referred to the constraints depending on the distribution of θ as **statistic constraints**.

4.2.4 A Framework of Solving Functional Optimization with Both Types of Constraints

A functional optimization problem with I instantaneous constraints and J statistic constraints can be expressed as follows,

$$\min_{\boldsymbol{x}(\theta)} \quad \mathbb{E}_{\theta}\{f\left[\boldsymbol{x}(\theta),\theta\right]\} \tag{4.4}$$

$$\text{s.t.} \quad C_i\left[\boldsymbol{x}(\theta),\theta\right] \leq 0, \ \forall \theta \in \mathcal{D}_{\theta}, i = 1, \ldots, I, \tag{4.4a}$$

$$\mathbb{E}_{\theta}\left\{C_j\left[\boldsymbol{x}(\theta),\theta\right]\right\} \leq 0, j = I+1, \ldots, I+J. \tag{4.4b}$$

To find the optimal solution of problem (4.4), we first define the Lagrangian of the problem as

$$L \triangleq \int_{\theta \in D_{\theta}} f\left[\boldsymbol{x}(\theta),\theta\right] p(\theta) \mathrm{d}\theta + \sum_{i=1}^{I} \int_{\theta \in D_{\theta}} \upsilon_i(\theta) C_i\left[\boldsymbol{x}(\theta),\theta\right] p(\theta) \mathrm{d}\theta +$$
$$\sum_{j=I+1}^{I+J} \lambda_j \int_{\theta \in D_{\theta}} C_j\left[\boldsymbol{x}(\theta),\theta\right] p(\theta) \mathrm{d}\theta,$$

where $\upsilon_i(\theta) \geq 0, \forall \theta \in \mathcal{D}_{\theta}$, and $\lambda_j \geq 0$ are the Lagrange multipliers. Noting that every Lagrange multiplier related to each instantaneous constraint in (4.4a) is a function of θ, because the constraint should be satisfied for all the possible values of θ.

4.2.4.1 Theoretical Approach

The theory of calculus of variations in [14] indicates that the optimal solution of problem (4.4) should satisfy the following conditions,

$$\frac{\delta L}{\delta \boldsymbol{x}(\boldsymbol{\theta})} = 0, \tag{4.5}$$

$$v_i(\boldsymbol{\theta}) C_i \left[\boldsymbol{x}(\boldsymbol{\theta}), \boldsymbol{\theta}\right] = 0, \ \forall \boldsymbol{\theta} \in \mathcal{D}_{\boldsymbol{\theta}}, \tag{4.6}$$

$$\lambda_j \mathbb{E}_{\boldsymbol{\theta}} \left\{C_j \left[\boldsymbol{x}(\boldsymbol{\theta}), \boldsymbol{\theta}\right]\right\} = 0, \tag{4.7}$$

$$v_i(\boldsymbol{\theta}) \geq 0, \ \forall \boldsymbol{\theta} \in \mathcal{D}_{\boldsymbol{\theta}}, \lambda_j \geq 0, \tag{4.8}$$

$$\text{(4.4a)}, \ \text{(4.4b)}, \ \forall \boldsymbol{\theta} \in \mathcal{D}_{\boldsymbol{\theta}}.$$

From the definition of the Lagrangian, (4.5) can be derived as follows,

$$\left\{ \frac{\partial f\left[\boldsymbol{x}(\boldsymbol{\theta}), \boldsymbol{\theta}\right]}{\partial \boldsymbol{x}(\boldsymbol{\theta})} + \sum_{i=1}^{I} v_i(\boldsymbol{\theta}) \frac{\partial C_i\left[\boldsymbol{x}(\boldsymbol{\theta}), \boldsymbol{\theta}\right]}{\partial \boldsymbol{x}(\boldsymbol{\theta})} + \sum_{j=I+1}^{I+J} \lambda_j \frac{\partial C_j\left[\boldsymbol{x}(\boldsymbol{\theta}), \boldsymbol{\theta}\right]}{\partial \boldsymbol{x}(\boldsymbol{\theta})} \right\} p(\boldsymbol{\theta}) = \mathbf{0}, \tag{4.9}$$

which is the simplified form of the Euler–Lagrange equation defined in [14].

These conditions are the first-order necessary conditions to achieve the optimality of functional optimization problems, like the KKT conditions of variable optimization problems [5]. However, the condition in (4.7) is an integral equation, which comes from the statistic constraints. This makes solving functional optimization problems rather challenging. In particular, even if the closed-form expression of $\boldsymbol{x}(\boldsymbol{\theta})$ can be obtained from (4.5), (4.6), (4.8), (4.4a) and (4.4b), the closed-form expressions of the Lagrange multiplier for the statistic constraints $\lambda_j, j = I+1, \ldots, I+J$ are hard to derive since integral equations are in general difficult to solve. For example, the optimal solution of problem (4.3) is the well-known "water-filling" policy, where the water level satisfying (4.7) and (4.4b) does not have a closed-form expression and has to be obtained from a binary search in [13]. On the other hand, if the closed-form expression of $\boldsymbol{x}(\boldsymbol{\theta})$ cannot be obtained, one has to employ the FEM with extremely high complexity for finding the numerical result of the integration in (4.7).

In what follows, we resort to unsupervised deep learning to solve problem (4.4).

4.2.4.2 Learning Approach

To deal with the constraints of a problem, one can solve its primal-dual problem. In particular, if problem (4.4) is convex and the Slater's condition holds, then it is equivalent to the following problem [5, 14],

$$\max_{v_i(\boldsymbol{\theta}), \lambda_j} \min_{\boldsymbol{x}(\boldsymbol{\theta})} L \tag{4.10}$$

$$\text{s.t. } (4.8)$$

The Slater's condition generally holds in optimization problems with continuous variables. However, the convexity does not hold in many cases. If the problem is non-convex, a local optimal solution of problem (4.4) can be obtained by solving problem (4.10) [22].

To find the solution with unsupervised deep learning, we approximate the two functions in L, $\boldsymbol{x}(\boldsymbol{\theta})$ and $\boldsymbol{v}(\boldsymbol{\theta}) \triangleq [v_1(\boldsymbol{\theta}), \ldots, v_I(\boldsymbol{\theta})]^{\mathrm{T}}$, by two DNNs denoted as $\mathcal{N}_x(\boldsymbol{\theta}; \boldsymbol{\omega}_x)$ and $\mathcal{N}_v(\boldsymbol{\theta}; \boldsymbol{\omega}_v)$ respectively with model parameters $\boldsymbol{\omega}_x$ and $\boldsymbol{\omega}_v$. According to the Universal Approximation Theory, a deterministic continuous function defined over a compact set can be approximated by a DNN, and the approximation can be arbitrarily accurate [16]. By replacing $\boldsymbol{x}(\boldsymbol{\theta})$ and $\boldsymbol{v}(\boldsymbol{\theta})$ with $\hat{\boldsymbol{x}}(\boldsymbol{\theta}) \triangleq \mathcal{N}_x(\boldsymbol{\theta}; \boldsymbol{\omega}_x)$ and $\hat{\boldsymbol{v}}(\boldsymbol{\theta}) \triangleq \mathcal{N}_v(\boldsymbol{\theta}; \boldsymbol{\omega}_v)$, problem (4.10) can be re-written as,

$$\max_{\omega_v, \lambda_j} \min_{\omega_x} \quad \hat{L} = \mathbb{E}_{\boldsymbol{\theta}} \left\{ f\left[\hat{\boldsymbol{x}}(\boldsymbol{\theta}), \boldsymbol{\theta}\right] + \sum_{i=1}^{I} \hat{v}_i(\boldsymbol{\theta}) C_i \left[\hat{\boldsymbol{x}}(\boldsymbol{\theta}), \boldsymbol{\theta}\right] + \sum_{j=I+1}^{I+J} \lambda_j \{C_j \left[\hat{\boldsymbol{x}}(\boldsymbol{\theta}), \boldsymbol{\theta}\right]\} \right\} \tag{4.11}$$

$$\text{s.t.} \quad \hat{v}_i(\boldsymbol{\theta}) \geq 0, \forall \boldsymbol{\theta} \in \mathcal{D}_{\boldsymbol{\theta}}, \lambda_j \geq 0. \tag{4.11a}$$

Then, the primal-dual method can be used to update primal variables $\boldsymbol{\omega}_x$, dual variables $\boldsymbol{\omega}_v$, and the Lagrange multiplier for statistical constraint λ_j iteratively to find a solution of problem (4.11). At the tth iteration, these variables can be updated by the Stochastic Gradient Descent (SGD) method and the Stochastic Gradient Ascent (SGA) method as

$$\boldsymbol{\omega}_x^{(t+1)} = \boldsymbol{\omega}_x^{(t)} - \phi_{\omega_x}(t) \nabla_{\omega_x} \hat{L}^{(t)}, \tag{4.12}$$

$$\boldsymbol{\omega}_v^{(t+1)} = \boldsymbol{\omega}_v^{(t)} + \phi_{\omega_v}(t) \nabla_{\omega_v} \hat{L}^{(t)}, \tag{4.13}$$

$$\lambda_j^{(t+1)} = \left(\lambda_j^{(t)} + \phi_{\lambda_j}(t) \frac{\partial \hat{L}^{(t)}}{\partial \lambda_j} \right)^+, \tag{4.14}$$

where $(x)^+ \triangleq \max\{x, 0\}$ ensures $\lambda_j^{(t+1)} > 0$, $\phi_{\omega_x}(t)$, $\phi_{\omega_v}(t)$ and $\phi_{\lambda_j}(t)$ are the learning rates for updating $\boldsymbol{\omega}_x$, $\boldsymbol{\omega}_v$ and λ_j, and $\nabla_{\omega_x} \hat{L}^{(t)}$ and $\nabla_{\omega_v} \hat{L}^{(t)}$ are the gradients of $\hat{L}$ w.r.t. ω_x and ω_v, respectively.[1] The method to compute the gradient and derivative is provided in Appendix A.2.

To guarantee $\hat{v}_i(\boldsymbol{\theta}) \geq 0, \forall \boldsymbol{\theta} \in \mathcal{D}_{\boldsymbol{\theta}}$, we need to choose a proper activation function in the output layer of $\mathcal{N}_v(\boldsymbol{\theta}; \boldsymbol{\omega}_v)$, e.g. `ReLU`$(x) \triangleq \max(x, 0)$ or `SoftPlus`$(x) \triangleq \ln[1 + \exp(x)]$ [12, 24].

The DNNs are trained by optimizing $\boldsymbol{\omega}_x$, $\boldsymbol{\omega}_v$, λ_j, $j = I+1, \ldots, I+J$ respectively with the SGD and SGA methods. As shown in [10], the primal-dual method converges at least to a local optimal solution of the primal-dual problem of the

1 The gradient of a scalar x w.r.t. to a vector $\boldsymbol{y}_{N_y \times 1}$ is defined as $\nabla_{\boldsymbol{y}} x \triangleq [\partial x / \partial y_1, \ldots, \partial x / \partial y_{N_y}]^{\mathrm{T}}$ and the gradient of a vector $\boldsymbol{x}_{N_x \times 1}$ w.r.t. to a vector $\boldsymbol{y}_{N_y \times 1}$ is defined as $\nabla_{\boldsymbol{y}} \boldsymbol{x} \triangleq [\nabla_{\boldsymbol{y}} x_1, \ldots, \nabla_{\boldsymbol{y}} x_{N_x}]$.

original problem. A local optimal solution is either at a stationary point of $\hat{L}$ or on the boundary of the feasible region. Thus, the following properties hold for the obtained solutions: $\nabla_{\omega_x}\hat{L} = \mathbf{0}$, $\nabla_{\omega_v}\hat{L} = \mathbf{0}$ (or $\hat{v}_i(\theta) = 0$) and $\partial\hat{L}/\partial\lambda_j = 0$ (or $\lambda_j = 0$), $j = I+1, \ldots, I+J$. These properties implicitly serve as the "supervised signal" of the DNNs. Since the DNNs are trained without labels, it belongs to unsupervised learning.

4.3 Resource Allocation with Unsupervised Learning in URLLC

In this section, we illustrate how to apply the framework presented in the previous section. To this end, we minimize the bandwidth required by satisfying the QoS of every user with URLLC by optimizing bandwidth allocation with or without dynamic power allocation.

For the policy without power allocation, the BS only allocates bandwidth among users according to their large-scale channel gains, which is formulated as a variable optimization problem. For the policy with power allocation, the BS also adjusts transmit power according to the small-scale channel gains of the users, which is formulated as a hybrid variable and functional optimization problem, where the "variables" are in two timescales.

4.3.1 System Model and QoS Constraints

4.3.1.1 System, Traffic, and Channel models

Consider a DL orthogonal frequency division multiple access system, where a BS with N_t antennas serves K single-antenna users. The maximal transmit power and the total bandwidth of the BS are denoted by P_{max} and W_{max}, respectively.

The packets for each user arrive at the buffer of the BS randomly. The inter-arrival time between packets could be shorter than the service time of each packet. Therefore, the packets may wait in the buffer of the BS. We consider a queueing model that the packets for different users wait in different queues and are served according to a first-come-first-serve order.

Time is discretized into slots, each with duration T_s. The duration for DL data transmission in one time slot is $\tau < T_s$. Since the E2E delay requirement in URLLC is typically shorter than the coherence time of the small-scale channel, the channel is quasi-static and time diversity cannot be exploited. To improve the transmission reliability within the delay bound, we consider frequency hopping, where each user is assigned with different subchannels in adjacent slots. When the frequency interval between adjacent subchannels is larger than the coherence bandwidth, the small-scale channel gains of a user among slots are mutually independent. Since the packet size u in URLLC is typically small (e.g. 20 bytes or 32 bytes [2]), the bandwidth required for transmitting each packet is less than the channel coherence bandwidth. Therefore, the small-scale channel is flat fading.

As shown in [13], the large-scale channel gain of a user varies when its moving distance is comparable to the decorrelation distance of shadowing, i.e. 50 ~ 100 m. Thus, the coherence time of the large-scale channel gain is around a few seconds, much longer than the delay bound $D_{\max}$ and the slot duration T_s (e.g. in 5G New Radio, T_s can be much shorter than 1 ms [3]). We assume that large-scale channel gains stay constant in each frame that consists of N_f time slots, and may vary in different frames. The relations among the timescales of the frames, slots, and the required delay bound are illustrated in Figure 4.1.

In URLLC, the blocklength of channel coding is short due to the short transmission duration, and hence the impact of decoding errors on reliability cannot be ignored. Since Shannon's capacity formula cannot be employed to characterize the probability of decoding errors [25], we consider the achievable rate in finite blocklength regime. In quasi-static flat fading channels, when small-scale channel gain is available at the transmitter and receiver, the achievable rate of the kth user can be accurately approximated by [36],

$$s_k \approx \frac{\tau W_k}{u \ln 2}\left[\ln\left(1+\frac{\alpha_k g_k P_k}{N_0 W_k}\right) - \sqrt{\frac{V_k}{\tau W_k}} Q_G^{-1}\left(\varepsilon_k^c\right)\right] \text{ (packets/slot)}, \tag{4.15}$$

where W_k and P_k are the bandwidth and the transmit power allocated to the kth user, respectively, ε_k^c is the decoding error probability of the kth user, α_k and g_k are the large-scale channel gain and small-scale channel gain of the kth user, respectively, $Q_G^{-1}(x)$ is the inverse of the Gaussian Q-function, and V_k is the channel dispersion given by $V_k = 1 - \frac{1}{\left[1+\frac{\alpha_k g_k P_k}{N_0 W_k}\right]^2}$ [36].

Although the achievable rate is in closed-form, it is still too complicated to obtain graceful results. As shown in [25], if the SNR $\frac{\alpha_k g_k P_k}{N_0 W_k} \geq 5$ dB, $V_k \approx 1$ is accurate. Since high SNR is required to ensure ultra-high reliability and ultra-low latency, such approximation is reasonable. Even when the SNR is not high, we can obtain a lower bound of the achievable rate by substituting $V_k \approx 1$ into s_k. Then, the required ε^c can be satisfied if the lower bound of (4.15) is used to characterize the achievable rate.

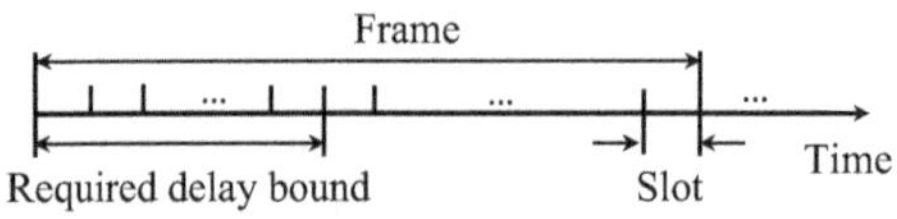

Figure 4.1 Large-scale channels change among frames, and small-scale channels change among time slots due to frequency hopping.

4.3.1.2 Reliability and Delay Constraints

The QoS requirement of each user can be characterized by an E2E delay bound $D_{\max}$ for each packet and the overall packet loss probability $\varepsilon_{\max}$.

The uplink transmission delay, backhaul delay and processing delay have been studied in [28], [11] and [23], respectively, and can be subtracted from the E2E delay. In this paper, $D_{\max}$ is the DL delay, which consists of the queueing delay (denoted as D_k^{q} for the kth user), transmission delay D^{t} (which equals to T_s, including the data transmission time τ and the channel training time) and decoding delay D^{c}.

D^{t} and D^{c} are constant values depending on the standardization and hardware [7]. Due to the random packet arrival, D_k^{q} is random. To ensure the delay requirement, D_k^{q} should be bounded by $D_{\max}^{\mathrm{q}} \triangleq D_{\max} - D^{\mathrm{t}} - D^{\mathrm{c}}$ with a very low probability, because a packet will be useless if the queueing delay of the packet exceeds $D_{\max}^{\mathrm{q}}$.

Denote $\varepsilon_k^{\mathrm{q}} \triangleq \Pr\{D_k^{\mathrm{q}} > D_{\max}^{\mathrm{q}}\}$ as the queueing delay violation probability. Then, the overall reliability requirement can be characterized by $1 - (1 - \varepsilon_k^{\mathrm{c}})(1 - \varepsilon_k^{\mathrm{q}}) \approx \varepsilon_k^{\mathrm{c}} + \varepsilon_k^{\mathrm{q}} \leq \varepsilon_{\max}$. This approximation is very accurate, because the values of ε^{c} and ε^{q} are very small in URLLC.

Effective bandwidth and effective capacity have been widely used to analyze the tail probability of queueing delay, i.e. $D_{\max}^{\mathrm{q}}$ is large or $\varepsilon_k^{\mathrm{q}}$ is extremely small [6, 35]. As analyzed in [27], if the slot duration is much shorter than the delay bound, which is true in URLLC, effective bandwidth can be used to analyze the queueing delay at the BS for Poisson, interrupted Poisson, and switched Poisson arrival processes.

We take the Poisson arrival process with the average packet arrival rate a_k packets/slot as an example, whose effective bandwidth can be expressed as [27]

$$B_k^{\mathrm{E}} = \frac{\ln(\varepsilon_{\max}/2)}{D_{\max}^{\mathrm{q}} \ln\left[1 - \frac{\ln(\varepsilon_{\max}/2)}{a_k D_{\max}^{\mathrm{q}}}\right]} \text{ (packets/slot).} \tag{4.16}$$

If the constant packet service rate (i.e. the achievable rate) of the kth user is no less than B_k^{E}, then we have $\Pr\{D_k^{\mathrm{q}} > D_{\max}^{\mathrm{q}}\} \leq \exp\{-\vartheta_k B_k^{\mathrm{E}} D_{\max}^{\mathrm{q}}\}$, where ϑ_k is the QoS component, which reflects the decay rate of the tail probability of the queueing delay. By setting the upper bound in the inequity equals to $\varepsilon_{\max}/2$, we can obtain

$$\vartheta_k = \ln\left[1 - \frac{\ln(\varepsilon_{\max}/2)}{a_k D_{\max}^{\mathrm{q}}}\right]. \tag{4.17}$$

Since the small-scale channel gains of a user are independent among slots owing to frequency hopping, the effective capacity of the kth user can be expressed as [34]

$$C_k^{\mathrm{E}} = -\frac{1}{\vartheta_k} \ln \mathbb{E}_{g_k} \left\{e^{-\vartheta_k s_k}\right\} \text{ (packets/slot).} \tag{4.18}$$

When both the packet arrival process and the packet service process are stochastic, $D^{\mathrm{q}}_{\max}$ and $\varepsilon^{\mathrm{q}}_k$ can be satisfied if [20]

$$C^{\mathrm{E}}_k \geq B^{\mathrm{E}}_k. \tag{4.19}$$

To simplify the optimization problem, we set $\varepsilon^{\mathrm{c}}_k = \varepsilon^{\mathrm{q}}_k = \varepsilon_{\max}/2$. The results in [27, 28] show that the optimal values of $\varepsilon^{\mathrm{c}}_k$ and $\varepsilon^{\mathrm{q}}_k$ are of the same order of magnitude, and the simplification will only lead to a negligible performance loss. Then, the QoS of each user, characterized by $D_{\max}$ and $\varepsilon_{\max}$, can be satisfied if (4.19) holds after substituting the expression of s_k in (4.15) into (4.18). Such a QoS constraint is complicated and may not be expressed in closed-form.

4.3.2 Bandwidth Allocation: A Variable Optimization Problem

In this subsection, we assume that the transmit power does not change according to small-scale channel gains, which is reasonable in practical cellular networks where modulation and coding schemes are adjusted according to channel realizations with fixed power allocation [1]. We optimize the bandwidth allocation policy according to the large-scale channel gains of users. Hence, the environmental parameters can be expressed as $\boldsymbol{\theta} = \boldsymbol{\alpha} \triangleq [\alpha_1, \cdots, \alpha_K]^{\mathrm{T}}$.

4.3.2.1 Problem Formulation

In particular, assume that $P_0 = P_{\max}/W_{\max}$. Then, by substituting $\varepsilon^{\mathrm{c}}_k = \varepsilon_{\max}/2$ into (4.15), the achievable rate of the kth user can be re-written as follows,

$$s_k = \frac{\tau W_k}{u \ln 2}\left[\ln\left(1 + \frac{\alpha_k g_k}{N_0} P_0\right) - \frac{Q_{\mathrm{G}}^{-1}(\varepsilon_{\max}/2)}{\sqrt{\tau W_k}}\right]. \tag{4.20}$$

The bandwidth allocation problem can be formulated as a variable optimization problem that minimizes the total bandwidth required to ensure the QoS of every user, i.e.

$$\min_{W_k, k=1,\dots,K} \quad \sum_{k=1}^{K} W_k \tag{4.21}$$

$$\text{s.t.} \quad \mathbb{E}_{g_k}\left\{e^{-\vartheta_k s_k}\right\} - e^{-\vartheta_k B^{\mathrm{E}}_k} \leq 0, k = 1, \dots, K \tag{4.21a}$$

$$\sum_{k=1}^{K} W_k \leq W_{\max}, \tag{4.21b}$$

$$W_k \geq 0, k = 1, \dots, K,$$

where (4.21a) is obtained by substituting (4.18) into (4.19), s_k is given by (4.20), and $W_{\max}$ is the maximal total bandwidth.

Since the left-hand side of the constraint in (4.21b) is the same as the objective function, we can remove it when solving problem (4.21). If the minimal bandwidth required to guarantee the QoS requirement of every user exceeds $W_{\max}$, problem (4.21) will be infeasible. After removing the constraint in (4.21b), the bandwidth allocation of every user is mutually independent of each other. Thus, problem (4.21) can be equivalently decomposed into K single-user problems,

$$\min_{W_k} \quad W_k \tag{4.22}$$

$$\text{s.t.} \quad (4.21\text{a}), W_k \geq 0.$$

In the remainder of this subsection, the index k is omitted for notational simplicity.

4.3.2.2 Optimizing W from the Variable Optimization Problem

To provide a baseline for the unsupervised deep learning method, we first find the optimal solution of problem in (4.22) for any given realizations of the environmental parameters.

Since s_k in (4.20) increases with W, the left-hand side of (4.21a) decreases with W, and the minimal bandwidth is obtained when the equality in (4.21a) holds. If effective capacity can be derived as a closed-form expression, say in large-scale antenna systems [26], then we can use binary search to find the minimal bandwidth. In general wireless systems, the effective capacity does not have closed-form expression, and hence (4.21a) cannot be expressed in closed form. To find the optimal bandwidth allocated to each user, one can use stochastic optimization through the following iterations,

$$W^{(t+1)} = \left[W^{(t)} + \phi(t)\left(e^{-\vartheta s^{(t)}} - e^{-\vartheta B^{\mathrm{E}}}\right)\right]^{+}, \tag{4.23}$$

where $\phi(t) > 0$ is the learning rate, $s^{(t)}$ is the achievable rate computed from (4.20) given the realization of g in the tth iteration, and one realization of g can be obtained in each slot. With $\phi(t) \sim \mathcal{O}\left(\frac{1}{t}\right)$, $\{W^{(t)}\}$ converges to the unique optimal bandwidth [4] thanks to the monotonicity of the function of the left-hand side of (4.21a).

4.3.2.3 Optimizing $W(\alpha)$ with Unsupervised Deep Learning

For the sake of learning to optimize problem (4.22), we first formulate a functional optimization problem of finding the mapping from α to the optimal solution of problem (4.22) as follows,

$$\min_{W(\alpha)} \quad \mathbb{E}_\alpha \{W(\alpha)\} \tag{4.24}$$

$$\text{s.t.} \quad \mathbb{E}_g \{e^{-\vartheta s[W(\alpha);\alpha]}\} - e^{-\vartheta B^{\mathrm{E}}} \leq 0, \tag{4.24a}$$

$$W(\alpha) \geq 0,$$

where $s[W(\alpha); \alpha] = \frac{\tau W(\alpha)}{u \ln 2}\left[\ln\left(1 + \frac{\alpha g}{N_0} P_0\right) - \frac{Q_{\mathrm{G}}^{-1}(\varepsilon_{\max}/2)}{\sqrt{\tau W(\alpha)}}\right]$ is the re-written expression of (4.20), and (4.24a) is an instantaneous constraint although it consists of expectation, because the expectation is taken over small-scale channel gains for a given realization of the environment parameter α. The constraint in (4.24a) is non-convex, hence problem (4.24) is non-convex. According to the discussion in Section 4.2.3, a local optimal solution of problem (4.24) can be found by solving its primal-dual problem,

$$\max_{\upsilon(\alpha)} \min_{W(\alpha)} L_1 \triangleq \mathbb{E}_\alpha \left\{W(\alpha) + \upsilon(\alpha)\left(\mathbb{E}_g\{e^{-\vartheta s[W(\alpha);\alpha]}\} - e^{-\vartheta B^{\mathrm{E}}}\right)\right\} \tag{4.25}$$

$$\text{s.t. } W(\alpha) \geq 0, \; \upsilon(\alpha) > 0, \forall \alpha > 0,$$

where $\upsilon(\alpha)$ is the Lagrange multiplier function. The constraint $W(\alpha) \geq 0$ and the corresponding Lagrange multiplier are not included in L_1, because the optimal bandwidth is always positive and the corresponding Lagrange multiplier is always zero.

To apply the framework in Section 4.2.4 to solve problem (4.25), we approximate the functions $W(\alpha)$ and $\upsilon(\alpha)$ by two DNNs, denoted as $\hat{W} \triangleq \mathcal{N}_W(\alpha; \omega_W)$ and $\hat{\upsilon} \triangleq \mathcal{N}_\upsilon(\alpha; \omega_\upsilon)$, respectively. By using appropriate activation function in the output layers of both DNNs, $\hat{W}$ and $\hat{\upsilon}$ are positive. The model parameters of the DNNs, ω_W and ω_υ, can be obtained iteratively as follows,

$$\begin{aligned}\omega_W^{(t+1)} &= \omega_W^{(t)} - \phi_{\omega_W}(t) \nabla_{\omega_W} \hat{L}_1^{(t)} \\ &= \omega_W^{(t)} - \frac{\phi_{\omega_W}(t)}{N_{\mathrm{b}}} \sum_{n=1}^{N_{\mathrm{b}}} \left[\nabla_{\omega_W} \mathcal{N}_W\left(\alpha^{(t,n)}; \omega_W^{(t)}\right) \frac{\mathrm{d}\hat{L}_1^{(t)}}{\mathrm{d}\hat{W}^{(t,n)}}\right],\end{aligned} \tag{4.26}$$

$$\begin{aligned}\omega_\upsilon^{(t+1)} &= \omega_\upsilon^{(t)} + \phi_{\omega_\upsilon}(t) \nabla_{\omega_\upsilon} \hat{L}_1^{(t)} \\ &= \omega_\upsilon^{(t)} + \frac{\phi_{\omega_\upsilon}(t)}{N_{\mathrm{b}}} \sum_{n=1}^{N_{\mathrm{b}}} \left[\nabla_{\omega_\upsilon} \mathcal{N}_\upsilon\left(\alpha^{(t,n)}; \omega_W^{(t)}\right) \frac{\mathrm{d}\hat{L}_1^{(t)}}{\mathrm{d}\hat{\upsilon}^{(t,n)}}\right],\end{aligned} \tag{4.27}$$

where $\hat{L}_1^{(t)} \triangleq \frac{1}{N_{\mathrm{b}}} \sum_{n=1}^{N_{\mathrm{b}}} \left[\hat{W}^{(t,n)} + \hat{\upsilon}^{(t,n)}\left(e^{-\vartheta \hat{s}^{(t,n)}} - e^{-\vartheta B^{\mathrm{E}}}\right)\right]$ is the estimated objective function in (4.25) with N_{b} realizations of large-scale channel gains while $\alpha^{(t,n)}$ and $\hat{s}^{(t,n)}$ are respectively the nth realizations of the large-scale channel gain and the achievable rate in the tth iteration, $\hat{W}^{(t,n)} \triangleq \mathcal{N}_W\left(\alpha^{(t,n)};\omega_W^{(t)}\right)$ and $\hat{\upsilon}^{(t,n)} \triangleq \mathcal{N}_\upsilon\left(\alpha^{(t,n)};\omega_\upsilon^{(t)}\right)$. In (4.26) and (4.26), the derivative of $\hat{L}_1^{(t)}$ w.r.t. $\hat{W}^{(t,n)}$ and $\hat{\upsilon}^{(t,n)}$ can be derived as follows,

$$\frac{\mathrm{d}\hat{L}_1^{(t)}}{\mathrm{d}\hat{W}^{(t,n)}} = 1 - \hat{\upsilon}^{(t,n)}\vartheta\frac{\partial \hat{s}^{(t,n)}}{\partial \hat{W}^{(t,n)}}e^{-\vartheta \hat{s}^{(t,n)}}, \quad \frac{\mathrm{d}\hat{L}_1^{(t)}}{\mathrm{d}\hat{\upsilon}^{(t,n)}} = e^{-\vartheta \hat{s}^{(t,n)}} - e^{-\vartheta B^{\mathrm{E}}},$$

where the values of B^{E} and ϑ are computed according to (4.16) and (4.17), respectively, and

$$\frac{\partial \hat{s}^{(t,n)}}{\partial \hat{W}^{(t,n)}} = \frac{1}{u\ln 2}\left[\tau\ln\left(1+\frac{\alpha^{(t,n)}gP_0}{N_0}\right) - \frac{Q_{\mathrm{G}}^{-1}\left(\varepsilon_{\max}/2\right)}{2}\sqrt{\frac{\tau}{\hat{W}^{(t,n)}}}\right].$$

The gradient matrices $\nabla_{\omega_W}\mathcal{N}_W\left(\alpha^{(t,n)};\omega_W^{(t)}\right)$ and $\nabla_{\omega_\upsilon}\mathcal{N}_\upsilon\left(\alpha^{(t,n)};\omega_\upsilon^{(t)}\right)$ can be computed by backward propagation.

After the convergence of iterations, we can obtain a well-trained DNN $\mathcal{N}_W\left(\alpha;\omega_W\right)$, which can approximate the optimal function of $W(\alpha)$. Then, the BS only needs to compute the bandwidth allocated to each user from $\mathcal{N}_W\left(\alpha;\omega_W\right)$ after obtaining the large-scale channel gain of each user at the beginning of each frame.

4.3.3 Bandwidth and Power Allocation: A Hybrid Variable and Functional Optimization Problem

In this subsection, we illustrate how to solve a functional optimization problem subject to both instantaneous and statistic constraints. Although the BSs in the fourth generation cellular systems do not adjust transmit power according to small-scale channel, the total bandwidth required by URLLC can be further reduced with dynamic power allocation. We optimize bandwidth allocation according to the large-scale channel gains of multiple users (i.e. $\boldsymbol{\theta} = \boldsymbol{\alpha}$) and power allocation according to their small-scale channel gains (i.e. $\boldsymbol{\theta} = \boldsymbol{g} \triangleq [g_1, \cdots, g_K]^{\mathrm{T}}$). Hence, the jointly optimized policy operates in two timescales.

4.3.3.1 Problem Formulation

To reflect the impact of the two-timescale resource allocation, we re-write the achievable rate of the kth user in (4.15) to satisfy $\varepsilon_k^{\mathrm{c}} = \varepsilon_{\max}/2$ as,

$$s_k[W_k, P_k(\boldsymbol{g}); g_k] = \frac{\tau W_k}{u\ln 2}\left[\ln\left(1+\frac{\alpha_k g_k P_k(\boldsymbol{g})}{N_0 W_k}\right) - \frac{Q_{\mathrm{G}}^{-1}(\varepsilon_{\max}/2)}{\sqrt{\tau W_k}}\right]. \tag{4.28}$$

The problem of joint bandwidth and power allocation that minimizes the total bandwidth required to ensure the QoS under the constraint of maximal power $P_{\max}$ can be formulated as,

$$\min_{W_k, P_k(\boldsymbol{g})} \quad \sum_{k=1}^{K} W_k \tag{4.29}$$

$$\text{s.t.} \quad \mathbb{E}_{\boldsymbol{g}}\left\{e^{-\vartheta_k s_k [W_k, P_k(\boldsymbol{g}); g_k]}\right\} - e^{-\vartheta_k B_k^{\mathrm{E}}} \leq 0, \; k = 1, \cdots, K \tag{4.29a}$$

$$\sum_{k=1}^{K} P_k(\boldsymbol{g}) \leq P_{\max}, \tag{4.29b}$$

$$W_k \geq 0, P_k(\boldsymbol{g}) \geq 0, \; k = 1, \cdots, K. \tag{4.29c}$$

The left-hand side of (4.29a) is a function of $P_k(\boldsymbol{g})$, which measures the QoS requirement in each frame and depends on the distribution of environmental parameters $\boldsymbol{g}$. Thus, (4.29a) are the statistic constraints for the functional optimization. The constraints in (4.29b) and (4.29c) only depend on specific realizations of environmental parameters, and hence are instantaneous constraints. The total bandwidth constraint is removed as explained in the previous subsection. If the minimal total bandwidth is higher than $W_{\max}$, then the problem is infeasible.

This is a generic functional optimization problem, including both functional optimization for $P_k(\boldsymbol{g})$ and variable optimization for W_k. In what follows, we apply the proposed framework to solve this hybrid variable and functional optimization problem.

4.3.3.2 Optimizing W_k and $P_k(g)$ from Necessary Conditions

To provide a baseline for the learning-based solution, we first derive the optimal solution of problem (4.29) from the necessary conditions. To simplify the notation, in the following we again use s_k to denote $s_k[W_k, P_k(\boldsymbol{g}); g_k]$ in (4.28).

The Lagrangian of problem (4.29) can be expressed as follows,

$$L_2 \triangleq \sum_{k=1}^{K} W_k + \sum_{k=1}^{K} \lambda_k \left(\mathbb{E}_{\boldsymbol{g}}\{e^{-\vartheta_k s_k}\} - e^{-\vartheta_k B_k^{\mathrm{E}}}\right) + \int_{\mathbb{R}_+^K} \left[h(\boldsymbol{g}) \left(\sum_{k=1}^{K} P_k(\boldsymbol{g}) - P_{\max} \right) - \sum_{k=1}^{K} \upsilon_k(\boldsymbol{g}) P_k(\boldsymbol{g}) \right] p(\boldsymbol{g}) \mathrm{d}\boldsymbol{g}$$

where λ_k, $h(\boldsymbol{g})$ and $\upsilon_k(\boldsymbol{g})$ are the Lagrange multipliers. Similar to the Lagrangian in (4.25), the constraint $W_k \geq 0$ and the corresponding Lagrange multiplier are omitted in L_2.

Then, the optimal solution of problem (4.29) should satisfy its first-order necessary conditions, which can be derived as [14],

$$\frac{\partial L_2}{\partial P_k(\boldsymbol{g})} = \left[h(\boldsymbol{g}) - \upsilon_k(\boldsymbol{g}) - \lambda_k \vartheta_k \frac{\partial s_k}{\partial P_k(\boldsymbol{g})} e^{-\vartheta s_k}\right] p(\boldsymbol{g}) = 0, \tag{4.30}$$

$$\frac{\partial L_2}{\partial W_k} = 1 - \lambda_k \vartheta_k \mathbb{E}_{\boldsymbol{g}} \left\{ \frac{\partial s_k}{\partial W_k} e^{-\vartheta s_k} \right\} = 0, \tag{4.31}$$

$$\lambda_k \left(\mathbb{E}_{\boldsymbol{g}} \{ e^{-\vartheta_k s_k} \} - e^{-\vartheta_k B_k^{\mathrm{E}}} \right) = 0, \tag{4.32}$$

$$h(\boldsymbol{g}) \left(\sum_{k=1}^{K} P_k(\boldsymbol{g}) - P_{\max} \right) = 0, \forall \boldsymbol{g} \in \mathbb{R}_+^K, \tag{4.33}$$

$$\upsilon_k(\boldsymbol{g}) P_k(\boldsymbol{g}) = 0, \forall \boldsymbol{g} \in \mathbb{R}_+^K, \tag{4.34}$$

$$W_k \geq 0, \lambda_k \geq 0, P_k(\boldsymbol{g}) \geq 0, h(\boldsymbol{g}) \geq 0, \upsilon_k(\boldsymbol{g}) \geq 0, \forall \boldsymbol{g} \in \mathbb{R}_+^K, \tag{4.35}$$

(4.29a) and (4.29b),

where $p(\boldsymbol{g})$ is the joint PDF of $\boldsymbol{g}$.

Optimal Power Allocation: Since $p(\boldsymbol{g}) > 0$, from (4.30) and (4.28) we have

$$\begin{aligned} h(\boldsymbol{g}) &= \lambda_k \vartheta_k \frac{\partial s_k}{\partial P_k(\boldsymbol{g})} e^{-\vartheta s_k} + \upsilon_k(\boldsymbol{g}) \\ &= \lambda_k \vartheta_k \frac{\tau W_k}{u \ln 2} \frac{\alpha_k g_k}{N_0 W_k} \frac{1}{(1+\gamma_k)} e^{-\vartheta s_k} + \upsilon_k(\boldsymbol{g}) \\ &= \frac{\lambda_k \vartheta_k \alpha_k g_k \tau}{N_0 u \ln 2 (1+\gamma_k)} (1+\gamma_k)^{-\frac{\vartheta_k \tau W_k}{u \ln 2}} e^{\frac{\vartheta_k \sqrt{\tau W_k} Q_G^{-1}(\varepsilon_{\max}/2)}{u \ln 2}} + \upsilon_k(\boldsymbol{g}) \\ &= \frac{\beta_k g_k}{(1+\gamma_k)^{\frac{1}{\eta_k}}} + \upsilon_k(\boldsymbol{g}), \end{aligned} \tag{4.36}$$

where $\gamma_k \triangleq \frac{\alpha_k g_k P_k(\boldsymbol{g})}{N_0 W_k}$ is the SNR of the kth user, $\beta_k \triangleq \frac{\lambda_k \vartheta_k \alpha_k \tau}{N_0 u \ln 2} e^{\frac{\vartheta_k \sqrt{\tau W_k} Q_G^{-1}(\varepsilon_{\max}/2)}{u \ln 2}}$, and $\eta_k \triangleq 1/\left(1 + \frac{\vartheta_k \tau W_k}{u \ln 2}\right)$.

From (4.36), we can see that if $h(\boldsymbol{g}) > \beta_k g_k \geq \beta_k g_k/(1+\gamma_k)^{1/\eta_k}$, then $\upsilon_k(\boldsymbol{g}) > 0$. To satisfy the condition in (4.34), we have $P_k(\boldsymbol{g}) = 0$. In the case that $h(\boldsymbol{g}) < \beta_k g_k$, if $P_k(\boldsymbol{g}) = 0$, then $\upsilon_k(\boldsymbol{g})$ will be negative, which contradicts the constraint $\upsilon_k(\boldsymbol{g}) \geq 0$ in (4.35). Thus, we have $P_k(\boldsymbol{g}) > 0$. To meet the constraint in (4.34), $\upsilon_k(\boldsymbol{g}) = 0$. It is not hard to see that when $h(\boldsymbol{g}) = \beta_k g_k$, the solution is $P_k(\boldsymbol{g}) = 0$ and $\upsilon_k(\boldsymbol{g}) = 0$. Given the solutions in the above cases, the optimal power allocation policy can be expressed as,

$$P_k(\boldsymbol{g}) = \frac{N_0 W_k}{\alpha_k g_k}\left[\left(\frac{g_k}{g_k^{\mathrm{th}}(\boldsymbol{g})}\right)^{\eta_k} - 1\right]^+, \tag{4.37}$$

where $g_k^{th}(\boldsymbol{g}) \triangleq \frac{h(\boldsymbol{g})}{\beta_k}$. Since the bandwidth required to guarantee the QoS of each user decreases with the transmit power allocated to the user, the optimal solution of problem (4.29) is obtained when the equality in (4.29b) holds. Substituting (4.37) into $\sum_{k=1}^{K} P_k(\boldsymbol{g}) = P_{\max}$, we have

$$\sum_{k=1}^{K} \frac{N_0 W_k}{\alpha_k g_k}\left[\left(\frac{g_k}{g_k^{\mathrm{th}}(\boldsymbol{g})}\right)^{\eta_k} - 1\right]^+ = \sum_{k\in\mathbb{K}^+} \frac{N_0 W_k}{\alpha_k g_k}\left[\left(\frac{g_k}{g_k^{\mathrm{th}}(\boldsymbol{g})}\right)^{\eta_k} - 1\right] = P_{\max}, \tag{4.38}$$

where $\mathbb{K}^+$ denotes the set of users with positive transmit power. Since the values of $\eta_k, k = 1, \cdots, K$, differ among users, $g_k^{\mathrm{th}}(\boldsymbol{g})$ cannot be obtained in a closed-form expression.

A Symmetric Case: when all users have identical large-scale channel gains (i.e. $\alpha_k = \alpha$) and have the same average packet arrival rate (i.e. $a_k = a$, and hence $\vartheta_k = \vartheta$), W_k and η_k are identical for different users (i.e. $W_k = W$ and $\eta_k = \eta$). In this case, $g_k^{\mathrm{th}}(\boldsymbol{g}) = g^{\mathrm{th}}(\boldsymbol{g})$, which can be derived as follows,

$$g^{\mathrm{th}}(\boldsymbol{g}) = \left(\frac{\frac{\alpha P_{\max}}{N_0 W} + \sum_{k\in\mathbb{K}^+} {g_k}^{-1}}{\sum_{k\in\mathbb{K}^+} {g_k}^{\eta-1}}\right)^{-\frac{1}{\eta}}. \tag{4.39}$$

Substituting (4.39) into (4.37), we can derive the optimal power allocation policy as

$$P_k(\boldsymbol{g}) = \frac{N_0 W}{\alpha g_k}\left(\frac{\frac{\alpha g_k P_{\max}}{N_0 W} + g_k \sum_{k\in\mathbb{K}^+} {g_i}^{-1}}{{g_k}^{1-\eta} \sum_{k\in\mathbb{K}^+} {g_i}^{\eta-1}} - 1\right)^+. \tag{4.40}$$

It is worth noting that the elements in $\mathbb{K}^+$ and the function $P_k(\boldsymbol{g})$ rely on each other. To find the solution, for each given realization of $\boldsymbol{g}$ we compute (4.40) and update $\mathbb{K}^+$ iteratively from the initial user set $\mathbb{K}_0^+ = \{1, 2, \cdots, K\}$. According to the results of $P_k(\boldsymbol{g})$ obtained from (4.40), the users with negative transmit power are removed from $\mathbb{K}^+$. By repeating this procedure until $P_k(\boldsymbol{g}) \geq 0$ for all $k \in \mathbb{K}^+$, we can obtain the optimal power allocation policy.

Optimal Bandwidth Allocation: due to the expectation in (4.29a) and the complex expression of s_k in (4.28), the optimal bandwidth allocation cannot be obtained in closed-form. The solution can be found with stochastic optimization using the iteration formula in (4.23), where $s_k^{(t)}$ is obtained by substituting the optimal power control policy in (4.40) into (4.28) (instead of (4.20) with $P_0 = P_{\max}/W_{\max}$ as in Section (4.3.2)).

Remark 1. Since the power allocation in (4.40) and the bandwidth allocation found with stochastic optimization yield the unique solution that satisfies the necessary conditions, the obtained solution is globally optimal to problem (4.29) in the symmetric scenario.

4.3.3.3 Optimizing W_k and $P_k(g)$ with Unsupervised Learning

Even in the symmetric scenario, the optimal power allocation policy in (4.40) is not in closed-form. In general asymmetric cases, the expression of $g_k^{\text{th}}(\boldsymbol{g})$ cannot be derived from (4.38), and again there is no closed-form solution of $P_k(\boldsymbol{g})$. To avoid using a high complexity numerical method such as FEM to find the solution of the functional optimization for $P_k(\boldsymbol{g})$, we apply the method in Section 4.2.4 to solve problem (4.29), i.e. we turn to solving the following problem,

$$\max_{\lambda_k, h(\boldsymbol{g}), \upsilon_k(\boldsymbol{g})} \min_{W_k, P_k(\boldsymbol{g})} L_2 \tag{4.41}$$
$$\text{s.t. } P_k(\boldsymbol{g}) \geq 0, h(\boldsymbol{g}) \geq 0, \lambda_k \geq 0.$$

We approximate $P_k(\boldsymbol{g})$ by $\mathcal{N}_P(\boldsymbol{g}; \boldsymbol{\omega}_P)$, which is a DNN with model parameters $\boldsymbol{\omega}_P$, input $\boldsymbol{g}$, and output $\left[\hat{P}_1(\boldsymbol{g}; \boldsymbol{\omega}_P), \cdots, \hat{P}_K(\boldsymbol{g}; \boldsymbol{\omega}_P)\right]^{\mathrm{T}}$. In order not to lose any information during the forward propagation, the dimension of each hidden layer is set to be the number of users, which is the same as the input and output dimensions. As mentioned before (4.38), the optimal solution of problem (4.29) is obtained when the equality in (4.29b) holds. By applying `Softmax` function as the activation function in the output layer, we can guarantee that $\hat{P}_k(\boldsymbol{g}; \boldsymbol{\omega}_P) \geq 0$ and $\sum_{k=1}^{K} \hat{P}_k(\boldsymbol{g}; \boldsymbol{\omega}_P) = P_{\max}$. Thereby, the term $\int_{\mathbb{R}_+^K} \left[h(\boldsymbol{g})\left(\sum_{k=1}^{K} P_k(\boldsymbol{g}) - P_{\max}\right) - \sum_{k=1}^{K} \upsilon_k(\boldsymbol{g}) P_k(\boldsymbol{g})\right] d\boldsymbol{g}$ can be removed from the Lagrangian, and the corresponding Lagrange multiplier functions $h(\boldsymbol{g})$ and $\upsilon_k(\boldsymbol{g})$ can also be removed. By replacing $P_k(\boldsymbol{g})$ in (4.41) with $\hat{P}_k(\boldsymbol{g}; \boldsymbol{\omega}_P)$, the joint power and bandwidth allocation optimization problem then becomes,

$$\max_{\lambda_k} \min_{W_k, \boldsymbol{\omega}_P} \hat{L}_2 \triangleq \sum_{k=1}^{K} \left[W_k + \lambda_k \left(\mathbb{E}_{\boldsymbol{g}}\{e^{-\vartheta_k \hat{s}_k}\} - e^{-\vartheta_k B_k^{\mathrm{E}}}\right)\right] \tag{4.42}$$
$$\text{s.t. } \lambda_k \geq 0,$$

where $\hat{s}_k = \frac{\tau W_k}{u \ln 2} \left[\ln\left(1 + \frac{\alpha_k g_k \hat{P}_k(\boldsymbol{g}; \boldsymbol{\omega}_P)}{N_0 W_k}\right) - \frac{Q_{\mathrm{G}}^{-1}(\varepsilon_k^{\mathrm{c}})}{\sqrt{\tau W_k}}\right]$.

The model parameters of the DNN ω_P, the allocated bandwidth W_k, $k = 1, \ldots, K$, and the Lagrange multipliers λ_k, $k = 1, \ldots, K$ can be obtained from the following iterations,

$$\omega_P^{(t+1)} = \omega_P^{(t)} - \phi_{\omega_P}(t) \nabla_{\omega_P} \hat{L}_2^{(t)} = \omega_P^{(t)} - \phi_{\omega_P}(t) P_{\max} \nabla_{\omega_P} \mathcal{N}\left(\boldsymbol{g}; \omega_P^{(t)}\right) \nabla_{\boldsymbol{p}} \hat{L}_2^{(t)}, \quad (4.43)$$

$$W_k^{(t+1)} = \left[W_k^{(t)} - \phi_W(t) \frac{\partial \hat{L}_2^{(t)}}{\partial W_k} \right]^+, \quad (4.44)$$

$$\lambda_k^{(t+1)} = \left[\lambda_k^{(t)} + \phi_\lambda(t) \frac{\partial \hat{L}_2^{(t)}}{\partial \lambda_k} \right]^+ = \left[\lambda_k^{(t)} + \phi_\lambda(t) \frac{1}{N_\mathrm{b}} \sum_{n=1}^{N_\mathrm{b}} \left(e^{-\vartheta_k \hat{s}_k^{(t,n)}} - e^{-\vartheta_k B_k^\mathrm{E}} \right) \right]^+, \quad (4.45)$$

where $\hat{L}_2^{(t)} \triangleq \frac{1}{N_\mathrm{b}} \sum_{n=1}^{N_\mathrm{b}} \sum_{k=1}^{K} \left[W_k + \lambda_k \left(e^{-\vartheta_k \hat{s}_k^{(t,n)}} - e^{-\vartheta_k B_k^\mathrm{E}} \right) \right]$, $\hat{s}_k^{(t,n)}$ is the nth realization of the achievable rate in the tth iteration, and N_b is number of realizations of small-scale channel gains in each iteration. The gradient matrix of the DNN *w.r.t.* the model parameters $\nabla_{\omega_P} \mathcal{N}\left(\boldsymbol{g}; \omega_P^{(t)}\right)$ can be computed through backward propagation, and the gradient vector $\nabla_{\boldsymbol{p}} \hat{L}_2^{(t)}$ is with the dimension of K and the kth element of $-\frac{1}{N_\mathrm{b}} \sum_{n=1}^{N_\mathrm{b}} \lambda_k^{(t)} \vartheta_k \frac{\partial \hat{s}_k^{(t,n)}}{\partial \hat{P}_k} e^{-\vartheta_k \hat{s}_k^{(t,n)}}$.

Remark 2. From (4.45), we can find that the iteration converges only if $\frac{1}{N_\mathrm{b}} \sum_{n=1}^{N_\mathrm{b}} \left(e^{-\vartheta_k \hat{s}_k^{(t,n)}} - e^{-\vartheta_k B_k^\mathrm{E}} \right) \to 0$. This means that the QoS requirement in (4.29a) can be ensured when the iterations converge, because the constraints are judiciously controlled by the unsupervised learning framework.

Remark 3. For mobile users, ω_P, W_k and λ_k, $k = 1, \ldots, K$ need to be found again whenever their large-scale channel gains vary, say from the iterations in (4.43), (4.44) and (4.45) with random initial values. To accelerate convergence, we can employ pre-training, where the well-trained values of ω_P, W_k and λ_k, $k = 1, \ldots, K$ with fixed user locations are used to initialize the iteration for re-training when the user locations change. Alternatively, we can also find the mapping from all environment parameters to the bandwidth and power allocation by further converting the variable optimization for W_k into a functional optimization problem as in Section 4.2.2.

4.4 Simulation Results

In this section, we evaluate the performance achieved by the unsupervised deep learning when solving the variable and functional optimization problems in the previous section. For the bandwidth allocation problem without power allocation, we compare the performance of unsupervised learning with supervised learning in terms of approximation accuracy of the policy solution and the QoS violation. For the two-timescale bandwidth and power allocation problem, we compare the unsupervised learning with the global optimal solution in the symmetric scenario, considering that obtaining the labels for supervised learning is prohibitive.

We consider multiple users in a cell with a radius of 250 m. At the beginning of each slot, the small-scale channel gains of all the users are randomly generated from Rayleigh distribution. Other simulation parameters are listed in Table 4.1, unless otherwise specified.

We apply fully-connected DNNs in learning algorithms, and use `TanH` in the input layer and the hidden layers as an example activation function, where similar results can be obtained with other activation functions. The activation functions for the output layers will be introduced later. The fine-tuned batch size for learning is $N_b = 100$.

4.4.1 Bandwidth Allocation without Power Allocation

The users uniformly distributed along a road, which is 50 m minimal distance away from the BS. Since the bandwidth allocation without power allocation is independent for each user, we only consider the bandwidth and QoS constraint of one user.

The two DNNs have six hidden layers, and each layer has 16 neurons. We use `Softplus` in the output layers in all DNNs to ensure that the outputs are positive. The learning rate is $\phi(t) = 0.5/(1+10^{-4}t)$. To evaluate the performance of learning in terms of the approximation accuracy to the optimal policy and the

Table 4.1 Simulation parameters.

Duration of each slot T_s	0.1 ms
Duration of DL transmission τ	0.05 ms
Transmission delay D^t	1 slot (0.1 ms) [7]
Decoding delay D^c	1 slot (0.1 ms) [7]
Overall packet loss probability ε_{max}	10^{-5}
DL delay bound D_{max}	10 slots (1 ms)
Maximal transmit power of BS P_{max}	43 dBm
Path loss model $-10\lg(\alpha)$	$35.3 + 37.6\lg(d_k)$
Number of antennas N_t	8
Single-sided noise spectral density N_0	−173 dBm/Hz
Packet size u	20 bytes (160 bits) [2]
Average packet arrival rate a	0.2 packets/slot

QoS guarantee, we define the relative error of the learnt bandwidth allocation to the optimal solution as $\sigma \triangleq \left|\hat{W}(\alpha)/W^*(\alpha)-1\right|$, and the QoS violation of the learnt solution as $\nu \triangleq \left(\mathbb{E}_{\mathrm{g}}\left\{e^{\vartheta(B^{\mathrm{E}}-\hat{s})}\right\}-1\right)^+$.

In Figure 4.2, we show the Complementary Cumulative Distribution Functions (CCDF) of σ and ν achieved by the unsupervised learning approach in Section 4.3.2.3, and those obtained by the supervised learning approach where a DNN is used to learn the optimal policy $W^*(\alpha)$ and is trained by taking the optimal solutions of problem in (4.22) as labels. The results are obtained through 100 trials, where in each trial the NNs are trained through 10 000 iterations and are tested on 1000 realizations of the large-scale channel gains. It is shown that the unsupervised learning approach outperforms the supervised learning approach. With unsupervised learning, the relative approximation error of the allocated bandwidth is less than 1% and the QoS violation probability is less than 2% with a probability of 99.999%.

4.4.2 Joint Bandwidth and Power Allocation

To show the performance gap of the solution obtained with unsupervised learning from the global optimal solution, we first consider a symmetric scenario, where all users are in the cell-edge, i.e. the user-BS distances are 250 m. Then, we evaluate the performance considering an asymmetric scenario, where the users are uniformly located in the road with 50 m minimal distance away from the BS, i.e. the user-BS distances are distributed from 50 m to 250 m. The DNN has two

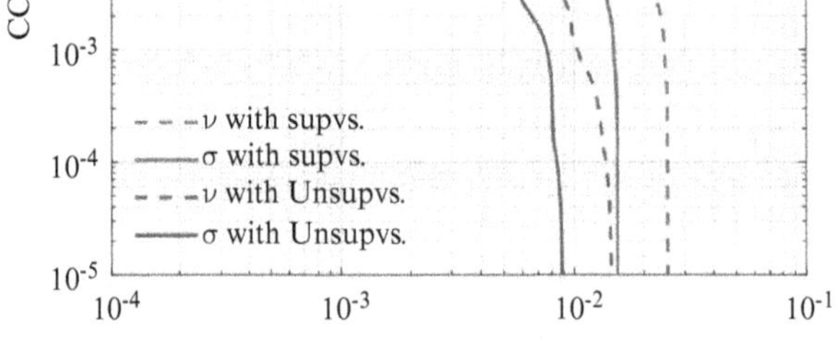

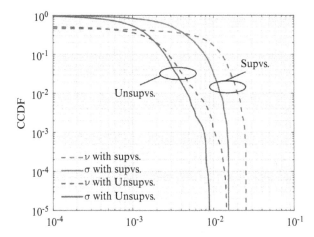

Figure 4.2 Complementary cumulative distributions of the relative approximation error of $\hat{W}$ and the QoS violation.

hidden layers, and the number of neurons in each layer is equal to the number of users. We use `Softmax` in the output layer to ensure the maximum transmit power constraint and $P_k(\boldsymbol{g}) \geq 0$ for all $k = 1, \cdots, K$. The learning rate is set to be $\phi(t) = 1/(1+0.1t)$, which turns out to be a good setting in our experience.

The joint optimal policy (with legend "w MUD w FD") is obtained from the method in Section 4.3.3.2 with around 200 iterations in the symmetric scenario, which exploits multi-user diversity by dynamically adjusting the transmit power according the small-scale channel gains of users, and exploits frequency diversity by frequency hopping. The learning-based bandwidth and power allocation policy (with legend "w MUD w FD (NN)") is obtained from the iterations in (4.43), (4.44) and (4.45) with random initial values. In each slot, the channel realizations in recent N_b slots are taken as a batch, which is used for 10 iterations in each slot. The training procedure converges after 100 slots, unless otherwise specified.

To show the gain from multi-user diversity, we compare the joint optimal policy with the optimal bandwidth allocation policy obtained through (4.23) after sufficient iterations, where the transmit power is equally allocated in the frequency domain without exploiting multi-user diversity (with legend "w/o MUD w FD"). To show the gain from frequency diversity, we compare with a heuristic policy in [30], which also exploits multi-user diversity by scheduling the users according to their small-scale channel gains but does not exploit frequency diversity (with legend "w MUD w/o FD"). Finally, we show the performance of the policy in [27] as a baseline, which optimizes the bandwidth allocation, but exploits neither multi-user diversity nor frequency diversity (with legend "w/o MUD w/o FD").

In Figure 4.1a, we provide the results in the symmetric scenario. It shows that the performance of learning-based policy (i.e. "w MUD w FD (NN)") is almost the same as the global optimal policy derived in (4.40) (i.e. "w MUD w FD"). In Figure 4.1b, we provide the results in the asymmetric scenario, where only the learning-based policy is simulated since the optimal solution is not available in this scenario. From both scenarios we can see that exploiting multi-user diversity or frequency diversity individually can significantly improve the bandwidth efficiency, while the gain from frequency diversity is larger. Once the frequency diversity is exploited, multi-user diversity only provides marginal performance gain.

To show the convergency of the learning-based solution, we consider the sum of the absolute values of average gradients $\zeta^{(t)} \triangleq \left\|\mathbb{E}_{\boldsymbol{g}}\{\nabla_{\omega_P}\hat{L}^{(t)}\}\right\|_1 + \sum_{k=1}^{K}\left|\mathbb{E}_{\boldsymbol{g}}\left\{\frac{\partial \hat{L}^{(t)}}{\partial W_k}\right\}\right| + \sum_{k=1}^{K}\left|\mathbb{E}_{\boldsymbol{g}}\left\{\frac{\partial \hat{L}^{(t)}}{\partial \lambda_k}\right\}\right|$ and the QoS constraint violation $\xi^{(t)} \triangleq \sum_{k=1}^{K}\left[\mathbb{E}_{\boldsymbol{g}}\left\{e^{\vartheta_k\left(B_k^{\mathrm{E}} - \mathcal{S}_k^{(t)}\right)}\right\} - 1\right]^+ \Big/ K$. The training algorithm in (4.43), (4.44) and (4.45) is considered to be converged at the tth slot if $\zeta^{(t)} < 1\%$ and $\xi^{(t)} < 1\%$.

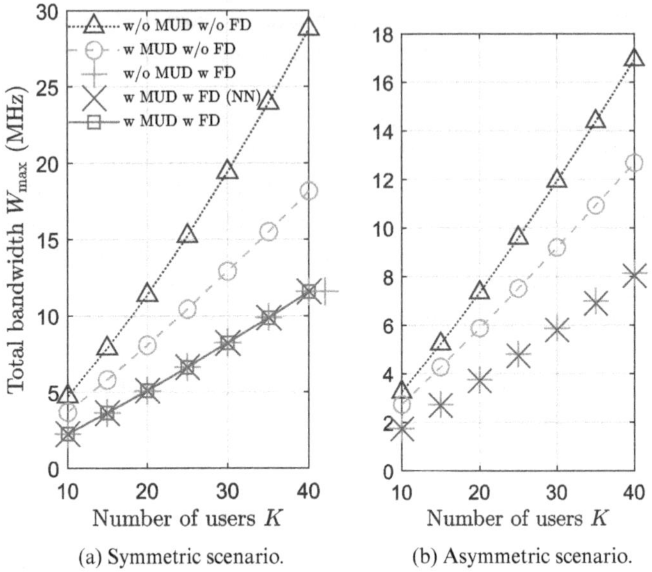

Figure 4.3 Total bandwidth required to support the QoS of each user. ©IEEE 2019. Reprinted with permission from [32].

The convergence speeds with and without pre-training are shown in Table 4.2, which are obtained from 100 000 trails. For the results without pre-training, 40 users are randomly dropped in the road in each trail and the realizations of their large- and small-scale channel gains are used to train ω_P, W_k and λ_k, $k=1,\cdots,K$, with random initializations. For the results with pre-training, all users move at the velocity of 72 kph along the road in the same direction. The well-trained values of ω_P, W_k and λ_k, $k = 1,\cdots,K$, are fine-tuned every 0.1 s using the channels at the new locations. Without pre-training, 10 000 time slots (i.e. 1 s) are required to achieve 99.99% convergence percentage, i.e. the QoS of each user is ensured with a probability of 99.99% according to Remark 2. We can see that the pre-training, which can be accomplished off-line, shortens the convergence time significantly.

Table 4.2 Number of time slots for convergence in an asymmetric scenario, where T_s =0.1 milliseconds.

Convergence percentage	99.9%	99.99%
w/o pre-training	5 000	> 10 000
w pre-training	3	1 000

The complexity of the training is low. A computer with Intel® Core™ i7-6700 CPU is able to finish around 1000 iterations in 0.1 s without using the acceleration from GPU.

4.5 Conclusion

In this chapter, we proved that the problem of finding the mapping from environment parameters to the solutions of constrained variable optimizations can be formulated as functional optimizations with instantaneous constraints, and established a unified unsupervised deep learning framework to solve functional optimizations with both instantaneous and statistic constraints. We considered two example problems in downlink URLLC to illustrate how to apply this framework. The first problem is variable optimization, where bandwidth allocation is optimized according to large-scale channel gains. The second problem is a hybrid variable and functional optimization with two types of constraints, where we jointly optimized bandwidth allocation according to large-scale channel gains and power allocation according to small-scale channel gains. Simulation results showed that, for the bandwidth allocation problem, unsupervised learning is superior to the supervised learning in both the accuracy of approximating the optimal solution and the guarantee of the QoS constraint. For the joint bandwidth allocation and power allocation problem, the performance of the learning-based solution is almost the same as the global optimal solution in a symmetric scenario. For both problems, the QoS achieved by the solution using unsupervised learning can be guaranteed with very high probability. The training algorithm converges rapidly with pre-training, and is with low computational complexity. As a byproduct, the optimization results also showed that the bandwidth utilization efficiency of URLLC can be improved more significantly by exploiting frequency diversity than by multi-user diversity.

Bibliography

[1] 3GPP. *LTE; E-UTRA; Physical layer procedures.* TS 36.213 v. 8.8.0 Release 8, Oct. 2009.

[2] 3GPP. *Study on Scenarios and Requirements for Next Generation Access Technologies.* Technical Specification Group Radio Access Network, Technical Report 38.913, Release 14, Oct. 2016.

[3] 3GPP. Study on New Radio (NR) access technology; physical layer aspects (Release 14). TR 38.802, 3GPP, 2017. v2.0.0.

[4] Léon Bottou. Online algorithms and stochastic approximations. In David Saad, editor, *Online Learning and Neural Networks*. Cambridge University Press, Cambridge, UK, 1998. URL http://leon.bottou.org/papers/bottou-98x. revised, Oct. 2012.

[5] S. Boyd and L. Vandanberghe. *Convex Optimization*. Cambridge University Press, 2004.

[6] C. Chang and Joy A. Thomas. Effective bandwidth in high-speed digital networks. *IEEE J. Sel. Areas Commun.*, 13(6):1091–1100, Aug. 1995.

[7] M. Condoluci, T. Mahmoodi, E. Steinbach, and M. Dohler. Soft resource reservation for low-delayed teleoperation over mobile networks. *IEEE Access*, 5:10445–10455, May 2017. ISSN 2169-3536. doi: 10.1109/ACCESS.2017.2707319.

[8] Wei Cui, Shen Kaiming, and Wei Yu. Spatial deep learning for wireless scheduling. *IEEE J. Sel. Areas Commun.*, 37(6):1248–1261, June 2019.

[9] Rui Dong, Changyang She, Wibowo Hardjawana, Yonghui Li, and Branka Vucetic. Deep learning for radio resource allocation with diverse quality-of-service requirements in 5G. *IEEE Transactions on Wireless Communications*, 20(4):2309–2324, 2021. doi: 10.1109/TWC.2020.3041319.

[10] Mark Eisen, Clark Zhang, Luiz F. O. Chamon, Daniel D Lee, and Alejandro Ribeiro. Learning optimal resource allocations in wireless systems. *IEEE Trans. Signal Process.*, 67(10):2775–2790, May 2019.

[11] G. Zhang, T. Q. S. Quek, M. Kountouris, *et al.* Fundamentals of heterogeneous backhaul design–analysis and optimization. *IEEE Trans. Commun.*, 64(2): 876–889, Feb. 2016.

[12] Xavier Glorot, Antoine Bordes, and Yoshua Bengio. Deep sparse rectifier neural networks. In *Proc. AISTATS*, pages 315–323, 2011.

[13] Andrea Goldsmith. *Wireless Communications*. Cambridge University Press, 2005.

[14] John Gregory. *Constrained optimization in the calculus of variations and optimal control theory*. Chapman and Hall/CRC, 2018.

[15] J. Guo and C. Yang. Predictive resource allocation with deep learning. In *Proc. IEEE VTC Fall*, 2018.

[16] Kurt Hornik, Maxwell B Stinchcombe, and Halbert White. Multilayer feedforward networks are universal approximators. *Neural Networks*, 2(5): 359–366, 1989.

[17] H. Lee, S. H. Lee, and T. Q. S. Quek. Deep learning for distributed optimization: Applications to wireless resource management. *IEEE J. Sel. Areas Commun.*, 37(10):2251–2266, 2019.

[18] Daniel Liberzon. *Calculus of Variations and Optimal Control Theory: A Concise Introduction*. Princeton University Press, 2012.

[19] Dong Liu, Chengjian Sun, Chenyang Yang, and Lajos Hanzo. Optimizing wireless systems using unsupervised and reinforced-unsupervised deep learning. *IEEE Network*, 34(4):270–277, July 2020.

[20] L. Liu, P. Parag, J. Tang, W. Y. Chen, and J. F. Chamberland. Resource allocation and quality of service evaluation for wireless communication systems using fluid models. *IEEE Trans. on Inf. Theory*, 53(5):1767–1777, May 2007. ISSN 0018-9448. doi: 10.1109/TIT.2006.894682.

[21] L. Liu, B. Yin, S. Zhang, X. Cao, and Y. Cheng. Deep learning meets wireless network optimization: Identify critical links. *IEEE Transactions on Network Science and Engineering*, 7(1):167–180, 2020.

[22] David G Luenberger. *Optimization by vector space methods*. John Wiley & Sons, 1997.

[23] B. Makki, T. Svensson, G. Caire, and M. Zorzi. Fast HARQ over finite blocklength codes: A technique for low-latency reliable communication. *IEEE Trans. Wireless Commun.*, 18(1):194–209, Jan 2019. ISSN 1536-1276. doi: 10.1109/TWC.2018.2878713.

[24] Vinod Nair and Geoffrey E Hinton. Rectified linear units improve restricted boltzmann machines. In *Proc. ICML*, pages 807–814, 2010.

[25] Sebastian Schiessl, James Gross, and Hussein Al-Zubaidy. Delay analysis for wireless fading channels with finite blocklength channel coding. In *Proc. ACM MSWiM*, pages 13–22, 2015. ISBN 9781450337625. doi: 10.1145/2811587.2811596. URL https://doi.org/10.1145/2811587.2811596.

[26] C. She, C. Yang, and L. Liu. Energy-efficient resource allocation for MIMO-OFDM systems serving random sources with statistical QoS requirement. *IEEE Trans. Commun.*, 63(11):4125–4141, Nov 2015. ISSN 1536-1276. doi: 10.1109/TWC.2018.2878713.

[27] C. She, C. Yang, and T. Q. S. Quek. Cross-layer optimization for ultra-reliable and low-latency radio access networks. *IEEE Trans. Wireless Commun.*, 17(1): 127–141, Jan 2018. ISSN 1536-1276. doi: 10.1109/TWC.2017.2762684.

[28] C. She, C. Yang, and T. Q. S. Quek. Joint uplink and downlink resource configuration for ultra-reliable and low-latency communications. *IEEE Trans. Commun.*, 66(5): 2266–2280, May 2018. ISSN 0090-6778. doi: 10.1109/TCOMM.2018.2791598.

[29] Changyang She, Rui Dong, Zhouyou Gu, Zhanwei Hou, Yonghui Li, Wibowo Hardjawana, Chenyang Yang, Lingyang Song, and Branka Vucetic. Deep learning for ultra-reliable and low-latency communications in 6G networks. *IEEE Network*, 34(5):219–225, 2020. doi: 10.1109/MNET.011.1900630.

[30] C. Sun, C. She, and C. Yang. Exploiting multi-user diversity for ultra-reliable and low-latency communications. In *Proc. IEEE Globecom Workshops*, 2017. doi: 10.1109/GLOCOMW.2017.8269133.

[31] Chengjian Sun and Chenyang Yang. Learning to optimize with unsupervised learning: Training deep neural networks for URLLC. In *Proc. IEEE PIMRC*, 2019.

[32] Chengjian Sun and Chenyang Yang. Unsupervised deep learning for ultra-reliable and low-latency communications. In *Proc. IEEE Globecom*, 2019.

[33] H. Sun, X. Chen, Q. Shi, M. Hong, X. Fu, and N. D. Sidiropoulos. Learning to optimize: Training deep neural networks for interference management. *IEEE Trans. on Signal Proc.*, 66(20):5348–5453, Oct. 2018. ISSN 1536-1276. doi: 10.1109/TWC.2018.2880907.

[34] J. Tang and X. Zhang. Quality-of-service driven power and rate adaptation over wireless links. *IEEE Trans. Wireless Commun.*, 6(8):3058–3068, August 2007. ISSN 1536-1276. doi: 10.1109/TWC.2007.051075.

[35] Dapeng Wu and Rohit Negi. Effective capacity: A wireless link model for support of quality of service. *IEEE Trans. Wireless Commun.*, 2(4):630–643, Jul. 2003.

[36] W. Yang, G. Durisi, T. Koch, and Y. Polyanskiy. Quasi-static multiple-antenna fading channels at finite blocklength. *IEEE Trans. Inf. Theory*, 60(7): 4232–4264, Jul. 2014.

[37] Olgierd Cecil Zienkiewicz, Robert Leroy Taylor, Perumal Nithiarasu, and JZ Zhu. *The finite element method*, volume 3. McGraw-hill London, 1977.

A

Appendix for Unsupervised Learning

A.1 Proof of Proposition 1

Proof. We first prove that $\boldsymbol{x}^*(\theta)$ is optimal for problem (4.2). Denote $\boldsymbol{x}^*(\theta_1)$ as an optimal solution of problem (4.1) given an arbitrary realization $\theta_1 \in \mathcal{D}_\theta$, and denote the objective function in problem (4.2) as

$$\mathcal{F}\left[\boldsymbol{x}(\theta)\right] \triangleq \int_{\theta \in \mathcal{D}_\theta} f\left[\boldsymbol{x}(\theta);\theta\right] p(\theta)\mathrm{d}\theta.$$

Let $\boldsymbol{x}_1(\theta), \theta \in \mathcal{D}_\theta$ be an arbitrary feasible solution of problem (4.2). Since problems (4.1) and (4.2) have the same constraints, they have the same feasible region. Thus, $\boldsymbol{x}_1(\theta_1)$ is a feasible solution of problem (4.1). Given the realization θ_1, the optimal solution of problem (4.1) is better than any feasible solutions of problem (4.1), i.e.

$$f\left[\boldsymbol{x}^*(\theta_1);\theta_1\right] - f\left[\boldsymbol{x}_1(\theta_1);\theta_1\right] \leq 0,\ \forall \theta_1 \in \mathcal{D}_\theta. \tag{A.1}$$

Since $p(\theta) \geq 0$, we further have,

$$\mathcal{F}\left[\boldsymbol{x}^*(\theta)\right] - \mathcal{F}\left[\boldsymbol{x}_0(\theta)\right] = \int_{\theta \in \mathcal{D}_\theta} \left[f\left[\boldsymbol{x}^*(\theta);\theta\right] - f\left[\boldsymbol{x}_0(\theta);\theta\right]\right] p(\theta)\mathrm{d}\theta \leq 0. \tag{A.2}$$

Since $\boldsymbol{x}^*(\theta), \theta \in \mathcal{D}_\theta$, satisfies all the constraints in problem (4.2), it is a feasible solution of problem (4.2). (A.2) indicates that $\boldsymbol{x}^*(\theta), \theta \in \mathcal{D}_\theta$ is better than an arbitrary solution of problem (4.2). Thus, it is optimal for problem (4.2).

In what follows, we prove that the value of $\boldsymbol{x}_{opt}(\theta)$ for arbitrary realization of θ is optimal for problem (4.1) with probability one. From the definition of $\boldsymbol{x}_{opt}(\theta)$

Ultra-Reliable and Low-Latency Communications (URLLC) Theory and Practice: Advances in 5G and Beyond, First Edition. Edited by Trung Q. Duong, Saeed R. Khosravirad, Changyang She, Petar Popovski, Mehdi Bennis and Tony Q.S. Quek.
© 2023 John Wiley & Sons Ltd. Published 2023 by John Wiley & Sons Ltd.

and $\boldsymbol{x}^*(\theta)$, we have

$$f\left[\boldsymbol{x}_{opt}(\theta);\theta\right] - f\left[\boldsymbol{x}^*(\theta),\theta\right] \geq 0,\ \forall\theta \in \mathcal{D}_\theta. \tag{A.3}$$

Suppose there exists a non-zero measure set, $\mathcal{D}_\theta^+$, such that for any $\theta' \in \mathcal{D}_\theta^+$, $\boldsymbol{x}_{opt}(\theta')$ is not optimal for problem (4.1). In other words,

$$f\left[\boldsymbol{x}_{opt}(\theta');\theta'\right] - f\left[\boldsymbol{x}^*(\theta'),\theta'\right] > 0, \forall\theta' \in \mathcal{D}_\theta^+. \tag{A.4}$$

Here, a non-zero measure set is a set that $\Pr\{\theta' \in \mathcal{D}_\theta^+\} > 0$.

From (A.3) and (A.4), we can derive that

$$\begin{aligned}
&\mathcal{F}\left[\boldsymbol{x}_{opt}(\theta)\right] - \mathcal{F}\left[\boldsymbol{x}^*(\theta)\right] \\
&= \int_{\theta\in\mathcal{D}_\theta} \left[f\left[\boldsymbol{x}_{opt}(\theta);\theta\right] - f\left[\boldsymbol{x}^*(\theta),\theta\right]\right] p(\theta)\mathrm{d}\theta \\
&\geq \int_{\theta\in\mathcal{D}_\theta^+} \left[f\left[\boldsymbol{x}_{opt}(\theta);\theta\right] - f\left[\boldsymbol{x}^*(\theta),\theta\right]\right] p(\theta)\mathrm{d}\theta > 0.
\end{aligned} \tag{A.5}$$

This is in contradiction with the fact that $\boldsymbol{x}_{opt}(\theta)$ is the optimal solution of problem (4.2).

This completes the proof.

A.2 The method to compute (4.12), (4.13) and (4.14)

Proof. For notational simplicity, we omitted the index of iteration t in this appendix. To compute (4.12), (4.13) and (4.14), we only need to compute $\nabla_{\omega_x}\hat{L}$, $\nabla_{\omega_v}\hat{L}$ and $\partial\hat{L}/\partial\lambda_j$.

The value of $\nabla_{\omega_x}\hat{L}$ can be obtained from the following expression,

$$\begin{aligned}
\nabla_{\omega_x}\hat{L} = &\int\limits_{\theta\in D_\theta} \left[\nabla_{\omega_x}\hat{\boldsymbol{x}}(\theta)\right]\left[\nabla_{\hat{\boldsymbol{x}}(\theta)} f\left(\hat{\boldsymbol{x}}(\theta);\theta\right)\right] p\left(\theta\right)\mathrm{d}\theta \\
&+ \sum_{j=I+1}^{J} \lambda_j \int\limits_{\theta\in D_\theta} \left[\nabla_{\omega_x}\hat{\boldsymbol{x}}(\theta)\right]\left[\nabla_{\hat{\boldsymbol{x}}(\theta)} C_j\left(\hat{\boldsymbol{x}}(\theta);\theta\right)\right] p\left(\theta\right)\mathrm{d}\theta \\
&+ \sum_{i=1}^{I} \int\limits_{\theta\in D_\theta} \hat{\lambda}_j\left(\theta\right)\left[\nabla_{\omega_x}\hat{\boldsymbol{x}}(\theta)\right]\left[\nabla_{\hat{\boldsymbol{x}}(\theta)} C_i\left(\hat{\boldsymbol{x}}(\theta);\theta\right)\right]\mathrm{d}\theta,
\end{aligned} \tag{B.1}$$

where $\nabla_{\omega_x}\hat{\boldsymbol{x}}(\theta) \triangleq [\nabla_{\omega_x}\hat{x}_1(\theta), ..., \nabla_{\omega_x}\hat{x}_{N_x}(\theta)]$ can be obtained via backward propagation.

The values of $\nabla_{\omega_v}\hat{L}$ and $\partial\hat{L}/\partial\lambda_j$ can be obtained from

$$\nabla_{\omega_v}\hat{L} = \sum_{i=1}^{I} \int_{\theta\in D_\theta} [\nabla_{\omega_v}\hat{v}_i(\theta)]\,[C_i\,(\hat{\boldsymbol{x}}(\theta);\theta)]\,\mathrm{d}\theta, \tag{B.2}$$

$$\frac{\partial\hat{L}}{\partial\lambda_j} = \int_{\theta\in D_\theta} C_j\,(\hat{\boldsymbol{x}}(\theta);\theta)\,p\,(\theta)\,\mathrm{d}\theta, \tag{B.3}$$

where $\nabla_{\omega_\lambda}\hat{\lambda}_i(\theta), i = 1, \ldots, I$, can be obtained via backward propagation.

5

Channel Coding and Decoding Schemes for URLLC

Chentao Yue, Mahyar Shirvanimoghaddam, Branka Vucetic, and Yonghui Li*

The School of Electrical and Information Engineering, The University of Sydney, 2006, NSW, Darlington, Australia
** Corresponding Author*

5.1 Short Block-length Codes for URLLC Communications

Since 1948, when Shannon introduced the notion of channel capacity [41], researchers have been looking for powerful channel codes that can approach this limit. Low Density Parity Check (LDPC) codes and turbo codes have been shown to perform very close to Shannon's limit at large block lengths and have been widely applied in the 3rd generation (3G) and 4th generation (4G) of mobile standards [31]. The polar code proposed by Arikan in 2009 [3] has attracted much attention in the last decade and has been chosen as the standard coding scheme for the 5th generation 5G-enhanced Mobile Broadband (eMBB) control channels and the physical broadcast channel. Polar codes take advantage of a simple successive cancellation decoder, which is optimal for asymptotically large block lengths [43, 44].

Recently, short code design and related decoding algorithms have attracted a great deal of interest among industry and academia [32, 52]. This interest was triggered by the stringent requirements of the new ultra-reliable and low-latency communications (URLLC) service for mission critical Internet of Things (IoT) services, including the hundreds-of-microsecond time-to-transmit latency, block error rates of 10^{-5}, and the bit-level granularity of codeword size and code rate. These URLLC service requirements mandate the use of short block-length codes. Shannon's formula, which was originally obtained for infinite block length, breaks down for the short block length regime. Polyanskiy et al. [39] showed that channel dispersion cannot be regarded as Gaussian for the Gaussian channel under

Ultra-Reliable and Low-Latency Communications (URLLC) Theory and Practice: Advances in 5G and Beyond, First Edition. Edited by Trung Q. Duong, Saeed R. Khosravirad, Changyang She, Petar Popovski, Mehdi Bennis and Tony Q.S. Quek.
© 2023 John Wiley & Sons Ltd. Published 2023 by John Wiley & Sons Ltd.

the finite block length regime, and derived new bounds for the achievable rate, referred to as the Normal Approximation (NA), which are tighter than Shannon's bounds. Following this theoretical milestone, researchers are committed to seeking practical coding schemes achieving the new capacity bound in the short block length regime to meet the requirements of URLLC.

Next, we focus on the key requirements and performance benchmarks for URLLC channel coding.

5.1.1 Key Requirements and Performance Benchmarks

The following three aspects are considered in the design of channel codes and decoding algorithms for URLLC.

Latency

In the physical layer, we mainly focus on user plane latency, which is defined as the time to successfully deliver a data block from the transmitter to the receiver via the radio interface in both uplink and downlink directions. User plane latency consists of the time-to-transmit latency, the propagation delay, the processing latency, and the re-transmission time. The time-to-transmit latency is required to be of the order of a hundred microseconds, which is much less than the 1 ms currently considered in 4G [36]. Propagation delay is typically defined as the delay of propagation through the transmission medium, and it depends on the distance between the transmitter and receiver. The processing delay includes the latency introduced by channel estimation, decoding, etc. Thus, in URLLC, fast decoding algorithms are needed to conserve the overall budget of latency.

Reliability

Reliability is defined as the success probability of transmitting k information bits within the desired user plane latency at a certain channel quality. The failure of transmission happens when the packet is lost, or it is received late, or it has residual errors. Therefore, it is essential to maximize the reliability of every packet in order to minimize the error rate, so as to minimize the number of re-transmissions [44]. In the context of channel coding, we use Block Error Rate (BLER) as a metric to compare different channel codes in terms of reliability. The lower the BLER, the more reliable the transmission is.

Flexibility

The flexibility of the channel coding scheme is an important aspect along with the evaluation of the coding performance. Bit-level granularity of the codeword size and code operating rate is desired for URLLC [2]. The actual coding rate used in transmission could not be restricted and optimized for specified ranges [2].

The channel codes, therefore, need to be flexible to enable Hybrid Automatic Repeat Request (HARQ). The number of re-transmissions, however, needs to be kept as low as possible to minimize the latency.

The above critical requirements put forward specific provisions on the design of codes and decoders for URLLC. Specifically, the latency requirement demands the design and application of low-complexity decoders. The decoders share the latency budget of the processing delay with other physically-layered components, which means that the channel decoder for URLLC must target the processing time as low as a few microseconds [42]. This ultra-low processing time can only be achieved by deploying decoders that fully utilize parallelism and other techniques resulting in low overall complexity.

Length-compatible codes should be designed to meet the requirements of bit-level flexibility in URLLC. For codes that naturally have only a few fixed rates or code lengths (e.g. Bose-Chaudhuri-Hocquenghem codes, polar codes, and LDPC codes), techniques such as extending, puncturing, and shortening are used to generalize the code parameters [42, 46]. These techniques are based on a mother code that can evolve to different code lengths and code rates. However, the intrinsic structural characteristics of the mother codes must be maintained during the generalization, to be compatible with the decoding approach.

Furthermore, the requirements of reliability place demands on both the code and decoder design. For the design of short codes, there is the issue of the gap to Shannon's limit. That is, if we decrease the block length, the coding gain will be reduced and the gap to Shannon's limit will increase. Thus, Shannon's theoretical model breaks down for short codes, as the channel capacity defined as the maximum possible rate at which reliable communications is possible, is only valid for infinite block length. As an excellent expansion of Shannon's limit, the finite block length bound was shown to be tight for moderate block lengths (>100 bits) [15]. Therefore, in a short-block length scenario, the NA of finite block length bound [15, 39] is considered as the performance benchmark for comparison. For the transmission of a coding block of size n over a channel with the Signal-to-Noise Ratio (SNR) γ, the NA of the BLER ϵ is given by [39]

$$\epsilon = Q\left(\sqrt{\frac{1}{nV(\gamma)}}\left(\frac{nC(\gamma)-k}{\log_2(e)}+\frac{\ln(n)}{2}\right)\right), \tag{5.1}$$

where $Q(x)$ is the standard Q-function, k is the message length, $C(\gamma)$ is the channel capacity, and $V(\gamma)$ is the channel dispersion coefficient. For a binary AWGN (BI-AWGN) channel at SNR γ, we have

$$C(\gamma) = 1 + \frac{H^{(1)}(0)}{\ln(2)}, \quad V(\gamma) = H^{(2)}(0) - \left(H^{(1)}(0)\right)^2, \tag{5.2}$$

and

$$H^{(\ell)}(0) = \frac{1}{\sqrt{2\pi\gamma}} \int_{-\infty}^{-\infty} e^{-\frac{(x-\gamma)^2}{2\gamma}} (-h(x))^{\ell} dx, \tag{5.3}$$

and $h(x) = \ln(1 + e^{-2x})$ [15].

Given the code parameters and the channel SNR, the design of short codes should target the BLER given by the NA. However, many well-known codes show considerable gaps to the NA under practical decoders. Figure 5.1 shows the normal approximation for different code rates and information block lengths. As can be seen, Long-Term Evolution (LTE) turbo and tail-biting convolutional code (TB-CC) codes show a considerable gap to the normal approximation at short blocks. However, when the block length of the turbo code increases, the gap to the normal approximation and Shannon's limit decreases. These gaps to NA at short block length is not only because of the codes, but is also because of the sub-optimal decoding approaches. It has been shown that the BLER gap of short LDPC codes to the NA can be reduced by using near-optimal decoders rather than iterative Belief Propagation (BP) decoders [32, 56].

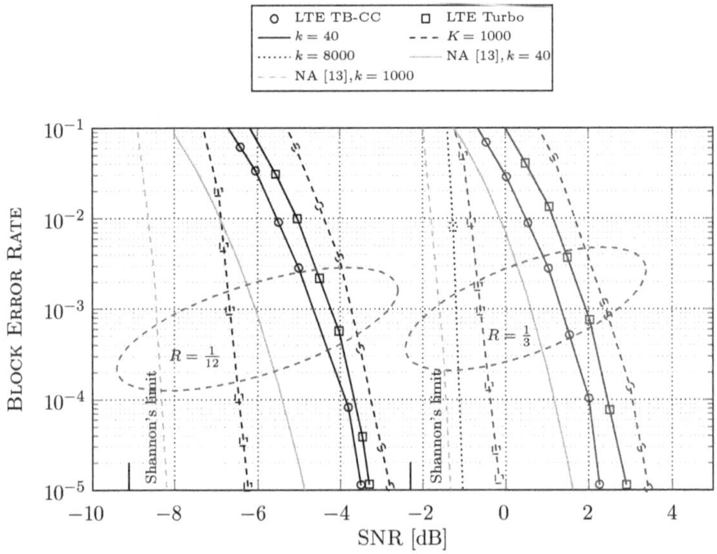

Figure 5.1 Comparison of error performance of LTE channel codes with different information block lengths, k. This is reused with permission from [44].

5.1.2 Candidate Short Block Length Codes for URLLC

Several candidate channel codes such as Bose–Chaudhuri–Hocquenghem (BCH), LDPC, polar, TB-CC, and turbo codes, have been considered for URLLC data channels [32, 44]. While some of these codes perform close to Shannon's limit at asymptotically long block lengths, they usually suffer from performance degradation if the code length is short, e.g. turbo codes with iterative decoding in short and moderate block lengths show a gap of more than 1 dB to the NA [31]. TB-CC can eliminate the rate loss of conventional convolutional codes due to the zero tail termination, but its decoding process is more complex than that of conventional codes [44]. Although LDPC codes have already been selected for eMBB data channels in 5G, recent investigations have shown that there exist error floors for LDPC codes constructed using the base graph at high SNRs [32, 44] at moderate and short block lengths; hardly satisfying ultra-reliability requirements. Polar codes outperform LDPC codes with no error floor at short block lengths, but for short codes, it still falls short of the finite block length capacity bound [44], i.e. the maximal channel coding rate achievable at a given block length and error probability [39]. BCH codes have large minimum distances and approximately approach the NA; however, their decoding complexity is a challenging issue.

Next, we briefly discuss these channel code candidates for URLLC. For block codes, we use k and n to denote the information block length and codeword length, respectively. For convolutional codes, we use k, n, and m to denote the bit input and bit output per time instant and memory order, respectively.

5.1.2.1 BCH Codes

BCH codes have attracted the interest of the research community recently [32, 44, 50, 55, 56]. As a class of powerful cyclic codes, BCH codes are usually constructed with the lowest-degree polynomials of 2^t roots over finite fields [31], where t is referred to as the error correction capability, and have large minimum distances and a high-density generator matrix. The decoding of BCH codes is a challenge and can hardly be accomplished by efficient modern decoders (e.g. iterative decoders). The first algebraic decoding algorithm specifically for BCH codes was proposed by Peterson [38], then it was further improved by Chieng [12] and Berlekamp [4]. However, these classical algebraic decoding algorithms were developed based on the use of syndromes to locate errors, and there is a gap in performance compared to maximum-likelihood (ML) decoding. As the ML decoding is usually highly complex and introduces a significant delay at the receiver, the complexity issue of its ML decoding needs to be addressed urgently to enable the usage of BCH codes in URLLC.

The ML decoding is defined as the decoder maximizing the *a posterior* probability (APP) $\Pr(\hat{\mathbf{c}} = \mathbf{c}|\mathbf{r})$, where $\mathbf{c}$ is the transmitted codeword, $\hat{\mathbf{c}}$ is the estimate of $\mathbf{c}$, and $\mathbf{r}$ is the received signal [31], $\Pr(\hat{\mathbf{c}} = \mathbf{c}|\mathbf{r})$ is also referred to as the likelihood of the estimate $\hat{\mathbf{c}}$. It can be shown that for a noisy channel, the likelihood can only be determined by using the soft information carried by the received signal. For a code with information length k and block length n, ML decoding needs to determine the likelihood for 2^k codeword estimates, and the practical complexity of this is prohibitive. The Ordered Statistics Decoder (OSD) is one of the possible ML decoding alternatives suitable for BCH codes [18]. OSD is a near ML soft decision decoding with the complexity of $\mathcal{O}(k^m)$, where m is the decoding order of OSD. In recent work [58], the complexity of OSD is significantly reduced which makes OSD a good choice for short block length codes.

It is worth noting that although BCH codes have large minimum distances which avoid flooring the performance at low BLER, they are not flexible as the block length and information length cannot be selected arbitrarily.

5.1.2.2 Low Density Parity Check (LDPC) Codes

LDPC codes were originally proposed by Gallager in the early 1960s and later rediscovered in the 1990s, when researchers began to investigate codes-on-graph based on Tanner's work in 1981 and iterative decoding [31]. LDPC codes with iterative BP decoding have been shown to perform very close to Shannon's limit with only a fraction of a decibel gap. Binary LDPC codes with iterative BP decoding, however, do not perform well at short to moderate block-lengths which is mainly due to the existence of many short cycles in the code's bipartite graph. Recently, protograph-based LDPC codes have been shown to perform well under belief propagation decoding at short-to-moderate block length [16], but their performance is not comparable with BCH codes under OSD or TB-CCs with large memory. However, they favor the very low decoding complexity under iterative decoding algorithms. Non-binary LDPC codes are also shown to perform very close to the finite length performance bound where the decoding complexity is the major drawback.

LDPC codes have been adopted for the eMBB data channel of 5G and it is a natural extension to apply LDPC codes for URLLC. However, recent investigations demonstrated error floors for LDPC codes at certain rates and block lengths [1]. Moreover, although the rate-compatible LDPC codes have been exhaustively studied by applying puncturing [14, 22], shortening [45], and extending [11, 30], the resulting codes are sub-optimal over a large range of rates with trying to keep the sparse feature of codes for enabling BP decoding. Recently, it has been shown that the decoding complexity of LDPC also increases with increasing flexibility [1].

5.1.2.3 Polar Codes

Polar codes as introduced in [3], are binary linear codes that can provably achieve the capacity of a binary-input discrete memoryless channel using low-complexity encoding and decoding in the asymptotically large block-length scenarios. The ingenious technique, namely channel polarization, facilitates the construction of these codes. In channel polarization, a length-n block code is translated to n independent and identical binary-input discrete memoryless channels. It is proved that when n is large, these synthesized channels have capacities of either (close to) zero or one. The message is only sent over the set of near-perfect channels, and the unreliable channels are unused. The unused channels are constantly transmitting bits '0', which are referred to as frozen bits. In practice, frozen bits are assigned as a priori knowledge for both the encoder and decoder.

Following the inverse process of channel polarization, the Successive Cancellation (SC) decodes polar codes with a complexity of $\mathcal{O}(n \log n)$ [3]. The SC decoder sequentially performs the inverse polarization by traversing a binary tree structure. However, SC has inferior error correction performance when the block length is finite. Also, its sequential nature introduces high processing latency on the implementation.

A significant improvement in the decoding performance is achieved by using Successive Cancellation List (SCL) decoding [10]. The SCL keeps a list of the L most likely decoding paths over the binary tree at all times, unlike the SC decoder, which keeps only one decoding path and results in only one decoding candidate. It has been shown that Cyclic Redundancy Check (CRC)-aided SCL (CA-SCL) has especial performance gain compared to SC decoders [34] at short-to-moderate block length, where the CRC is served as an outer code of the polar coding. In this way, the CRC checksum is used at the decoder side to pick the right decoding path in the list, even if it is not the most probable path.

Polar codes have been selected for short blocks for control channels in eMBB [1], and they outperform LDPC codes in short block lengths and low code rates without any sign of error floor; therefore, more suitable for URLLC [1]. In terms of designing length-flexible polar codes, the puncturing technique was first investigated in [35] and further studied in [6]. The shortening of polar codes was presented in [6, 51], and extending methods are introduced in [24, 29]. Because of these techniques, 1-bit granularity can be achieved for all coding rates and for a full range of block sizes for polar codes. However, due to the channel polarization, the design of length-flexible polar codes must consider the frozen bits to allow SCL to work efficiently, which somehow results in sub-optimal codes. Furthermore, the implementation complexity of SCL increases with increasing list size, which also hinders the application of polar codes in URLLC to some extent.

5.1.2.4 Convolutional Codes

Convolutional Codes (CC) were first introduced by Elias in 1955 [31]. They differ from block codes as the encoder contains memory. Generally, a rate $R = k/n$ convolutional encoder with memory order m can be realized as a linear sequential circuit with input memory m, k inputs, and n outputs, where inputs remain in the encoder for m time units after entering. Large minimum distances and low error probabilities for convolutional codes are achieved by not only increasing k and n, but also by increasing the memory order. The decoding complexity however scales in general exponentially with the memory order in both Viterbi and Bahl, Cocke, Jelinek, and Raviv (BCJR) algorithms [31].

When short packets have to be transmitted, terminated convolutional codes represent a promising candidate solution, although the rate loss due to a zero tail termination at short block lengths may be unacceptable. A tail-biting approach [21] eliminates the rate loss and hence it deserves particular attention when comparing channel codes for short blocks. For these reasons, TB-CCs are currently considered within the 5G standardization for URLLC. It is worth mentioning that the decoders for TB-CCs are more complex than those for convolutional codes. TB-CC was used in LTE for the broadcast channel and downlink/uplink control information.

5.1.2.5 Turbo Codes

In 1993, Berrou, Glavieux, and Thitimajshima, introduced turbo coding, which combines a parallel concatenation of two convolutional encoder and iterative maximum a posteriori probability (MAP) decoding [5]. Turbo codes have been extensively used for the data channel in LTE. For large blocks, turbo codes are capable of performing within a few tenths of dB from Shannon's limit. Unfortunately, turbo codes with iterative decoding in short and moderate block lengths show a gap of more than 1 dB to the finite-length performance benchmark. LTE turbo code is known to be well designed for medium block length and code rate $\geq 1/3$. When the code rate and block length are small, LTE-turbo code performance is degraded. For turbo codes, 1-bit granularity is feasible for all coding rates and for full range of block size, and the ability of turbo codes to support both Chase combining and incremental redundancy HARQ is well known [1].

5.1.3 Comparison Between Channel Codes for URLLC

In this section, different channel code candidates for URLLC are compared in terms of reliability, and algorithmic complexity. We consider a binary input AWGN channel, where unit power binary phase shift keying (BPSK) signals are sent over a channel which are subject to the AWGN of variance σ_n^2. The SNR is then defined as

$\frac{1}{\sigma_n^2}$. For each SNR point and code rate, the simulation is run to obtain 100 codeword errors at the decoder output.

5.1.3.1 Reliability

Figure 5.2 shows the BLER versus the SNR for different candidate channel codes at rate $R = \frac{1}{2}$ and block length $n = 128$ under the Maximum Likelihood Decoding (MLD) [23]. By using an optimal decoder in the (ML sense), the plot gives insights into the code performance itself. As can be seen in this figure, the extended BCH (eBCH) code closely approaches the NA benchmark over the whole SNR region and can provide a very low BLER as small as 10^{-7} with only 0.1 dB gap to the NA bound [15, 39]. Another competitive code is the TB-CC code with $m = 14$, which can provide a BLER of 10^{-5} with only 0.1 dB gap to the NA benchmark, however, when it goes to a lower BLER of 10^{-7} the gap increases to 0.3 dB. Decreasing the memory to 11, TB-CC still gives a performance within 0.1 dB gap from the normal approximation at a BLER of 10^{-5}. Other competitive codes are LDPC codes designed over a large Galois field (here $\mathbb{F}_{256}$), which have almost the same performance as the TB-CC code with $m = 14$. The Circular Viterbi Algorithm (CVA) has been used for decoding of TB-CCs [21].

The BCH code outperforms all other existing codes owing to its better distance spectrum. Other codes are mainly designed to provide good performance while maintaining the decoding complexity at a reasonable order. We will discuss the trade-off between complexity and performance in more detail in later sections.

Comparison of different channel codes with different rates and information block lengths is depicted in Figure 5.1. Rates $R = 1/3$, $1/6$, and $1/12$ are considered for CRC-aided polar codes (CA-polar), TB-CC, turbo, and LDPC codes. As shown, by fixing the code rates, all codes exhibit better BLER performance when the block length is large ($k = 4000$) than the short block length cases ($k = 40$). The CA-polar has the best BLER among all the counterparts, showing acceptable gaps to the NA bounds. It is worth noting that LDPC codes can still offer good performance at very low code rates (e.g. $R = 1/12$) with the low complexity min-sum decoding algorithm.

5.1.3.2 Complexity Versus Performance

Turbo and LDPC codes have shown to provide near-capacity performance at large block-lengths with reasonable complexity due to the iterative nature of the decoders, and the fact that most of the calculations can be done in parallel. The complexity of such decoders, for example belief propagation, scale linearly with the block-length. In fact, in most of the complexity analyses of such codes, the complexity is usually characterized in terms of the block length. However, for

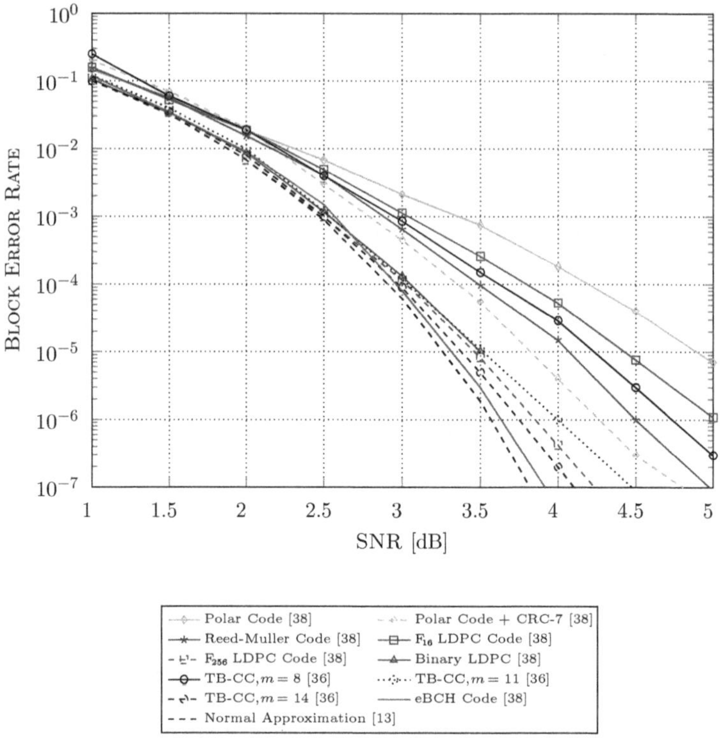

Figure 5.2 Comparison of error performance of different rate $R = 1/2$ channel codes with codeword length of $n = 128$ under MLD [23]. This figure is reused with permission from [44].

short block lengths other code parameters have significant impact on the decoding complexity. Here, we only focus on the algorithmic complexity, which can be represented in terms of the number of binary operations. For example, the decoding complexity of a TB-CC code using the Viterbi decoder is mainly dominated by the memory order, in short block lengths, as the memory order should be usually large to guarantee the performance.

Figure 5.4 shows the complexity versus the performance of different channel codes. As can be seen, polar codes with the SCL decoder achieve the error rate of 10^{-4} at only 0.5dB gap to the normal approximation benchmark with the complexity of the order of 10^3 operations per bit. The complexity can be reduced by reducing the list size, which however degrades the performance. TB-CC codes have huge complexity, which significantly increases with the memory order. The original OSD decoder has the complexity of the order of $\mathcal{O}(k^m)$, for m being the order number. An order-5 OSD generates $\sum_{i=1}^{5} \binom{k}{i}$ codeword candidates during the decoding. The black dash curve in Figure 5.4 is obtained with purely reducing the maximum allowed number of codeword candidates of the order-5 OSD.

5.1.4 Recommendations of Coding and Decoding in URLLC

As has been shown by comparison in Section 5.1.3, each type of code has its advantages and drawbacks. Some of them are superior to others in terms of exploiting fast decoding algorithms, while some of them could achieve narrow gaps to the NA bound at the expense of decoding complexity.

It can be first concluded that it is almost impractical to select TB-CC as the encoding scheme of URLLC. From Figure 5.4, it can be seen that the TB-CC provides the inferior trade-offs between the decoding complexity and the reliability, which is even worse than the BCH codes decoded by OSD. Second, turbo codes will be scarcely considered for URLLC applications because of its performance degradation in short/moderate block lengths.

Polar codes, LDPC codes, and the BCH codes are conceivable coding scheme candidates for the URLLC. It has been shown that the CRC-aided polar codes offer a desirable trade-off between decoding complexity and the gap to NA (see Figure 5.3). Nevertheless, the SCL decoding with list size $L = 32, 64$ still draws a relatively high decoding complexity. Thus, one can further design the low-complexity SCL decoding algorithms to empower polar codes of low-latency communications. Moreover, the rate matching (i.e. shortening and puncturing) of

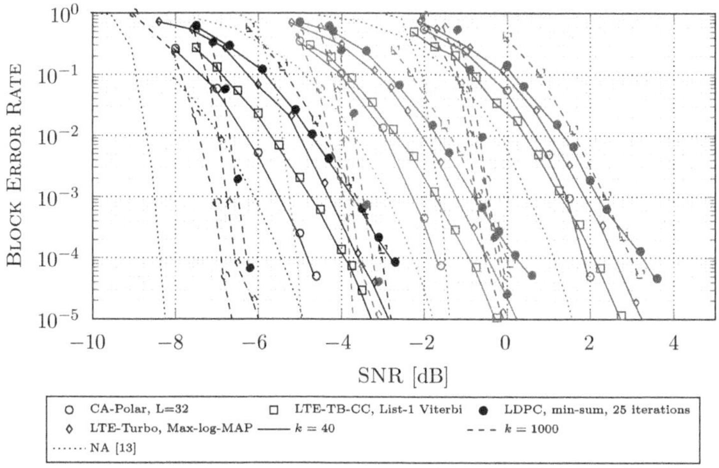

Figure 5.3 Comparison of different channel codes with different rates and information block lengths. Rates $R = 1/3$, $1/6$, and $1/12$, are respectively shown from the left-hand side to the right-hand side of the graph. This figure is reused with permission from [44].

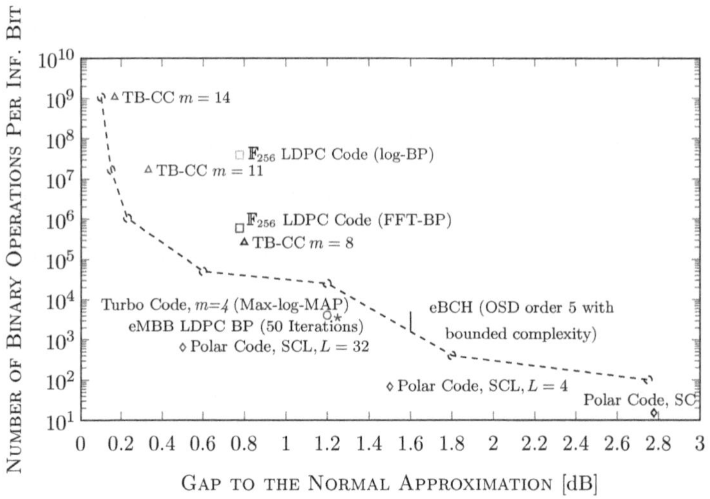

Figure 5.4 Algorithmic complexity versus performance for different rate-1/2 channel codes with block length $N = 128$ at BLER $= 10^{-4}$. This figure is reused with permission from [44].

polar codes still needs to be thoroughly studied. It has been demonstrated that in shortened and punctured patterns, the code bits must rely on the frozen bits, forcing the freezing of some of the most reliable synthesized channels and lowering the error correction probability [7].

Although LDPC codes show a considerable gap with the NA bound, it has the opportunity to be utilized in URLLC because of its low-complexity iterative BP decoding. As shown in Figure 5.3, the LDPC codes can provide an acceptable gap to the NA bound for low-rate short code, exhibiting a similar BLER performance compared to polar codes. However, the BLER performance of the LDPC code should be improved for a wider range of code rates for its comprehensive application in URLLC. The improvement of the BLER performance can be achieved by employing near-ML decoders. As reported by [32, Figure 1], when the LDPC codes are decoded by OSD, its gap to the NA bound can be shortened by 0.5 dB approaching similar BLER performance to the BCH codes (but slightly worse). Thus, one can design near-ML decoding algorithms that can take advantage of the sparse property of LDPC codes to improve the BLER performance while maintaining the low decoding complexity of LDPC codes.

As shown in Figure 5.2, under the OSD, BCH codes outperform other existing channel codes, including polar codes and LDPC codes. However, the algorithm complexity of OSD must be significantly reduced to enable the fast decoding of

BCH codes. Besides, the inflexibility of BCH codes will limit their employment in URLLC, as only a fixed range of code rates and code lengths of BCH codes is available.

It can be seen that for LDPC codes, polar codes, and BCH codes, they need superior decoders, which should be advanced in terms of both complexity and error correction capability. The OSD, as a universal decoder for block codes, rekindled interest recently [13, 50, 57, 58]. Compared to the specially designed decoder for different codes, OSD has the following advantages. First, OSD is universal and the required inputs of the decoding are the generator matrix and the transmitted signal only. This feature significantly benefits the implementation of decoders in devices and receivers. That is the receiver is equipped with the OSD decoder which is compatible with any utilized block codes. On the other hand, OSD does not depend on any unique structure of the codes. In fact, with OSD, the design of rate-compatible codes can evade obeying the unique structure of the generator matrix (e.g. frozen bits of polar and sparsity of LDPC) without degrading to sub-optimal performance. Finally, as proved in [18], OSD is near ML in terms of BLER, which guarantees the high reliability of decoding results. OSD can also be implemented using parallel structures. It can be designed as a branches-based decoder, with each branch generating codeword candidates separately.

Despite the above advantages, currently, the high computational complexity of OSD prohibits its application in URLLC. In the following sections of this chapter, we will introduce the algorithm of OSD and approaches to reduce its complexity.

5.2 Ordered Statistics Decoding

The ordered statistics decoding was proposed in 1995, as an approximation of the ML decoder for linear block codes [18] to reduce the decoding complexity. For a linear block code $\mathcal{C}(n, k)$, with minimum distance d_{H}, it has been proven that an OSD with the order of $m = \lceil d_{\mathrm{H}}/4 - 1 \rceil$ is asymptotically optimum approaching the same performance as the ML decoding [18]. However, the decoding complexity of an order-m OSD can be as high as $\mathcal{O}(k^m)$[18].

In OSD, the bit-wise Log-Likelihood Ratios (LLRs) of the received symbols are sorted in descending order, and the order of the received symbols and the columns of the generator matrix are permuted accordingly. Gaussian Elimination (GE) over the permuted generator matrix is performed to transform it to a systematic form. Then, the first k positions, referred to as the Most Reliable Basis (MRB), will be XORed with a set of the Test Error Patterns (TEP) with the Hamming weight up to a certain degree, where the maximum Hamming weight of TEPs is referred to as the decoding order. Then the vectors obtained by XORing the MRB are re-encoded using the permuted generator matrix to generate candidate codeword estimates. This is referred to as the reprocessing and will continue until all the TEPs with the Hamming weights up to the decoding order are processed. Finally, the codeword estimate with the minimum distance from the received signal is selected as the decoding output.

Next, we introduce the algorithm of the original OSD [18], and give the analysis of its error rate performance and the computational complexity. The frequently used notations in the remainder of this chapter are summarized in Table 5.1.

5.2.1 Algorithm of OSD

We consider a binary linear block code $\mathcal{C}(n,k)$ with BPSK modulation over an AWGN channel, where k and n denote the information block and codeword

Table 5.1 Table of notations.

Symbols	Definition
X	a random variable
$[X]_u^v$	a sequence of random variables $[X]_u^v = [X_u, X_{u+1}, \dots, X_v]$
x	values of scalar variables or the sample of random variables
$[x]_u^v$	a sequence of scalar variables or samples $[x]_u^v = [x_u, \dots, x_v]$
$f_X(x)$	the probability density function (pdf) of a continuous X
$f_X(x\|Z=z)$	the pdf of a continuous X conditioning on $\{Z=z\}$
$F_X(x)$	the cumulative distribution function (cdf) of a continuous X
$F_X(x\|Z=z)$	the cdf of a continuous X conditioning on $\{Z=z\}$
$p_Y(y)$	the probability mass function (pmf) of a discrete Y
$p_Y(y\|Z=z)$	the pmf of a discrete Y conditioning on $\{Z=z\}$
$\mathbb{E}[X]$	the mean of X
$\mathbb{E}[X\|Z=z]$	the mean of X conditioning on $\{Z=z\}$
σ_X^2	the variance of X
$\sigma_{X\|Z=z}^2$	the variance of X conditioning on $\{Z=z\}$
$\mathbf{A}$	a matrix
$\mathbf{A}^{\mathrm{T}}$	the transposition of matrix $\mathbf{A}$
$\mathbf{a}$	a row vector
$\mathcal{A}$	a set
$\mathcal{O}(x)$	bounded above by x with up to constant factor asymptotically
$\mathcal{N}(\cdot,\cdot)$	the Gaussian distribution
$\mathcal{B}(\cdot,\cdot)$	the binomial distribution

length, respectively. Let $\mathbf{b} = [b]_1^k$ and $\mathbf{c} = [c]_1^n$ denote the information sequence and codeword, respectively. Given the generator matrix $\mathbf{G}$ of code $\mathcal{C}(n,k)$, the encoding operation can be described as $\mathbf{c} = \mathbf{bG}$. At the channel output, the received signal (also referred to as the noisy signal) is given by $\mathbf{r} = \mathbf{s} + \mathbf{w}$, where $\mathbf{s} = [s]_1^n$ denotes the sequence of modulated symbols with $s_u = (-1)^{c_u} \in \{\pm 1\}$, $1 \leq u \leq n$, and $\mathbf{w} = [w]_1^n$ is the AWGN vector with zero mean and variance $\sigma_n^2 = N_0/2$, for N_0 being the single side-band power spectrum density. The SNR is then given by $\gamma = 2/N_0$.

At the receiver, the bit-wise hard decision vector $\mathbf{y} = [y]_1^n$ can be obtained according to the following rule:

$$y_u = \begin{cases} 1, & \text{for } r_u < 0, 1 \leq u \leq n \\ 0, & \text{for } r_u \geq 0, 1 \leq u \leq n \end{cases} \tag{5.4}$$

where y_u is the hard-decision estimation of codeword bit c_u.

In general, if the codewords in $\mathcal{C}(n,k)$ have equal transmission probability, the Log-Likelihood-Ratio (LLR) of the uth symbol of the received signal can be calculated as $\ell_u \triangleq \ln \frac{\Pr(c_u=0|r_u)}{\Pr(c_u=1|r_u)}$, which can be further simplified to $\ell_u = 4r_u/N_0$ if BPSK symbols are transmitted. We consider the scaled magnitude of LLR as the reliability corresponding to bitwise decision, defined by $\alpha_u = |r_u|$, where $|\cdot|$ is the absolute operation. Utilizing the bit reliability, the soft-decision decoding can be effectively conducted using the OSD algorithm [18]. In OSD, a permutation π_1 is performed to sort the received signal $\mathbf{r}$ and the corresponding columns of the generator matrix in descending order of their reliabilities. The sorted received symbols and the sorted hard-decision vector are denoted by $\mathbf{r}^{(1)} = \pi_1(\mathbf{r})$ and $\mathbf{y}^{(1)} = \pi_1(\mathbf{y})$, respectively, and the corresponding reliability vector and permuted generator matrix are denoted by $\boldsymbol{\alpha}^{(1)} = \pi_1(\boldsymbol{\alpha})$ and $\mathbf{G}^{(1)} = \pi_1(\mathbf{G})$, respectively.

Next, the systematic form matrix $\widetilde{\mathbf{G}} = [\mathbf{I}_k \ \widetilde{\mathbf{P}}]$ is obtained by performing GE on $\mathbf{G}^{(1)}$, where $\mathbf{I}_k$ is a k-dimensional identity matrix and $\widetilde{\mathbf{P}}$ is the parity sub-matrix. An additional permutation π_2 may be performed during GE to ensure that the first k columns are linearly independent. The permutation π_2 will inevitably disrupt the descending order property of $\boldsymbol{\alpha}^{(1)}$ to some extent; nevertheless, it has been shown that the disruption is minor[18]. Accordingly, the received symbols, the hard-decision vector, the reliability vector, and the generator matrix are sorted to $\widetilde{\mathbf{r}} = \pi_2(\pi_1(\mathbf{r}))$, $\widetilde{\mathbf{y}} = \pi_2(\pi_1(\mathbf{y}))$, $\widetilde{\boldsymbol{\alpha}} = \pi_2(\pi_1(\boldsymbol{\alpha}))$, and $\widetilde{\mathbf{G}} = \pi_2(\pi_1(\mathbf{G}))$, respectively.

After the GE and permutations, the first k index positions of $\widetilde{\mathbf{y}}$ are associated with the MRB [18], which is denoted by $\widetilde{\mathbf{y}}_{\mathrm{B}} = [\widetilde{y}]_1^k$, and the remaining positions are associated with the redundancy part. A test error pattern $\mathbf{e} = [e]_1^k$ is added to $\widetilde{\mathbf{y}}_{\mathrm{B}}$ to obtain one codeword estimate by re-encoding as follows.

$$\widetilde{\mathbf{c}}_{\mathbf{e}} = (\widetilde{\mathbf{y}}_{\mathrm{B}} \oplus \mathbf{e})\,\widetilde{\mathbf{G}} = \left[\widetilde{\mathbf{y}}_{\mathrm{B}} \oplus \mathbf{e} \ \ (\widetilde{\mathbf{y}}_{\mathrm{B}} \oplus \mathbf{e})\,\widetilde{\mathbf{P}}\right], \tag{5.5}$$

where $\widetilde{\mathbf{c}}_{\mathbf{e}} = [\widetilde{c}_{\mathbf{e}}]_1^n$ is the ordered codeword estimate with respect to TEP $\mathbf{e}$.

In OSD, TEPs are checked in increasing order of their Hamming weights; that is, in the i-reprocessing, all TEPs of Hamming weight i will be generated and re-encoded. The maximum Hamming weight of TEPs is limited to m, which is referred to as the decoding order of OSD. Thus, for an order-m decoding, maximum $\sum_{i=0}^{m} \binom{k}{i}$ TEPs will be re-encoded to find the best codeword estimate. For BPSK modulation, finding the best ordered codeword estimate $\widetilde{\mathbf{c}}_{\text{opt}}$ is equivalent to minimizing the Weighted Hamming Distance (WHD) between $\widetilde{\mathbf{c}}_{\mathbf{e}}$ and $\widetilde{\mathbf{y}}$, which is defined as [49]

$$d^{(\mathrm{W})}(\widetilde{\mathbf{c}}_{\mathbf{e}}, \widetilde{\mathbf{y}}) \triangleq \sum_{\substack{1 \le u \le n \\ \widetilde{c}_{\mathbf{e},u} \neq \widetilde{y}_u}} \widetilde{\alpha}_u. \tag{5.6}$$

Here, we also define the Hamming distance between $\widetilde{\mathbf{c}}_{\mathbf{e}}$ and $\widetilde{\mathbf{y}}$ as

$$d^{(\mathrm{H})}(\widetilde{\mathbf{c}}_{\mathbf{e}}, \widetilde{\mathbf{y}}) \triangleq ||\widetilde{\mathbf{c}}_{\mathbf{e}} \oplus \widetilde{\mathbf{y}}||, \tag{5.7}$$

where $||\cdot||$ is the ℓ_1-norm. For simplicity of notations, we denote the WHD and Hamming distance between $\widetilde{\mathbf{c}}_{\mathbf{e}}$ and $\widetilde{\mathbf{y}}$ by $d_{\mathbf{e}}^{(\mathrm{W})} = d^{(\mathrm{W})}(\widetilde{\mathbf{c}}_{\mathbf{e}}, \widetilde{\mathbf{y}})$ and $d_{\mathbf{e}}^{(\mathrm{H})} = d^{(\mathrm{H})}(\widetilde{\mathbf{c}}_{\mathbf{e}}, \widetilde{\mathbf{y}})$, respectively. Furthermore, we alternatively use $w(\mathbf{e})$ to denote the Hamming weight of a binary vector $\mathbf{e}$, e.g. $w(\mathbf{e}) = ||\mathbf{e}||$. Finally, the estimate $\hat{\mathbf{c}}_{\text{opt}}$ corresponding to the initial received sequence $\mathbf{r}$, is obtained by performing inverse permutations over $\widetilde{\mathbf{c}}_{\text{opt}}$, i.e. $\hat{\mathbf{c}}_{\text{opt}} = \pi_1^{-1}(\pi_2^{-1}(\widetilde{\mathbf{c}}_{\text{opt}}))$. The algorithm of the original OSD is summarized in Algorithm 1.

5.2.2 Error Rate Performance

5.2.2.1 Distributions of the Received Signals

For simplicity and without loss of generality, we assume an all-zero codeword from $\mathcal{C}(n,k)$ is transmitted. Thus, the uth symbol of the AWGN channel output $\mathbf{r}$ is given by $r_u = 1 + w_u$, $1 \le u \le n$. Channel output $\mathbf{r}$ is observed by the receiver and the bit-wise reliability is then calculated as $\alpha_u = |1 + w_u|$, $1 \le u \le n$. Let us consider the uth reliability as a random variable denoted by A_u, and thus the sequence of random variables representing the reliabilities is denoted by $[A]_1^n$. Correspondingly after the permutations, the random variables of ordered reliabilities $\boldsymbol{\alpha} = [\widetilde{\alpha}]_1^n$ are denoted by $[\widetilde{A}]_1^n$. Similarly, let $[R]_1^n$ and $[\widetilde{R}]_1^n$ denote the random variable sequences of the received signals before and after permutations, respectively. Note that $[A]_1^n$ and $[R]_1^n$ are two sequences of independent and identically distributed (i.i.d.) random variables. Thus, the pdf of R_u, $1 \le u \le n$, is given by

$$f_R(r) = \frac{1}{\sqrt{\pi N_0}} e^{-\frac{(r-1)^2}{N_0}}, \tag{5.8}$$

Algorithm 1 Original OSD [18]

Require:
Generator matrix $\mathbf{G}$, input signal $\mathbf{r}$, decoding order m,
Ensure:
The decoding output $\hat{\mathbf{c}}_{\text{opt}}$

//**Preprocessing**
Calculate reliability value $\alpha_i = |r_i|$
First permutation: $\boldsymbol{\alpha}' = \pi_1(\boldsymbol{\alpha})$, $\mathbf{r}' = \pi_1(\mathbf{r})$, $\mathbf{G}' = \pi_1(\mathbf{G})$
Gaussian elimination and second permutation: $\widetilde{\boldsymbol{\alpha}} = \pi_2(\boldsymbol{\alpha}')$, $\widetilde{\mathbf{r}} = \pi_2(\mathbf{r}')$, $\widetilde{\mathbf{G}} = \pi_2(\mathbf{G}')$
Perform hard-decision: $\widetilde{y}_i = \begin{cases} 1 & \text{for } \widetilde{r}_i < 0 \\ 0 & \text{for } \widetilde{r}_i \geq 0 \end{cases}$
Initialize $d_{\min}^{(\mathrm{W})} = \infty$, $\widetilde{\mathbf{c}}_{\text{opt}} = \widetilde{\mathbf{y}}$
//**Reprocessing**
for $\ell = 0 : m$ **do**
  for $j = 1 : \binom{k}{\ell}$ **do**
    Take an unprocessed weight-ℓ TEP $\mathbf{e}_j$.
    Calculate $\widetilde{\mathbf{c}}_{\mathbf{e}_j} = (\widetilde{\mathbf{y}}_{\mathrm{B}} \oplus \mathbf{e})\widetilde{\mathbf{G}}$
    Calculate the WHD $d_{\mathbf{e}_j}^{(\mathrm{W})}$
    if $d_{\mathbf{e}_j}^{(\mathrm{W})} < d_{\min}^{(\mathrm{W})}$ **then**
      Declaim $\widetilde{\mathbf{c}}_{\text{opt}} = \widetilde{\mathbf{c}}_{\mathbf{e}_j}$ and $d_{\min}^{(\mathrm{W})} = d_{\mathbf{e}_j}^{(\mathrm{W})}$
    end if
  end for
end for
return $\hat{\mathbf{c}}_{\text{opt}} = \pi_1^{-1}(\pi_2^{-1}(\widetilde{\mathbf{c}}_{\text{opt}}))$

and the pdf of A_u, $1 \leq u \leq n$, is given by

$$f_A(\alpha) = \begin{cases} 0, & \text{if } \alpha < 0, \\ \dfrac{e^{-\frac{(\alpha+1)^2}{N_0}}}{\sqrt{\pi N_0}} + \dfrac{e^{-\frac{(\alpha-1)^2}{N_0}}}{\sqrt{\pi N_0}}, & \text{if } \alpha \geq 0. \end{cases} \tag{5.9}$$

Given the Q-function defined by $Q(x) = \frac{1}{\sqrt{2\pi}} \int_x^{\infty} \exp(-\frac{u^2}{2})du$, the cdf of A_u can be derived as

$$F_A(\alpha) = \begin{cases} 0, & \text{if } \alpha < 0, \\ 1 - Q(\frac{\alpha+1}{\sqrt{N_0/2}}) - Q(\frac{\alpha-1}{\sqrt{N_0/2}}), & \text{if } \alpha \geq 0. \end{cases} \tag{5.10}$$

By omitting the second permutation in GE[1], the pdf of the uth order reliability $\widetilde{A}_u$ can be derived as [37]

$$f_{\widetilde{A}_u}(\widetilde{\alpha}_u) = \frac{n!}{(u-1)!(n-u)!}(1 - F_A(\widetilde{\alpha}_u))^{u-1} F_A(\widetilde{\alpha}_u)^{n-u} f_A(\widetilde{\alpha}_u). \tag{5.11}$$

Similarly, the joint pdf of $\widetilde{A}_u$ and $\widetilde{A}_v$, $1 \leq u < v \leq n$, can be derived as follows.

$$\begin{aligned} f_{\widetilde{A}_u,\widetilde{A}_v}(\widetilde{\alpha}_u, \widetilde{\alpha}_v) = & \frac{n!}{(u-1)!(v-u-1)!(n-v)!} \\ & \cdot (1 - F_A(\widetilde{\alpha}_u))^{u-1} \left(F_A(\widetilde{\alpha}_u) - F_A(\widetilde{\alpha}_v)\right)^{v-u-1} \\ & \cdot F_A(\widetilde{\alpha}_v)^{n-v} f_A(\alpha_u) f_A(\widetilde{\alpha}_v) \mathbf{1}_{[0,\widetilde{\alpha}_u]}(\widetilde{\alpha}_v), \end{aligned} \tag{5.12}$$

where $\mathbf{1}_{\mathcal{X}}(x) = 1$ if $x \in \mathcal{X}$ and $\mathbf{1}_{\mathcal{X}}(x) = 0$, otherwise. For the random variable sequence $[\widetilde{R}]_1^n$ of ordered received signal, the pdf of $\widetilde{R}_u$ and the joint pdf of $\widetilde{R}_u$ and $\widetilde{R}_v$, $0 \leq u < v \leq n$, are respectively given by

$$f_{\widetilde{R}_u}(\widetilde{r}_u) = \frac{n!}{(u-1)!(n-u)!}(1 - F_A(|\widetilde{r}_u|))^{u-1} F_A(|\widetilde{r}_u|)^{n-u} f_R(\widetilde{r}_u) \tag{5.13}$$

and

$$\begin{aligned} f_{\widetilde{R}_u,\widetilde{R}_u}(\widetilde{r}_u, \widetilde{r}_v) = & \frac{n!}{(u-1)!(v-u-1)!(n-v)!} \\ & \cdot (1 - F_A(|\widetilde{r}_u|))^{u-1} (F_A(|\widetilde{r}_u|) - F_A(|\widetilde{r}_v|))^{v-u-1} \\ & \cdot F_A(|\widetilde{r}_v|)^{n-v} f_R(\widetilde{r}_u) f_R(\widetilde{r}_v) \mathbf{1}_{[0,|\widetilde{r}_u|]}(|\widetilde{r}_v|). \end{aligned} \tag{5.14}$$

These results of distributions of ordered random variables, from (5.8) to (5.14), are frequently used in the analysis of OSD.

5.2.2.2 Error Probabilities

For the original OSD, the bits of MRB $\widetilde{\mathbf{y}}_\mathrm{B}$ are flipped by XORing $\widetilde{\mathbf{y}}_\mathrm{B}$ with a TEP. Thus, the error rate performance of original OSD is determined by the error probabilities of the bits of MRB $\widetilde{\mathbf{y}}_\mathrm{B}$. According to (5.13), the bit-wise error

1 The second permutation π_2 occurs only when the first k columns of $\pi_1(G)$ are not linearly independent. As shown in Eq. (59) of [18], the probability of permutation π_2 occurring is very small. Also, even if π_2 occurs, the number of operations of π_2 is much smaller than the number of operations of π_1 [18]. Therefore, we omit the π_2 in the following analyses for the simplicity.

probability $\mathrm{Pe}(\widetilde{u})$ of the uth ordered received symbol $\widetilde{R}_u$ is given by

$$\mathrm{Pe}(u) = \int_{-\infty}^{0} f_{\widetilde{R}_i}(\widetilde{r}) d\widetilde{r}, \tag{5.15}$$

and the joint error probability $\mathrm{Pe}(\widetilde{u}, \widetilde{v})$ of the uth and vth ordered received symbol is given by

$$\mathrm{Pe}(u, v) = \int_{-\infty}^{0} f_{\widetilde{R}_i,\widetilde{R}_j}(\widetilde{r}_u, \widetilde{r}_v) d\widetilde{r}_u d\widetilde{r}_v. \tag{5.16}$$

Note that the error probabilities given by (5.15) and (5.16) are conditional on the unknown instantaneous received signal. Let us assume that at the receiver, a sequence of the samples of $[\widetilde{A}]_1^n$ is given by $\widetilde{\boldsymbol{\alpha}} = [\widetilde{\alpha}]_1^n$, i.e. the receiver knows a signal sequence $\mathbf{r}$ with reliabilities $\widetilde{\boldsymbol{\alpha}}$. Thus, conditioning on $\widetilde{A}_u = \widetilde{\alpha}_u$, the error probability of the uth $(1 \leq u \leq n)$ bit of $\widetilde{\mathbf{y}}$ can be obtained as

$$\mathrm{Pe}(u|\widetilde{A}_u = \widetilde{\alpha}_u) = \frac{f_R(-\widetilde{\alpha}_u)}{f_R(-\widetilde{\alpha}_u) + f_R(\widetilde{\alpha}_u)} = \frac{1}{1 + \exp(4\alpha_u/N_0)}. \tag{5.17}$$

Next, we introduce an important Lemma that describes the number of errors over a specific range of the ordered received signals.

Lemma 5.1: Let random variable E_a^b denote the number of errors in the positions from a to b, $1 \leq a < b \leq n$ over the ordered hard-decision vector $\widetilde{\mathbf{y}}$. The probability mass function $p_{E_a^b}(j)$ of E_a^b, for $0 \leq j \leq b - a + 1$, is given by (5.19) on the top of the next page, where $f_{\widetilde{A}_a}(x)$ and $f_{\widetilde{A}_a,\widetilde{A}_b}(x, y)$ are given by (5.11) and (5.12), respectively, and $p(x, y)$ is given by

$$p(x, y) = \frac{Q(\frac{-2x-2}{\sqrt{2N_0}}) - Q(\frac{-2y-2}{\sqrt{2N_0}})}{Q(\frac{-2x-2}{\sqrt{2N_0}}) - Q(\frac{-2y-2}{\sqrt{2N_0}}) + Q(\frac{2y-2}{\sqrt{2N_0}}) - Q(\frac{2x-2}{\sqrt{2N_0}})}. \tag{5.18}$$

Proof. We refer interested readers to [58, Lemma 1] for the proof.

Lemma 5.1 can be used to derive many properties of OSD. For example, the number of errors over MRB in fact decides the error performance of OSD. If there are $m+1$ errors over the MRB, the OSD will output an incorrect codeword estimate as it can correct maximum m MRB errors. The number of MRB errors is simply described by E_1^k introduced in Lemma 5.1.

$$p_{E_a^b}(j) = \begin{cases} \int_0^{\infty}\int_0^{\infty}\binom{b-a+1}{j}p(x,y)^j(1-p(x,y))^{b-a+1-j}f_{\tilde{A}_{a-1},\tilde{A}_{b+1}}(x,y)dydx, \\ \qquad \text{for } a>1 \text{ and } b<n \\ \int_0^{\infty}\binom{n-a+1}{j}p(x,0)^j(1-p(x,0))^{n-a+1-j}f_{\tilde{A}_{a-1}}(x)dx, \\ \qquad \text{for } a>1 \text{ and } b=n \\ \int_0^{\infty}\binom{b}{j}p(\infty,y)^j(1-p(\infty,y))^{b-j}f_{\tilde{A}_{b+1}}(y)dy, \\ \qquad \text{for } a=1 \text{ and } b<n \\ \binom{n}{j}\left(1-Q\left(\frac{-2}{\sqrt{2N_0}}\right)\right)^j Q\left(\frac{-2}{\sqrt{2N_0}}\right)^{n-j}, \\ \qquad \text{for } a=1 \text{ and } b=n \end{cases} \tag{5.19}$$

Generally, the overall decoding error probability of an order-m OSD can be upper bounded by [13, 18]

$$\epsilon_e(m) \leq \mathrm{P}_{\mathrm{list}}(m) + \mathrm{P}_{\mathrm{ML}}, \tag{5.20}$$

where P_{ML} is the error rate of lMLD of the code $\mathcal{C}(n,k)$, and $\mathrm{P}_{\mathrm{list}}$ is the probability that the errors in the MRB, i.e. $\tilde{\mathbf{y}}_{\mathrm{B}}$, cannot be eliminated by any TEPs with Hamming weights less than m. It is worth noting that P_{ML} is decided by the code structure of $\mathcal{C}(n,k)$ (known as the error correction capability of the code itself), which can be determined by computer search.

From Lemma 5.1, it can be seen that for an order-m OSD, $\mathrm{P}_{\mathrm{list}}(m)$ is conveniently given by $1-\sum_{i=0}^{m} p_{E_1^k}(i)$. Thus, the error rate upper-bound $\epsilon_e(m)$ of an order-m OSD can be derived as

$$\epsilon_e(m) \leq 1 - \sum_{i=0}^{m} p_{E_1^k}(i) + \mathrm{P}_{\mathrm{ML}}, \tag{5.21}$$

where $p_{E_1^k}(i)$ is given by (5.19).

Furthermore, if the permutation π_2 is considered, a tighter error rate upper-bound can be given by [13, Eq. (24)]

$$\epsilon_e = \sum_{d=0}^{n-k-d_{\mathrm{H}}+1} \mathrm{P}_{\pi_2}(d)\left(1 - \sum_{i=0}^{m} p_{E_1^{k+d}}(i)\right) + \mathrm{P}_{\mathrm{ML}}, \tag{5.22}$$

where $P_{\pi_2}(d)$ [18, Equ. 59] is the probability that d dependent columns are permuted in the second permutation.

We depict the BLER performance of decoding (128, 64, 22) eBCH code with OSD in Figure 5.5. As shown, the order-4 decoding could result in a BLER performance close to the NA bound [15], despite that the optimal order is 5 for this block code with $d_{\mathrm{H}} = 22$.

5.2.3 The Computational Complexity

This section evaluates the overall computational complexity of OSD algorithms. Let the C_{total} denote the computational complexity of an OSD algorithm, then C_{total} can be derived as

$$\begin{aligned} C_{\mathrm{OSD}} = \mathcal{O}(n) + & \underbrace{\mathcal{O}(n\log n)}_{\text{Sorting (FLOP)}} + \underbrace{\mathcal{O}(n\min(k,n-k)^2)}_{\text{Gaussian elimination (BOP)}} \\ & + N_{\max} \underbrace{\mathcal{O}(k + k(n-k))}_{\text{re-encoding (BOP)}}. \end{aligned} \tag{5.23}$$

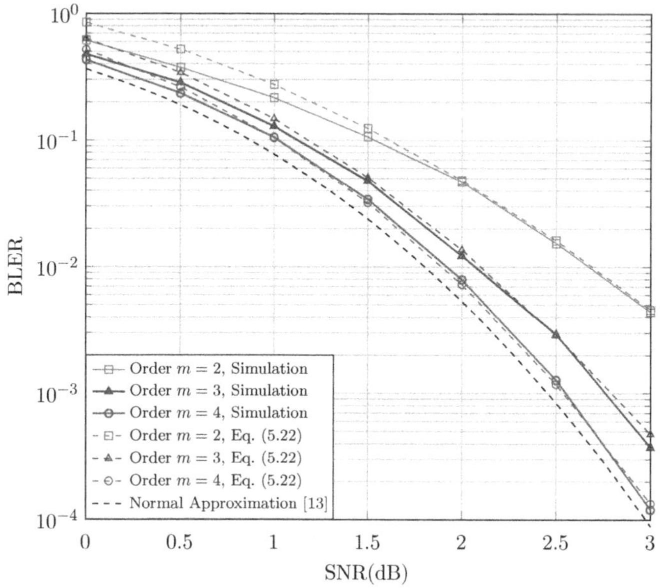

Figure 5.5 The BLER performance of OSD decoding (128,64,22) eBCH code.

where $N_{\max}$ is the number of re-encoded TEPs, and each term represents the complexity of various stages in the original OSD [18]. Specifically, the computational complexity of OSD depends on the following factors.

- Sorting (the first permutation π_1): merge sort algorithm can efficiently generate and perform the first permutation with average complexity of $\mathcal{O}(n \log n)$ FLOPs[18].
- Gaussian elimination: the operation to obtain systematic generation matrix $\widetilde{\mathbf{G}}$ from $\mathbf{G}$ can be done with $\mathcal{O}(n \min(k, n-k)^2)$ BOPs [18].
- Re-encoding: re-encoding $\widetilde{\mathbf{c}}_{\mathbf{e}} = (\widetilde{\mathbf{y}}_{\mathrm{B}} \oplus \mathbf{e})\widetilde{\mathbf{G}}$ uses k sign operations and $n-k$ parallel k XOR operations [18], which can be represented as $\mathcal{O}(k + k(n-k))$ BOPs.
- Number of TEPs $N_{\max}$: for OSD-based decoding, the total number of checked TEPs, $N_{\max}$, greatly affects the complexity since $N_{\max}$ times of re-encoding is required.
- Other components: the complexity of other components (e.g. the second permutation π_2) are summarized in the linear complexity term $\mathcal{O}(n)$

Equation (5.23) identifies the approaches to reduce the decoding complexity of OSD. For example, one can design some metrics to indicate the qualities of TEPs, and discard some TEPs with low qualities to reduce $N_{\max}$. Also, the complexity of GE takes a significant portion of the overall complexity. The overhead of GE can be relatively reduced by introducing partial GE techniques. Furthermore, a potential paralleled implementation can be used in the re-processing. That is, to use multiple branches re-encoding TEPs simultaneously to reduce the time complexity.

We will introduce several techniques to reduce the decoding complexity in the next section.

5.3 Fast OSD Approaches for URLLC

The original OSD has been in development for over 20 years. Over the decades, some significant improvements have been achieved. Some published papers have considered the information given by the most reliable bits in the MRB to design decoding techniques [17, 27, 48, 53, 54], while some studies also utilized the information outside the MRB to obtain further refinements [17, 48].

As significant results, skipping and stopping rules were introduced in [54] and [53] to prevent unpromising TEPs, which are unlikely to result in the correct output. The BMA approach proposed in [48] can significantly reduce the decoding complexity by using the "match" procedure, which defines a control band (CB) and identifies each TEP based on CB. The matching of candidates are implemented by memory spaces called "boxes", which however introduces a considerable number of extra overheads on computations and memory. The Iterative Information Set Reduction (IISR) technique was proposed in [17] to reduce the complexity of OSD. IISR applies a permutation over the positions around the boundary of MRB and generates a new MRB after each reprocessing. A fast OSD algorithm was proposed in [50], which combines the discarding rules from [53], the stopping criterion from [27], and the multibasis scheme in [25], reducing the complexity from $\mathcal{O}(k^m)$ to $\mathcal{O}(k^{m-2})$ at high SNRs. The complexity of OSD was further reduced by applying a segmentation technique [56], which adaptively divides the list of TEPs into segments and reduces the overhead of stopping and skipping rules. Most recently, deep learning has been applied to help OSD identify the required decoding order to achieve the optimal performance [8].

Some attempts were also made to analyze the error performance of the OSD algorithm and its alternatives [13, 18–20]. Most recently, the evolutions of the Hamming distance and weighted Hamming distance in the reprocessing stage of OSD were studied by [58], which introduces insights into how the techniques (e.g. stopping and skipping rules) could improve the performance of OSD. Based on the theoretical results provided in [58], a novel Probability-Based OSD (PB-OSD) was proposed in [57], which could significantly reduce the required number of TEPs with maintaining the similar BLER performance of the original OSD.

Towards the application of OSD in the URLLC systems, the end-to-end latency of applying the OSD was analyzed and optimized in [9]. In [9], the latency of transmission in using OSD is represented as a function of the decoding complexity. Then, the overall transmission latency is minimized under the power constraint, and information rate is maximized with the latency constraints.

In this section, we will introduce some properties of OSD and some of the techniques devised to reduce the complexity of OSD. Most of the introduced results are based on the latest study [58]. For other improved approaches of OSD, we refer the readers to the literature reviewed.

5.3.1 Some Properties of OSD

5.3.1.1 Approximations of Ordered Reliability

For the ordered reliability random variables $[\widetilde{A}]_1^n$, the pdf of $\widetilde{A}_u$, $1 \le u \le n$, is given by (5.11). In [58], it is shown that the pdf of $\widetilde{A}_u$ can be approximated by a normal distribution $\mathcal{N}(\mathbb{E}[\widetilde{A}_u], \sigma^2_{\widetilde{A}_u})$ with the pdf given by

$$f_{\widetilde{A}_u}(\widetilde{\alpha}_u) \approx \frac{1}{\sqrt{2\pi\sigma^2_{\widetilde{A}_u}}} \exp\left(-\frac{(\widetilde{\alpha}_u - \mathbb{E}[\widetilde{A}_u])^2}{2\sigma^2_{\widetilde{A}_u}} \right), \tag{5.24}$$

where

$$\mathbb{E}[\widetilde{A}_u] = F_A^{-1}(1 - \frac{u}{n}) \tag{5.25}$$

and

$$\sigma^2_{\widetilde{A}_u} = \pi N_0 \frac{(n-u)u}{n^3} \cdot \left(\exp\left(-\frac{(\mathbb{E}[\widetilde{A}_u]+1)^2}{N_0}\right) + \exp\left(-\frac{(\mathbb{E}[\widetilde{A}_u]-1)^2}{N_0}\right)\right)^{-2}. \tag{5.26}$$

Details of the approximation can be found in [58, Appendix A]. Similarly, the joint distribution of $\widetilde{A}_u$ and $\widetilde{A}_v$, $0 \leq u < v \leq n$, can be approximated to a bivariate normal distribution with the following joint pdf

$$f_{\widetilde{A}_u,\widetilde{A}_v}(\widetilde{\alpha}_u, \widetilde{\alpha}_v) \approx \frac{1}{2\pi\sigma_{\widetilde{A}_u}\sigma_{\widetilde{A}_v|\widetilde{A}_u=\widetilde{\alpha}_u}} \cdot \exp\left(-\frac{(\widetilde{\alpha}_u - \mathbb{E}[\widetilde{A}_u])^2}{2\sigma^2_{\widetilde{A}_u}} - \frac{(\widetilde{\alpha}_v - \mathbb{E}[\widetilde{A}_v|\widetilde{A}_u=\widetilde{\alpha}_u])^2}{2\sigma^2_{\widetilde{A}_v|\widetilde{A}_u=\widetilde{\alpha}_u}}\right), \tag{5.27}$$

where

$$\mathbb{E}[\widetilde{A}_v|\widetilde{A}_u = \widetilde{\alpha}_u] = \gamma^{-1}_{\widetilde{\alpha}_u}\left(\frac{v-u}{n-u}\right), \tag{5.28}$$

and

$$\sigma^2_{\widetilde{A}_v|\widetilde{A}_u=\widetilde{\alpha}_u} = \pi N_0 \frac{(n-v)(v-u)}{(n-u)^3} \cdot \left(\frac{\exp\left(\frac{-(\mathbb{E}[\widetilde{A}_v|\widetilde{A}_u=\widetilde{\alpha}_u]-1)^2}{N_0}\right) + \exp\left(\frac{-(\mathbb{E}[\widetilde{A}_v|\widetilde{A}_u=\widetilde{\alpha}_u]+1)^2}{N_0}\right)}{F_A(\widetilde{\alpha}_u)}\right)^{-2}. \tag{5.29}$$

In (5.28), $\gamma_{\widetilde{\alpha}_u}(t)$ is defined as follows

$$\gamma_{\widetilde{\alpha}_u}(t) = \frac{F_A(\widetilde{\alpha}_u) - F_A(t)}{F_A(\widetilde{\alpha}_u)}. \tag{5.30}$$

We show the distributions of ordered reliabilities in the decoding of a (128, 64, 22) eBCH code in Figure 5.6. As can be seen, the normal distribution $\mathcal{N}(\mathbb{E}[\widetilde{A}_u], \sigma^2_{\widetilde{A}_u})$ with the mean and variance given by (5.25) and (5.26), respectively, provides a good approximation to (5.11) for a wide range of u. Particularly, the approximation of the distribution of the uth reliability $\widetilde{A}_u$ is tight when u is not close to 1 or n. Specifically, when $u = n/2$ (by assuming n is even, similar analysis

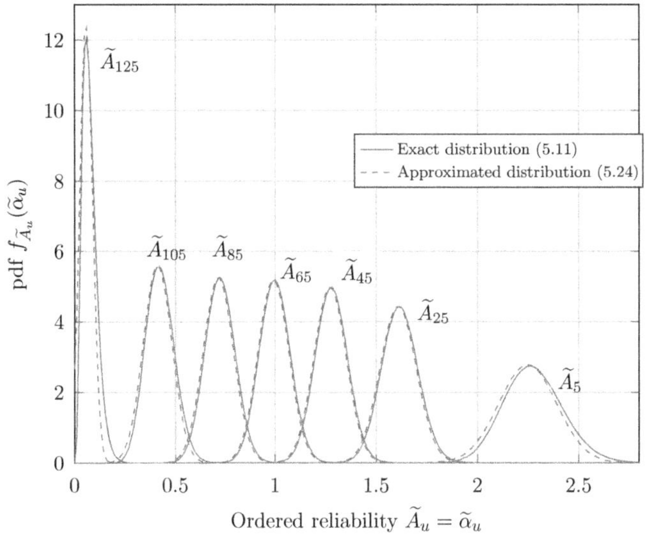

Figure 5.6 The approximation of the distribution of the uth ordered reliability in decoding a (128,64,22) eBCH code when SNR = 3 dB. This figure is reused with permission from [58].

can be drawn for $u = \lfloor n/2 \rfloor$ if n is odd), it can be seen that $\widetilde{A}_{\frac{n}{2}}$ is the median of the n samples $[\alpha_1, \alpha_2, \ldots, \alpha_n]$ of random variable A. Thus, when n is large, $\widetilde{A}_{\frac{n}{2}}$ is asymptotically normal with mean m_A and variance $\frac{1}{4nf_A(m_A)^2}$ [40], where m_A is the median of the distribution of A, defined as a real number satisfying

$$\int_{-\infty}^{m_A} f_A(x)dx \geq \frac{1}{2} \text{ and } \int_{m_A}^{\infty} f_A(x)dx \geq \frac{1}{2}. \tag{5.31}$$

Because $f_A(x)$ is a continuous pdf, it can be directly obtained that $m_A = F_A^{-1}(\frac{1}{2})$ from (5.31), that is, m_A is also given by (5.25) when $u = n/2$. Then, substituting $u = n/2$ and $m_A = F_A^{-1}(\frac{1}{2}) = \mathbb{E}[\widetilde{A}_{\frac{n}{2}}]$ into (5.26), it can be obtained that

$$\begin{aligned} \sigma^2_{\widetilde{A}_{\frac{n}{2}}} &= \frac{\pi N_0}{4n}\left(\exp\left(-\frac{(m_A+1)^2}{N_0}\right)+\exp\left(-\frac{(m_A-1)^2}{N_0}\right)\right)^{-2} \\ &= \frac{1}{4nf_A(m_A)^2}. \end{aligned} \tag{5.32}$$

Therefore, it can be concluded that (5.24) with mean (5.25) and variance (5.26) provides a tight approximation for $\widetilde{A}_{\frac{n}{2}}$, which is consistent with the results given in [40].

The introduced approximations of ordered reliabilities will be especially helpful in the analysis regarding OSD. For example, many variables (e.g. error probabilities, distance, etc.) in OSD rely on ordered reliabilities, and their characters can be conveniently analyzed with normal-approximated distributions of reliabilities.

5.3.1.2 Distribution of Distances from Estimates to Signals

This section introduces the distributions of the distance in OSD, where the "distance" is the distance from codeword estimates to the received signal. The properties of these distributions can be used to design tools to reduce the decoding complexity.

In [58], it has been verified that the distribution of the distance in OSD can be described by a mixed model of two random variables related to the number of channel errors and the code weight enumerator, respectively, and the weight of the mixture is determined by the channel condition in terms of SNR.

Let us take the 0-reprocessing Hamming distance $d_0^{(\mathrm{H})} = d^{(\mathrm{H})}(\widetilde{\mathbf{c}}_0, \widetilde{\mathbf{y}})$ as an example. In the 0-reprocessing where no TEP is added to MRB positions before re-encoding, $\widetilde{\mathbf{c}}_0$ is obtained as $\widetilde{\mathbf{c}}_0 = \widetilde{\mathbf{y}}_{\mathrm{B}}\widetilde{\mathbf{G}}$. To find the distribution of 0-reprocessing Hamming distance, we now regard it as a random variable denoted by $D_0^{(\mathrm{H})}$, and accordingly $d_0^{(\mathrm{H})}$ is the sample of $D_0^{(\mathrm{H})}$.

Let us re-write $\widetilde{\mathbf{y}}$ and $\widetilde{\mathbf{c}}_0$ as $\widetilde{\mathbf{y}} = [\widetilde{\mathbf{y}}_{\mathrm{B}} \ \ \widetilde{\mathbf{y}}_{\mathrm{P}}]$ and $\widetilde{\mathbf{c}}_0 = [\widetilde{\mathbf{c}}_{0,\mathrm{B}} \ \ \widetilde{\mathbf{c}}_{0,\mathrm{P}}]$, respectively, where subscript B and P denote the first k positions and the remaining positions of a length-n vector, respectively. Also, let us define $\widetilde{\mathbf{c}} = \pi_2(\pi_1(\mathbf{c})) = [\widetilde{\mathbf{c}}_{\mathrm{B}} \ \ \widetilde{\mathbf{c}}_{\mathrm{P}}]$ representing the transmitted codeword after permutations, which is unknown to the decoder but useful in the analysis later. Accordingly, we define $\widetilde{\mathbf{e}} = [\widetilde{\mathbf{e}}_{\mathrm{B}} \ \ \widetilde{\mathbf{e}}_{\mathrm{P}}]$ as the permuted hard-decision error, i.e. $\widetilde{\mathbf{e}} = \widetilde{\mathbf{c}} \oplus \widetilde{\mathbf{y}}$. For an arbitrary permuted codeword $\widetilde{\mathbf{c}}' = [\widetilde{\mathbf{c}}'_{\mathrm{B}} \ \ \widetilde{\mathbf{c}}'_{\mathrm{P}}]$ from $\mathcal{C}(n,k)$, where $\widetilde{\mathbf{c}}'$ is generated by an information vector $\mathbf{b}'$ with Hamming weight $w(\mathbf{b}') = q$ and the permuted generator matrix $\widetilde{\mathbf{G}}$, i.e. $\widetilde{\mathbf{c}}' = \mathbf{b}'\widetilde{\mathbf{G}}$, we further define $p_{\mathbf{c}_{\mathrm{P}}}(u,q)$ as the probability of $w(\widetilde{\mathbf{c}}'_{\mathrm{P}}) = u$ when $w(\mathbf{b}') = q$ i.e. $p_{\mathbf{c}_{\mathrm{P}}}(u,q) = \Pr(w(\widetilde{\mathbf{c}}'_{\mathrm{P}}) = u | w(\mathbf{b}') = q)$. It can be seen that $p_{\mathbf{c}_{\mathrm{P}}}(u,q)$ is characterized by the structure of the generator matrix $\mathbf{G}$ of $\mathcal{C}(n,k)$, which is independent of the channel conditions.

In the 0-reprocessing, the Hamming distance $D_0^{(\mathrm{H})}$ is affected by both the number of errors in $\widetilde{\mathbf{y}}_{\mathrm{P}}$ and also the Hamming weights of the parity part $\widetilde{\mathbf{c}}'_{\mathrm{P}}$ of permuted codewords $\widetilde{\mathbf{c}}'$ from $\mathcal{C}(n,k)$ simultaneously, which is explained in the following Lemma.

Lemma 5.2: After the 0-reprocessing of decoding a linear block code $\mathcal{C}(n,k)$, the Hamming distance $D_0^{(\mathrm{H})}$ between $\widetilde{\mathbf{y}}$ and $\widetilde{\mathbf{c}}_0$ is given by [58]

$$D_0^{(\mathrm{H})} = \begin{cases} E_{k+1}^n, & \text{w.p. } p_{E_1^k}(0), \\ W_{\mathbf{c}_\mathrm{P}}, & \text{w.p. } 1 - p_{E_1^k}(0), \end{cases} \tag{5.33}$$

where E_{k+1}^n is the random variable defined in Lemma (5.1) and $p_{E_1^k}(0)$ is given by (5.19). $W_{\mathbf{c}_\mathrm{P}}$ is a discrete random variable whose pmf is given by

$$p_{W_{\mathbf{c}_\mathrm{P}}}(j) = \sum_{u=0}^{n-k}\sum_{v=0}^{n-k} \frac{\binom{u}{\delta}\binom{n-k-u}{v-\delta}}{\binom{n-k}{v}} \cdot p_d(u) \cdot p_{E_{k+1}^n}(v) \cdot \mathbf{1}_{\mathbb{N}\bigcap[0,\min(u,v)]}(\delta), \tag{5.34}$$

where $\delta = (u+v-j)/2$, and

$$p_d(u) = \frac{1}{1 - p_{E_1^k}(0)} \sum_{q=1}^{k} p_{E_1^k}(q) p_{\mathbf{c}_\mathrm{P}}(u,q). \tag{5.35}$$

Proof. We refer interested readers to [58, Lemma 3].

From (5.19), we can see that the probability $p_{E_1^k}(0)$ is a function of k, n, and the noise power N_0. If k and n are fixed, $p_{E_1^k}(0)$ is a monotonically increasing function of SNR. This implies that the channel condition determines the weight of the composition of the Hamming distance. Based on Lemma 5.2, the distribution of $D_0^{(\mathrm{H})}$ is directly obtained in the following Theorem.

Theorem 5.1: Given a linear block code $\mathcal{C}(n,k)$, the pmf of the Hamming distance between $\widetilde{\mathbf{y}}$ and $\widetilde{\mathbf{c}}_0$, $D_0^{(\mathrm{H})}$, is given by [58]

$$p_{D_0^{(\mathrm{H})}}(j) = p_{E_1^k}(0) p_{E_{k+1}^n}(j) + \left(1 - p_{E_1^k}(0)\right) p_{W_{\mathbf{c}_\mathrm{P}}}(j), \tag{5.36}$$

where $p_{E_1^k}(0)$ is given by (5.19), and $p_{E_{k+1}^n}(j)$ and $p_{W_{\mathbf{c}_\mathrm{P}}}(j)$ are the pmfs of random variables E_{k+1}^n and $W_{\mathbf{c}_\mathrm{P}}$ given by (5.19) and (5.34), respectively.

Proof. The pmf of $D_0^{(\mathrm{H})}$ can be derived in the form of conditional probability as

$$p_{D_0^{(\mathrm{H})}}(j) = \Pr(\widetilde{\mathbf{e}}_\mathrm{B} = \mathbf{0}) p_{D_0^{(\mathrm{H})}}(j|\widetilde{\mathbf{e}}_\mathrm{B} = \mathbf{0}) + \Pr(\widetilde{\mathbf{e}}_\mathrm{B} \neq \mathbf{0}) p_{D_0^{(\mathrm{H})}}(j|\widetilde{\mathbf{e}}_\mathrm{B} \neq \mathbf{0}). \tag{5.37}$$

From the Lemma 5.2, we can see that $\Pr(\widetilde{\mathbf{e}}_\mathrm{B} = \mathbf{0})$ and $\Pr(\widetilde{\mathbf{e}}_\mathrm{B} \neq \mathbf{0})$ are given by $p_{E_1^k}(0)$ and $1 - p_{E_1^k}(0)$, respectively, and the conditional pmf $p_{D_0^{(\mathrm{H})}}(j|\widetilde{\mathbf{e}}_\mathrm{B} = \mathbf{0})$ and

$p_{D_0^{(\mathrm{H})}}(j|\widetilde{\mathbf{e}}_{\mathrm{B}} \neq \mathbf{0})$ are given by $p_{E_{k+1}^n}(j)$ and $p_{W_{\mathbf{c}_\mathrm{P}}}(j)$, respectively. Therefore, the pmf of $D_0^{(\mathrm{H})}$ can be obtained as (5.36).

It is important to note that in (5.36), $p_{W_{\mathbf{c}_\mathrm{P}}}(j)$ is affected by $p_{\mathbf{c}_\mathrm{P}}(j,q)$, and $p_{\mathbf{c}_\mathrm{P}}(j,q)$ is determined by the code structure and weight enumerator. One can find $p_{\mathbf{c}_\mathrm{P}}(j,q)$ if the codebook of $\mathcal{C}(n,k)$ is known or via computer search. It is beyond the scope of this chapter to theoretically determine $p_{\mathbf{c}_\mathrm{P}}(j,q)$ for a specific code; nevertheless, for a number of families of codes with binomial-like weight spectrum [33], e.g. BCH codes etc., $p_{\mathbf{c}_\mathrm{P}}(j,q)$ can be simply approximated to [58]

$$p_{\mathbf{c}_\mathrm{P}}(u,q) \approx \frac{1}{2^{n-k}}\binom{n-k}{u}. \tag{5.38}$$

This approximation can significantly simplify the pdf given by (5.36).

We take the decoding of eBCH codes as an example to verify the accuracy of Hamming distance distributions (5.36). We show the distribution of $D_0^{(\mathrm{H})}$ in decoding (128, 64, 22) eBCH code in Figure 5.7. As the SNR increases, it can be seen that the distribution will concentrate towards the left (i.e. $D_0^{(\mathrm{H})}$ becomes smaller), which indicates that the decoding error decreases as well.

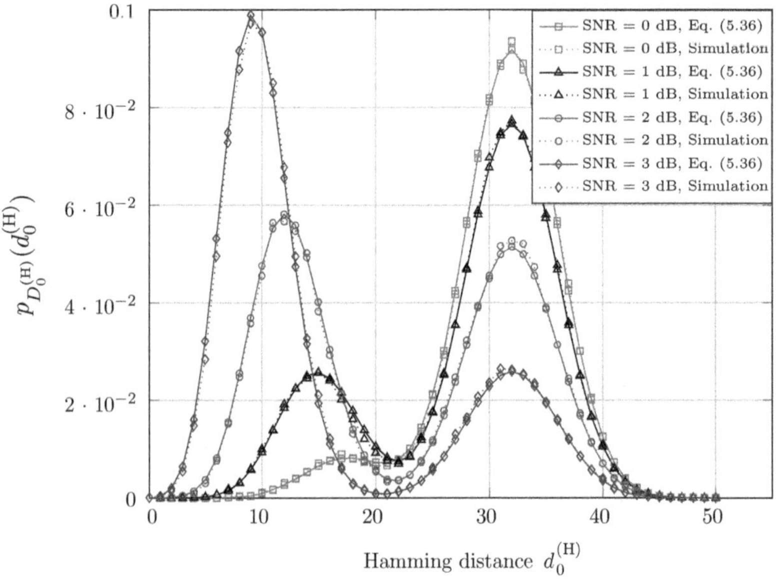

Figure 5.7 The distributions of $D_0^{(\mathrm{H})}$ in decoding (128,64,22) eBCH code at different SNRs. This figure is reused with permission from [58].

The results of the distribution of 0-reprocessing Hamming distance can be extended to any orders of reprocessing. In [58], the distributions of the WHD in OSD have also been investigated. We recommend readers to read [58] for details.

5.3.2 Decoding Rules to Reduce Complexity

The decoding complexity of OSD can be reduced by applying the Discarding Rule (DR) and Stopping Rule (SR). Given a TEP list, DRs are usually designed to identify and discard the unpromising TEPs, while SRs are typically designed to determine whether the best decoding result has been found and terminate the decoding process in advance. In this section, we introduce several notable SRs and DRs to reduce the decoding complexity.

5.3.2.1 Stopping Rules

Hard Individual Stopping Rule (HISR) [58]

Recalling the statistics of the Hamming distance $D_0^{(\mathrm{H})}$ proposed in Theorem 5.1, the pmf of Hamming distance $D_0^{(H)}$ is a mixture of two random variables E_{k+1}^{n} and $W_{\mathbf{c}_\mathrm{P}}$ which represent the number of errors in redundant positions and the Hamming weight of the redundant part of a codeword from $\mathcal{C}(n,k)$, respectively.

In fact, in (5.36), $p_{E_1^k}(0)$ is the probability that MRB has no errors, and we can regard $p_{E_1^k}(0)$ as the *a priori correct probability* of the codeword estimate $\widetilde{\mathbf{c}}_0$. Nevertheless, based on (5.36) we can further find the probability that MRB has no errors when given the Hamming distance $d_0^{(\mathrm{H})}$ (a sample of $D_0^{(\mathrm{H})}$), which is referred to as the *hard success probability* of $\widetilde{\mathbf{c}}_0$. The hard success probability can be regarded as the a posterior correct probability of $\widetilde{\mathbf{c}}_0$, given the value of $D_0^{(\mathrm{H})}$.

Furthermore, it is known that 0-reprocessing of OSD can be regarded as the reprocessing of a special all-zero TEP $\mathbf{0}$, where $\widetilde{\mathbf{y}}_\mathrm{B} \oplus \mathbf{0}$ is re-encoded. Thus, Eq. (5.36) in Theorem 5.1 is in fact the Hamming distance between $\hat{\mathbf{c}}_\mathbf{e}$ and $\mathbf{y}$ in the special case that $\mathbf{e} = \mathbf{0}$. Hence, for an arbitrary TEP $\mathbf{e}$, we can extend the results in Theorem 5.1 to obtain the hard success probability of an arbitrary estimate $\widetilde{\mathbf{c}}_\mathbf{e}$ with respect to $\mathbf{e}$. We introduce the hard success probability of the estimate $\widetilde{\mathbf{c}}_\mathbf{e}$ in the following Corollary.

Corollary 5.1: Given a linear block code $\mathcal{C}(n,k)$ and TEP $\mathbf{e}$, if the Hamming distance between $\widetilde{\mathbf{c}}_\mathbf{e}$ and $\widetilde{\mathbf{y}}$ is calculated as $d_\mathbf{e}^{(\mathrm{H})}$, the probability that the errors in MRB are eliminated by TEP $\mathbf{e}$ is given by

$$\mathrm{P}_\mathbf{e}^{\mathrm{suc}}(d_\mathbf{e}^{(\mathrm{H})}) = \mathrm{Pe}(\mathbf{e})\frac{p_{E_{k+1}^{n}}\left(d_\mathbf{e}^{(\mathrm{H})} - w(\mathbf{e})\right)}{p_{D_\mathbf{e}^{(\mathrm{H})}}\left(d_\mathbf{e}^{(\mathrm{H})} - w(\mathbf{e})\right)}, \tag{5.39}$$

where $p_{D_{\mathbf{e}}^{(\mathrm{H})}}(j)$ is the pmf given by

$$\begin{aligned} p_{D_{\mathbf{e}}^{(\mathrm{H})}}(j) &= \mathrm{Pe}(\mathbf{e}) p_{E_{k+1}^{n}}(j-v) \\ &\quad + (1-\mathrm{Pe}(\mathbf{e})) p_{W_{\mathbf{e},\mathrm{cp}}}(j-v|w(\mathbf{e})=v), \end{aligned} \tag{5.40}$$

for $j \geq w(\mathbf{e})$, and $\mathrm{Pe}(\mathbf{e})$ is given by

$$\begin{aligned} \mathrm{Pe}(\mathbf{e}) = &\underbrace{\int_{0}^{\infty} \cdots}_{k-w(\mathbf{e})} \underbrace{\int_{-\infty}^{0} \cdots}_{w(\mathbf{e})} \left(\frac{n!}{(n-k)!} F_A(|x_k|) \prod_{\ell=1}^{k} f_R(x_\ell) \prod_{\ell=2}^{k} \mathbf{1}_{[0,|x_{\ell-1}|]}(|x_\ell|) \right) \\ &\cdot \prod_{\substack{0<\ell\leq k \\ \mathbf{e}_\ell \neq 0}} dx_\ell \prod_{\substack{0<\ell\leq k \\ \mathbf{e}_\ell = 0}} dx_\ell . \end{aligned} \tag{5.41}$$

In (5.39), $p_{E_{k+1}^{n}}(j)$ is the pmf of random variable E_{k+1}^{n} given by (5.19), $p_{W_{\mathbf{e},\mathrm{cp}}}(j|w(\mathbf{e})=v)$ is a conditional pmf given by

$$\begin{aligned} &p_{W_{\mathbf{e},\mathrm{cp}}}(j|w(\mathbf{e})=v) \\ &= \sum_{\ell=0}^{n-k} \sum_{u=0}^{n-k} \frac{\binom{u}{\delta}\binom{n-k-u}{\ell-\delta}}{\binom{n-k}{\ell}} \sum_{q=0}^{k} \left(p_{W_{\mathbf{e},\tilde{\mathbf{e}}_{\mathrm{B}}}}(q|w(\mathbf{e})=v) p_{\mathbf{c}_{\mathrm{P}}}(\ell, q) \right) \\ &\cdot p_{E_{k+1}^{n}}(u) \cdot \mathbf{1}_{\mathbb{N} \cap [0,\min(u,\ell)]}(\delta), \end{aligned} \tag{5.42}$$

where $\delta = \frac{\ell+u-j}{2}$, and $p_{W_{\mathbf{e},\tilde{\mathbf{e}}_{\mathrm{B}}}}(q|w(\mathbf{e})=v)$ is a conditional pmf given by

$$p_{W_{\mathbf{e},\tilde{\mathbf{e}}_{\mathrm{B}}}}(q|w(\mathbf{e})=v) = \sum_{u=0}^{k} \frac{\binom{u}{\delta'}\binom{k-u}{v-\delta'}}{\binom{k}{v}} p_{E_1^k}(u) \cdot \mathbf{1}_{\mathbb{N},[0,\min(u,v)]}(\delta'), \tag{5.43}$$

for $\delta' = \frac{u+v-q}{2}$.

Proof. We refer interested readers to [58, Corollary 1, Corollary 2].

We show $\mathrm{P}_{\mathbf{e}}^{\mathrm{suc}}(d_{\mathbf{e}}^{(\mathrm{H})})$ as a function of $d_{\mathbf{e}}^{(\mathrm{H})}$ for TEP $\mathbf{e} = [0, \dots, 0, 1, 1, 0]$ in decoding (128, 64, 22) eBCH code in Figure 5.8. As can be seen, when $d_{\mathbf{e}}^{(\mathrm{H})}$ decreases, the probability that errors in MRB are eliminated increases rapidly. In other words, the a posterior correct probability of $\tilde{\mathbf{c}}_{\mathbf{e}}$ increases as $d_{\mathbf{e}}^{(\mathrm{H})}$ decreases. It is of interest that although the WHD usually measures the likelihood of a codeword estimate to the received signal, the Hamming distance can also represent the likelihood.

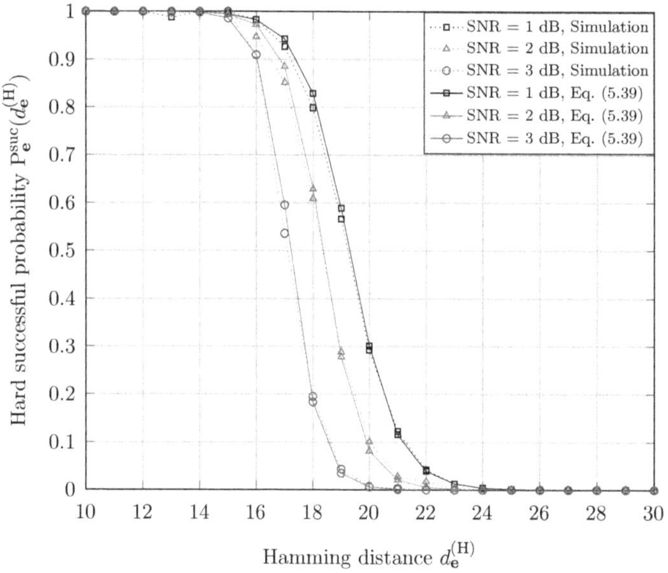

Figure 5.8 $P_{\mathbf{e}}^{suc}(d_{\mathbf{e}}^{(H)})$ in decoding (128,64,22) eBCH code at different SNR, for TEP $\mathbf{e} = [0, \ldots, 0,1,1,0]$. This figure is reused with permission from [58].

Despite the complicated form of (5.39), [58] shows that the hard success probability can be computed with $\mathcal{O}(n)$ Floating-Pointing Operations (FLOPs) on-the-fly when $\mathcal{C}(n,k)$ has the binomial-like weight spectrum. Specifically, when the reliability vector $[\widetilde{A}]_1^n = [\widetilde{\alpha}]_1^n$ is given, the hard success probability represented by $P_{\mathbf{e}}^{suc}(d_{\mathbf{e}}^{(H)}|\widetilde{\boldsymbol{\alpha}})$, can be approximated as [58]

$$P_{\mathbf{e}}^{suc}(d_{\mathbf{e}}^{(H)}|\widetilde{\boldsymbol{\alpha}}) \approx \left(1+\left(\frac{1-\mathrm{Pe}(\mathbf{e}|\widetilde{\boldsymbol{\alpha}})}{\mathrm{Pe}(\mathbf{e}|\widetilde{\boldsymbol{\alpha}})}\right)\left(\frac{2^{k-n}}{p^{(d_{\mathbf{e}}^{(H)}-w(\mathbf{e}))}(1-p)^{(n-k-d_{\mathbf{e}}^{(H)}+w(\mathbf{e}))}}\right)\right)^{-1}, \tag{5.44}$$

where $p = \frac{1}{n-k}\sum_{u=k+1}^{n} \mathrm{Pe}(u|\widetilde{\alpha}_u)$ is the arithmetic mean of the bit-wise error probabilities of $\widetilde{\mathbf{y}}_{\mathrm{P}}$, and $\mathrm{Pe}(u|\widetilde{\alpha}_u)$ is given by(5.17). Moreover, $\mathrm{Pe}(\mathbf{e}|\widetilde{\boldsymbol{\alpha}})$ is computed as

$$\mathrm{Pe}(\mathbf{e}|\widetilde{\boldsymbol{\alpha}}) = \prod_{\substack{0<u\le k\\ e_u\neq 0}} \mathrm{Pe}(u|\widetilde{\alpha}_u) \prod_{\substack{0<u\le k\\ e_u=0}} (1-\mathrm{Pe}(u|\widetilde{\alpha}_u)). \tag{5.45}$$

We now introduce the HISR. Given a predetermined threshold success probability $P_t^{suc} \in [0,1]$, if the Hamming distance $d_{\mathbf{e}}^{(H)}$ between $\widetilde{\mathbf{c}}_{\mathbf{e}}$ and $\widetilde{\mathbf{y}}$ satisfies the following condition

$$P_{\mathbf{e}}^{suc}(d_{\mathbf{e}}^{(H)}|\widetilde{\boldsymbol{\alpha}}) \ge P_t^{suc}, \tag{5.46}$$

the codeword $\hat{\mathbf{c}}_{\mathbf{e}} = \pi_1^{-1}(\pi_2^{-1}(\widetilde{\mathbf{c}}_{\mathbf{e}}))$ is selected as the decoding output, and the decoding is terminated. Therefore, the probability that errors in MRB are eliminated is lower bounded by $\mathrm{P}_t^{\mathrm{suc}}$ because of (5.46).

Soft Individual Stopping Rule (SISR) [58]

In [58], another stopping rule based on WHD is introduced, namely SISR. Different from the HISR introduce in the previous section, the SISR can make better use of the a priori soft information from the channel.

Similar to the HISR, the SISR also calculates the *a posterior* probability that a codeword estimate $\widetilde{\mathbf{c}}_{\mathbf{e}}$ is the correct estimate based on its difference pattern $\mathbf{d}_{\mathbf{e}} = \widetilde{\mathbf{c}}\oplus\widetilde{\mathbf{y}}$ from $\widetilde{\mathbf{y}}$. This a posterior probability is referred to as the *soft success probability*, denoted by $\widetilde{\mathrm{P}}_{\mathbf{e}}^{\mathrm{suc}}(\mathbf{d}_{\mathbf{e}})$.

It has been shown that when $\mathcal{C}(n,k)$ has a binomial-like weight spectrum, the soft success probability can be computed on-the-fly with the complexity $\mathcal{O}(n)$ FLOPs. Precisely, if the reliabilities of received signals are given as $[\widetilde{A}]_1^n = [\widetilde{\alpha}]_1^n$, the soft success probability of a codeword estimate $\widetilde{\mathbf{c}}_{\mathbf{e}}$, denoted by $\widetilde{\mathrm{P}}_{\mathbf{e}}^{\mathrm{suc}}(\mathbf{d}_{\mathbf{e}}|\widetilde{\alpha})$, can be approximately computed as

$$\widetilde{\mathrm{P}}_{\mathbf{e}}^{\mathrm{suc}}(\mathbf{d}_{\mathbf{e}}|\widetilde{\alpha}) \approx \left(1+\frac{1-\mathrm{Pe}(\mathbf{e}|\widetilde{\alpha})}{\mathrm{Pe}(\mathbf{e}|\widetilde{\alpha})}\prod_{\substack{k<u\leq n\\ d_{\mathbf{e},u}\neq 0}}\frac{1}{2\mathrm{Pe}(u|\widetilde{\alpha}_u)}\prod_{\substack{k<u\leq n\\ d_{\mathbf{e},u}=0}}\frac{1}{2-2\mathrm{Pe}(u|\widetilde{\alpha}_u)}\right)^{-1}, \tag{5.47}$$

where $d_{\mathbf{e},u}$ is the uth bit of $\mathbf{d}_{\mathbf{e}}$.

The SISR is described as follows. After each re-encoding, given a success probability threshold $\mathrm{P}_t^{\mathrm{suc}} \in [0,1]$, if the difference pattern $\mathbf{d}_{\mathbf{e}} = \widetilde{\mathbf{c}}_{\mathbf{e}} \oplus \widetilde{\mathbf{y}}$ between the generated codeword $\widetilde{\mathbf{c}}_{\mathbf{e}}$ and $\widetilde{\mathbf{y}}$ satisfies the following condition

$$\widetilde{\mathrm{P}}_{\mathbf{e}}^{\mathrm{suc}}(\mathbf{d}_{\mathbf{e}}|\widetilde{\alpha}) \geq \mathrm{P}_t^{\mathrm{suc}}, \tag{5.48}$$

the decoding is terminated and the codeword estimate $\hat{\mathbf{c}}_{\mathbf{e}} = \pi_1^{-1}(\pi_2^{-1}(\widetilde{\mathbf{c}}_{\mathbf{e}}))$ is selected as the decoding output.

Compared with the HISR, SISR terminates the decoding based on the difference pattern, rather than the number of different positions (Hamming distance). It is worth noting that if $\mathcal{C}(n,k)$ has a weight spectrum far from the binomial distribution, (5.47) and (5.44) need to be modified accordingly[58].

Decoding Optimality Condition (DOC) [31]

In [31, Theorem 10.1], a DOC was proposed to terminate the decoding early. Specifically, it has been proved that for a codeword estimate $\widetilde{\mathbf{c}}_{\mathbf{e}}$ in OSD, if the following condition

$$d_{\mathbf{e}}^{(\mathrm{W})} \leq g(\widetilde{\mathbf{c}}_{\mathbf{e}}, d_{\mathrm{H}}), \tag{5.49}$$

is satisfied, $\hat{\mathbf{c}}_{\mathbf{e}} = \pi_1^{-1}(\pi_2^{-1}(\widetilde{\mathbf{c}}_{\mathbf{e}}))$ is the maximum-likelihood estimate of the received sequence, where d_{H} is the minimum distance of $\mathcal{C}(n,k)$, and $g(\widetilde{\mathbf{c}}_{\mathbf{e}}, d_{\mathrm{H}})$ is given by [31, Eq.(10.31)]. It has been proved that (5.49) is a rigorous sufficient condition of the maximum-likelihood decoding [31]. On the other hand, the trade-off between complexity and error rate cannot be tuned as no parameters are introduced.

Probabilistic Sufficient Condition (PSC) [27]

In [27], a PSC on optimality for reliability based decoding was proposed. The PSC was also integrated with the decoder proposed in [50]. In the PSC, a syndrome-like index is calculated as

$$p_{sc} = [\widetilde{\mathbf{y}}_{\mathrm{B}} \oplus \mathbf{e} \;\; \widetilde{\mathbf{y}}_{\mathrm{P}}]\widetilde{\mathbf{H}}^{\mathrm{T}}, \tag{5.50}$$

where $\widetilde{\mathbf{H}}$ is the ordered parity matrix corresponding to $\widetilde{\mathbf{G}}$. Then, p_{sc} is compared with a parameter τ, and the decoding is terminated if $w(p_{sc}) \leq \tau$. Authors of [27] have shown that the probability of the "false alarm" of PSC can be negligible when τ is carefully selected. Furthermore, τ provides the flexibility between the complexity and error rate.

5.3.2.2 Discarding Rules

Soft Discarding Rule (SDR) [58]

In [58], an SDR technique is proposed to discard unpromising TEPs. The unpromising TEPs are defined as those TEPs with a very low probability of resulting in a correct codeword estimate. A *promising probability* is defined in [58] representing the probability that a TEP $\mathbf{e}$ can generate an estimate $\widetilde{\mathbf{c}}_{\mathbf{e}}$ having a less distance (to the signal) than the minimum distance $d_{\min}^{(\mathrm{W})}$ recorded so far. Let $\mathrm{P}_{\mathbf{e}}^{\mathrm{Pro}}(d_{\min}^{(\mathrm{W})})$ denote the promising probability with respect to $\mathbf{e}$, then $\mathrm{P}_{\mathbf{e}}^{\mathrm{Pro}}(d_{\min}^{(\mathrm{W})})$ is defined as

$$\mathrm{P}_{\mathbf{e}}^{\mathrm{Pro}}(d_{\min}^{(\mathrm{W})}) = \mathrm{Pr}(d_{\mathbf{e}}^{(\mathrm{W})} \leq d_{\min}^{(\mathrm{W})}). \tag{5.51}$$

It is shown that $\mathrm{P}_{\mathbf{e}}^{\mathrm{Pro}}$ can be efficiently evaluated based on Gaussian approximations when $\mathcal{C}(n,k)$ has a binomial-like weight spectrum[58]. Conditioning on $[\widetilde{A}]_1^n = [\widetilde{\alpha}]_1^n$, the promising probability $\mathrm{P}_{\mathbf{e}}^{\mathrm{Pro}}(d_{\min}^{(\mathrm{W})}|\widetilde{\alpha})$ can be approximated by

$$\begin{aligned}\mathrm{P}_{\mathbf{e}}^{\mathrm{pro}}(d_{\min}^{(\mathrm{W})}|\widetilde{\alpha}) &= \mathrm{Pe}(\mathbf{e}|\widetilde{\alpha})\left(1 - Q\left(\frac{d_{\min}^{(\mathrm{W})} - \mu_1}{\sigma_1}\right)\right) \\ &\quad + (1 - \mathrm{Pe}(\mathbf{e}|\widetilde{\alpha}))\left(1 - Q\left(\frac{d_{\min}^{(\mathrm{W})} - \mu_2}{\sigma_2}\right)\right),\end{aligned} \tag{5.52}$$

where

$$\mu_1 = \sum_{\substack{1\le u\le k\\ e_u\neq 0}} \widetilde{\alpha}_u + \sum_{u=k+1}^{n} \mathrm{Pe}(u|\widetilde{\alpha}_u)\widetilde{\alpha}_u, \tag{5.53}$$

$$\mu_2 = \sum_{\substack{1\le u\le k\\ e_u\neq 0}} \widetilde{\alpha}_u + \sum_{u=k+1}^{n} \frac{\widetilde{\alpha}_u}{2}, \tag{5.54}$$

$$\sigma_1^2 = \sum_{u=k+1}^{n}\sum_{v=k+1}^{n} \mathrm{Pe}(u|\widetilde{\alpha}_u)\mathrm{Pe}(v|\widetilde{\alpha}_v)\widetilde{\alpha}_u\widetilde{\alpha}_v - \left(\sum_{u=k+1}^{n} \mathrm{Pe}(u|\widetilde{\alpha}_u)\widetilde{\alpha}_u\right)^2, \tag{5.55}$$

$$\sigma_2^2 = \sum_{u=k+1}^{n-1}\sum_{v=u}^{n} \frac{\widetilde{\alpha}_u\widetilde{\alpha}_v}{2} - \left(\sum_{u=k+1}^{n} \frac{\widetilde{\alpha}_u}{2}\right)^2. \tag{5.56}$$

The SDR is described as follows. Given the threshold promising probability $\mathrm{P}_t^{\mathrm{pro}} \in [0,1]$ and the current recorded minimum WHD $d_{\mathrm{min}}^{(\mathrm{W})}$, if the soft promising probability of $\mathbf{e}$ satisfies

$$\mathrm{P}_{\mathbf{e}}^{\mathrm{pro}}(d_{\mathrm{min}}^{(\mathrm{W})}|\widetilde{\boldsymbol{\alpha}}) < \mathrm{P}_t^{\mathrm{pro}}, \tag{5.57}$$

the TEP $\mathbf{e}$ can be discarded without reprocessing.

Decoding Necessary Condition (DNC) [53]

In [53], a DNC was proposed as follows. A lower bound of the reliabilities of the TEPs is first estimated based on the so-far recorded WHD $d_{\mathrm{min}}^{(\mathrm{W})}$, i.e.

$$\ell^* = \frac{d_{\mathrm{min}}^{(\mathrm{W})} \sum_{u=1}^{k} \widetilde{\alpha}_u}{\sum_{u=1}^{k} \widetilde{\alpha}_u + \lambda \sum_{u=k+1}^{n} \widetilde{\alpha}_u}, \tag{5.58}$$

where λ is a parameter to be chosen. Then, for an arbitrary TEP $\mathbf{e}$, if the reliability of $\mathbf{e}$, defined as $\ell(\mathbf{e}) = \sum_{\substack{1\le u\le k\\ e_u\neq 0}} \widetilde{\alpha}_u$, satisfies $\ell(\mathbf{e}) \geq \ell^*$, $\mathbf{e}$ is discarded without re-encoding. Compared to the SDR proposed in [58], the DNC is more of heuristic.

5.3.2.3 Comparisons Between Decoding Rules

We compare the complexity of decoders with different stopping rules. The DOC [31], PSC [27], the HISR, and the SISR [58] are compared. We consider the order-3 decoding of (64, 30, 14) eBCH codes, which reaches the near-maximum-likelihood error performance [18]. All decoders are fine-tuned to reach the same error performance as the original OSD [18] which applies no stopping conditions. The average number of processed TEPs are compared in Figure 5.10a. As can be seen, the required number of TEPs of OSD can be significantly reduced by using stopping rules. Among these SRs, the HISR and SISR can result in lower numbers of required TEPs compared to the DOC [31] and PSC [27]. Furthermore, the soft condition (i.e. SISR) outperforms the hard condition (i.e. HIHR). The average decoding times for decoding a single codeword are further compared using MATLAB implementation on a 3.0 GHz CPU, as depicted in Figure 5.9b. It can be seen that the SISR can reduce the decoding time to less than 10 ms. It is worth noting that the DOC and PSC require a longer time to decode a codeword than the original OSD at low SNRs.

Next, we compare the complexity of decoders with different discarding rules. The DNC [53] and the SDR [58] are compared. We consider the order-3 decoding of (64, 30, 14) eBCH codes. All parameters in the simulated decoder are carefully selected to ensure that they can reach the same error rate as the original OSD [18], and the sequence of TEPs are ordered in descending order of the reliabilities, i.e. the *a priori* probabilities of TEPs eliminating the MRB errors.

The average numbers of re-encoded TEPs are compared in Figure 5.10a. It can be seen that the SDR can significantly reduce the number of re-encoded TEPs, and a notable improvement is shown compared to the DNC [53], especially at low SNRs. In addition, the average decoding times of decoding a single codeword are compared in Figure 5.10b. As shown, each simulated approach can significantly reduce the decoding time compared to the original OSD in both low and high SNR regimes. The main reason is that as shown in Section 5.3.2.2, the DDNC and SDR can be efficiently implemented with $O(n)$ FLOPs. We can also conclude that the SDR and DNC have similar decoding times at high SNRs, close to 1 ms; nevertheless, the SDR outperforms at low SNRs.

5.3.3 Probability-based OSD

As introduced in section 5.3.2, several techniques, including SRs and DRs, have been proposed in the literature. These techniques can significantly reduce the time complexity of OSD. Therefore, one can combine those best available techniques to design a fast OSD decoder as one of the possible decoder solutions of URLLC. In this section, we introduce the latest approach of OSD, namely Probability-Based OSD (PB-OSD) [57]. PB-OSD is devised based on the SISR and SDR, requiring much fewer TEPs than the original OSD.

5.3.3.1 Simple PB-OSD

The simple PB-OSD is a variant of the original OSD algorithm [18], where the SISR (5.48) and SDR (5.57) are applied to reduce the decoding complexity. In particular, the SDR is applied before each re-encoding to discard unpromising TEPs. Furthermore, SISR is applied to decide if the decoding can be terminated.

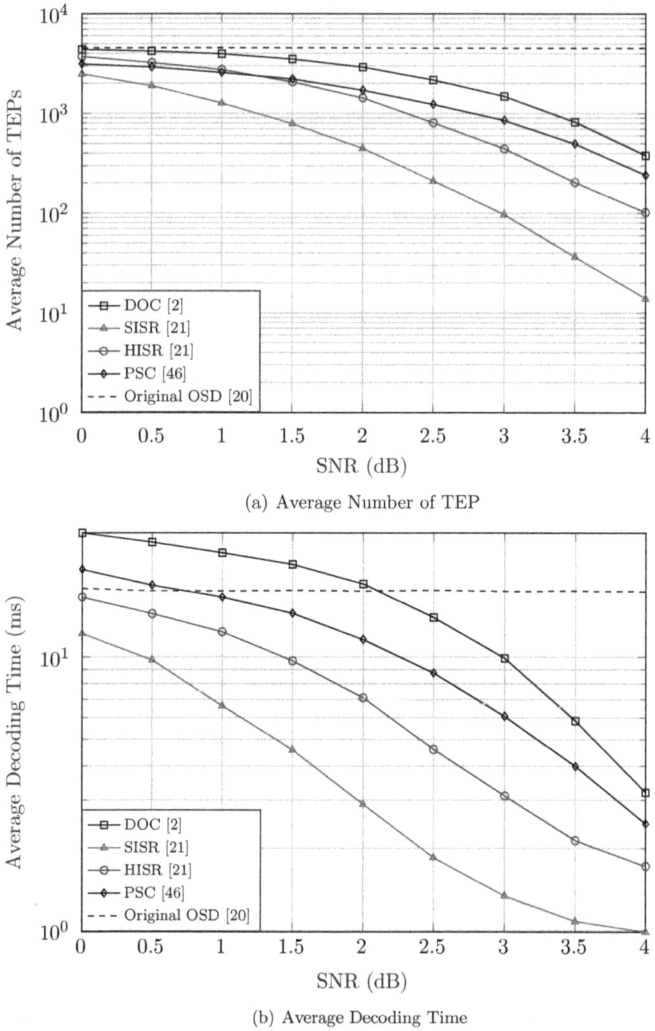

Figure 5.9 Decoding (64,30,14) eBCH code with order-3 OSD algorithms applying different stopping rules. This figure is reused with permission from [58].

The simple PB-OSD is conveniently implemented by slightly modifying the original OSD [18], and overhead of employing SDR and SISR has the complexity $\mathcal{O}(n)$ according to (5.52) and (5.47). We summarize the algorithm of simple PB-OSD in Algorithm 2.

Although the simple PB-OSD can be easily carried out, it has an obvious disadvantage. It can be seen that the SDR is checked for each TEP to discard the

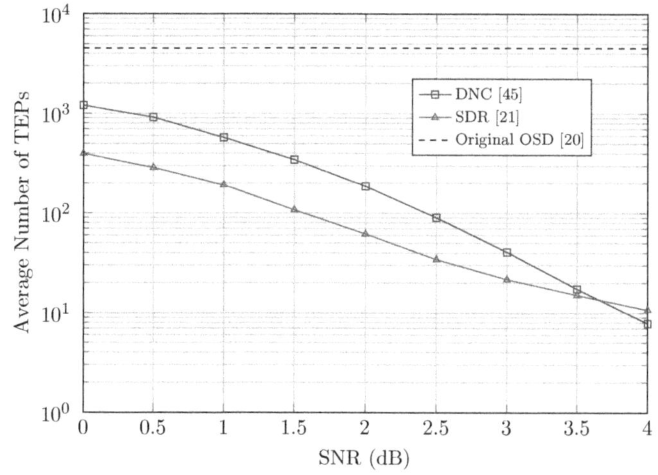

(a) Average Number of TEP

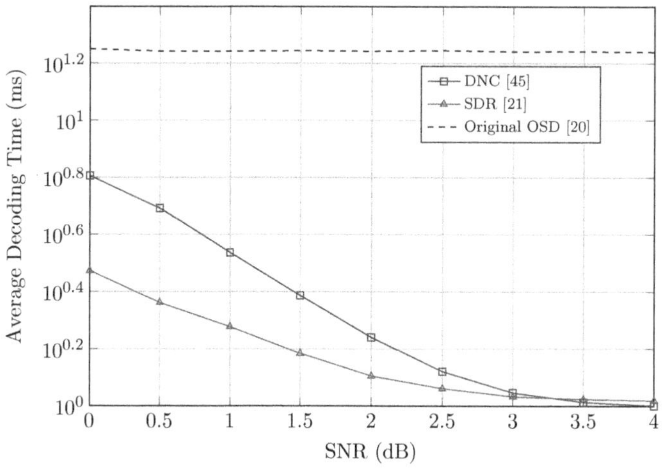

(b) Average Decoding Time

Figure 5.10 Decoding (64,30,14) eBCH code with order-3 OSD algorithms different discarding rules. This figure is reused with permission from [58].

unpromising ones. However, because SDR also has an overhead of processing, it might not reduce the actual time complexity of decoding. This issue of the simple PB-OSD can be addressed by carefully organizing the TEP sequence, which will be discussed in the following subsection.

Algorithm 2 Simple PB-OSD [57]

Require:

Matrix $\mathbf{G}$, input signal $\mathbf{r}$, decoding order m, parameters P_t^{suc}, P_t^{pro}

Ensure:

The decoding output $\hat{\mathbf{c}}_{opt}$

//**Preprocessing**

1: Calculate reliability value $\alpha_i = |r_i|$

2: First permutation: $\boldsymbol{\alpha}' = \pi_1(\boldsymbol{\alpha})$, $\mathbf{r}' = \pi_1(\mathbf{r})$, $\mathbf{G}' = \pi_1(\mathbf{G})$

3: Gaussian elimination and second permutation: $\tilde{\boldsymbol{\alpha}} = \pi_2(\boldsymbol{\alpha}')$, $\tilde{\mathbf{r}} = \pi_2(\mathbf{r}')$, $\tilde{\mathbf{G}} = \pi_2(\mathbf{G}')$

4: Perform hard-decision: $\tilde{y}_i = \begin{cases} 1 & \text{for } \tilde{r}_i < 0 \\ 0 & \text{for } \tilde{r}_i \geq 0 \end{cases}$

5: Calculate $Pe(u|\tilde{\alpha}_u)$ for $1 \leq u \leq n$.

6: Initialize $d_{min}^{(W)} = \infty$, $\tilde{\mathbf{c}}_{opt} = \tilde{\mathbf{y}}$

//**Reprocessing**

7: **for** $\ell = 0 : m$ **do**

8:   **for** $j = 1 : \binom{k}{\ell}$ **do**

9:     Take an unprocessed weight-ℓ TEP $\mathbf{e}_j$.

10:     Calculate $P_{\mathbf{e}}^{pro}(d_{min}^{(W)}|\tilde{\alpha})$ according to (5.52).

11:     **if** $P_{\mathbf{e}}^{pro}(d_{min}^{(W)}|\tilde{\alpha}) < P_t^{pro}$ **then**

12:       **Continue**

13:     **end if**

14:     Calculate $\tilde{\mathbf{c}}_{\mathbf{e}_j} = (\tilde{\mathbf{y}}_B \oplus \mathbf{e})\tilde{\mathbf{G}}$

15:     Calculate the WHD $d_{\mathbf{e}_j}^{(W)}$

16:     **if** $d_{\mathbf{e}_j}^{(W)} < d_{min}^{(W)}$ **then**

17:       Declaim $\tilde{\mathbf{c}}_{opt} = \tilde{\mathbf{c}}_{\mathbf{e}_j}$ and $d_{min}^{(W)} = d_{\mathbf{e}_j}^{(W)}$

18:       Calculate $\tilde{P}_{\mathbf{e}}^{suc}(\mathbf{d}_{\mathbf{e}}|\tilde{\alpha})$ according to (5.47).

19:       **if** $\tilde{P}_{\mathbf{e}}^{suc}(\mathbf{d}_{\mathbf{e}}|\tilde{\alpha} \geq P_t^{suc})$ **then**

20:         **return** $\hat{\mathbf{c}}_{opt} = \pi_1^{-1}(\pi_2^{-1}(\tilde{\mathbf{c}}_{opt}))$

21:       **end if**

22:     **end if**

23:   **end for**

24: **end for**

25: **return** $\hat{\mathbf{c}}_{opt} = \pi_1^{-1}(\pi_2^{-1}(\tilde{\mathbf{c}}_{opt}))$

5.3.3.2 The Optimal TEP Sequence

For a PB-OSD (which applies the SISR and SDR) with a maximum number of TEPs $N_{\max}$, let $\mathcal{E} = \{\mathbf{e}_1, \mathbf{e}_1, \dots, \mathbf{e}_{N_{\max}}\}$ denote a TEP sequence arranged in the descending order of $\mathrm{Pe}(\mathbf{e}|\widetilde{\boldsymbol{\alpha}})$, i.e. $\mathrm{Pe}(\mathbf{e}_1|\widetilde{\boldsymbol{\alpha}}) \geq \dots \geq \mathrm{Pe}(\mathbf{e}_{N_{\max}}|\widetilde{\boldsymbol{\alpha}})$. We define $\mathcal{E}$ as the optimal TEP sequence of the PB-OSD, by which the average number of required TEP is minimized for fixed parameters $\mathrm{P}_t^{\mathrm{suc}}$ and $\mathrm{P}_t^{\mathrm{pro}}$. In addition, the overhead of computing promising probability can be reduced by $\mathcal{E}$.

From (5.47), it can be seen that the success probability $\widetilde{\mathrm{P}}_{\mathbf{e}}^{\mathrm{suc}}(\mathbf{d}_{\mathbf{e}}|\widetilde{\boldsymbol{\alpha}})$ is a monotonically increasing function of $\mathrm{Pe}(\mathbf{e}|\widetilde{\boldsymbol{\alpha}})$ for specific $\mathbf{d}_{\mathbf{e}}$ and $\widetilde{\boldsymbol{\alpha}}$. We omit the proof of the monotonicity of $\widetilde{\mathrm{P}}_{\mathbf{e}}^{\mathrm{suc}}(\mathbf{d}_{\mathbf{e}}|\widetilde{\boldsymbol{\alpha}})$. Thus, if the channel input is unknown, the probability of $\widetilde{\mathbf{c}}_{\mathbf{e}}$ satisfying the SR, i.e. $\mathrm{Pr}(\widetilde{\mathrm{P}}_{\mathbf{e}}^{\mathrm{suc}}(\mathbf{d}_{\mathbf{e}}|\widetilde{\boldsymbol{\alpha}}) \geq \mathrm{P}_t^{\mathrm{suc}})$ is also an increasing function of $\mathrm{Pe}(\mathbf{e}|\widetilde{\boldsymbol{\alpha}})$. Therefore, if the decoder re-encodes TEPs according to $\mathcal{E}$, the sum probability $\sum_{i=1}^{\delta} \mathrm{Pr}(\widetilde{\mathrm{P}}_{\mathbf{e}_i}^{\mathrm{suc}}(\mathbf{d}_{\mathbf{e}_i}|\widetilde{\boldsymbol{\alpha}}) \geq \mathrm{P}_t^{\mathrm{suc}})$ can be maximized for any given δ, $1 \leq \delta \leq N_{\max}$. Specifically, considering an arbitrary TEP sequence $\mathcal{E}' = \{\mathbf{e}'_1, \mathbf{e}'_2, \dots, \mathbf{e}'_{N_{\max}}\}$, the probability that the decoding is terminated after re-encoding the jth TEP $\mathbf{e}'_j$, denoted by P_j, $1 < j < N_{\max}$, can be derived as

$$\mathrm{P}_j = \left(\prod_{u=1}^{j-1} \left(1 - \mathrm{Pr}(\widetilde{\mathrm{P}}_{\mathbf{e}'_u}^{\mathrm{suc}}(\mathbf{d}_{\mathbf{e}'_u}|\widetilde{\boldsymbol{\alpha}}) \geq \mathrm{P}_t^{\mathrm{suc}}) \right) \right) \cdot \mathrm{Pr}(\widetilde{\mathrm{P}}_{\mathbf{e}'_j}^{\mathrm{suc}}(\mathbf{d}_{\mathbf{e}'_j}|\widetilde{\boldsymbol{\alpha}}) \geq \mathrm{P}_t^{\mathrm{suc}}). \tag{5.59}$$

Particularly,

$$\mathrm{P}_1 = \mathrm{Pr}(\widetilde{\mathrm{P}}_{\mathbf{e}'_1}^{\mathrm{suc}}(\mathbf{d}_{\mathbf{e}'_1}|\widetilde{\boldsymbol{\alpha}}) \geq \mathrm{P}_t^{\mathrm{suc}})$$

and

$$\mathrm{P}_{N_{\max}} = \prod_{u=1}^{N_{\max}-1} \left(1 - \mathrm{Pr}(\widetilde{\mathrm{P}}_{\mathbf{e}}^{\mathrm{suc}}(\mathbf{d}_{\mathbf{e}'_u}|\widetilde{\boldsymbol{\alpha}}) \geq \mathrm{P}_t^{\mathrm{suc}})\right).$$

Therefore, the average number, N_a, of re-encoded TEP can be derived as

$$N_a = \sum_{j=1}^{N_{\max}} (j \cdot \mathrm{P}_j) \tag{5.60}$$

Then, it can be proved that (5.60) is minimized when $\mathcal{E}'$ is the optimal TEP sequence $\mathcal{E}$, summarized in the following Proposition.

Proposition 5.1: The average number of TEPs, N_a, given by (5.60) is minimized when $\mathcal{E}'$ is the optimal TEP sequence.

Proof. Assume that the TEP sequence is given by $\mathcal{E}' = \{\mathbf{e}'_1, \mathbf{e}'_2, \dots, \mathbf{e}'_{N_{\max}}\}$. Let us re-write (5.60) as

$$\begin{aligned} N_a = &\sum_{j=1}^{N_{\max}} \left(j\delta_j \Pr(\widetilde{\mathrm{P}}^{\mathrm{suc}}_{\mathbf{e}'_j}(\mathbf{d}_{\mathbf{e}'_j}|\widetilde{\boldsymbol{\alpha}}) \geq \mathrm{P}^{\mathrm{suc}}_t) \right) \\ &+ \prod_{u=1}^{N_{\max}} \left(1 - \Pr(\widetilde{\mathrm{P}}^{\mathrm{suc}}_{\mathbf{e}'_u}(\mathbf{d}_{\mathbf{e}'_u}|\widetilde{\boldsymbol{\alpha}}) \geq \mathrm{P}^{\mathrm{suc}}_t)\right) N_{\max}, \end{aligned} \tag{5.61}$$

where

$$\delta_j = \prod_{u=1}^{j-1} \left(1 - \Pr(\widetilde{\mathrm{P}}^{\mathrm{suc}}_{\mathbf{e}'_u}(\mathbf{d}_{\mathbf{e}'_u}|\widetilde{\boldsymbol{\alpha}}) \geq \mathrm{P}^{\mathrm{suc}}_t)\right). \tag{5.62}$$

In particular, $\delta_1 = 1$. Note that the second term of (5.61) is independent of the order of $\mathcal{E}'$. Because $\Pr(\widetilde{\mathrm{P}}^{\mathrm{suc}}_{\mathbf{e}}(\mathbf{d}_{\mathbf{e}}|\widetilde{\boldsymbol{\alpha}}) \geq \mathrm{P}^{\mathrm{suc}}_t)$ increases as $\mathrm{Pe}(\mathbf{e}|\widetilde{\boldsymbol{\alpha}})$ increases for an arbitrary TEP $\mathbf{e}$, for an arbitrary coefficient δ_j, δ_j is minimized when the following inequality holds:

$$\mathrm{Pe}(\mathbf{e}'_u|\widetilde{\boldsymbol{\alpha}}) \geq \mathrm{Pe}(\mathbf{e}'_v|\widetilde{\boldsymbol{\alpha}}), \quad \text{for} \quad 1 \leq u < j \leq v \leq N_{\max}. \tag{5.63}$$

Minimizing N_a by minimizing coefficients δ_j for all possible j at the same time, the following set of inequalities is obtained, i.e.,

$$\begin{cases} \mathrm{Pe}(\mathbf{e}'_u|\widetilde{\boldsymbol{\alpha}}) \geq \mathrm{Pe}(\mathbf{e}'_v|\widetilde{\boldsymbol{\alpha}}), & \text{for} \quad 1 \leq u < 2 \leq v \leq N_{\max}, \\ \mathrm{Pe}(\mathbf{e}'_u|\widetilde{\boldsymbol{\alpha}}) \geq \mathrm{Pe}(\mathbf{e}'_v|\widetilde{\boldsymbol{\alpha}}), & \text{for} \quad 1 \leq u < 3 \leq v \leq N_{\max}, \\ & \vdots \\ \mathrm{Pe}(\mathbf{e}'_u|\widetilde{\boldsymbol{\alpha}}) \geq \mathrm{Pe}(\mathbf{e}'_v|\widetilde{\boldsymbol{\alpha}}), & \text{for} \quad 1 \leq u < N_{\max} \leq v \leq N_{\max}, \end{cases} \tag{5.64}$$

which is equivalent to

$$\mathrm{Pe}(\mathbf{e}'_1|\widetilde{\boldsymbol{\alpha}}) \geq \mathrm{Pe}(\mathbf{e}'_2|\widetilde{\boldsymbol{\alpha}}), \dots, \geq \mathrm{Pe}(\mathbf{e}'_{N_{\max}}|\widetilde{\boldsymbol{\alpha}}). \tag{5.65}$$

Thus, $\mathcal{E}'$ minimizes N_a when it is an optimal TEP sequence.

On the other hand, the SDR is checked for each TEP individually, and the TEP sequence does not affect the number of discarded TEPs. Nevertheless, the overhead of computing the promising probability in SDR can be reduced by applying $\mathcal{E}$. To see this, we first prove that promising probability $\mathrm{P}^{\mathrm{pro}}_{\mathbf{e}}(d^{(\mathrm{W})}_{\min}|\widetilde{\boldsymbol{\alpha}})$ given by (5.52) is also a monotonically increasing function of $\mathrm{Pe}(\mathbf{e})$ when $d^{(\mathrm{W})}_{\min} \leq \frac{1}{2}\sum_{u=k+1}^{n} \widetilde{\alpha}_u$.

Proposition 5.2: For a given $d^{(\mathrm{W})}_{\min}$, the promising probability $\mathrm{P}^{\mathrm{pro}}_{\mathbf{e}}(d^{(\mathrm{W})}_{\min}|\widetilde{\boldsymbol{\alpha}})$ of $\mathbf{e}$, given by (5.52), is a monotonically increasing function of $\mathrm{Pe}(\mathbf{e}|\widetilde{\boldsymbol{\alpha}})$ if $d^{(\mathrm{W})}_{\min} \leq \frac{1}{2}\sum_{u=k+1}^{n} \widetilde{\alpha}_u$.

Proof. The derivative of $\mathrm{P}^{\mathrm{pro}}_{\mathbf{e}}(d^{(\mathrm{W})}_{\min}|\widetilde{\boldsymbol{\alpha}})$ with respect to $\mathrm{Pe}(\mathbf{e}|\widetilde{\boldsymbol{\alpha}})$ can be represented as

$$\begin{aligned}\frac{\partial\, \mathrm{P}^{\mathrm{pro}}_{\mathbf{e}}(d^{(\mathrm{W})}_{\min}|\widetilde{\boldsymbol{\alpha}})}{\partial\, \mathrm{Pe}(\mathbf{e}|\widetilde{\boldsymbol{\alpha}})} &= Q\left(\frac{d^{(\mathrm{W})}_{\min}-\mu_2}{\sigma_2}\right) - Q\left(\frac{d^{(\mathrm{W})}_{\min}-\mu_1}{\sigma_1}\right)\\ &\overset{(a)}{>} Q\left(\frac{d^{(\mathrm{W})}_{\min}-\mu_2}{\sigma_1}\right) - Q\left(\frac{d^{(\mathrm{W})}_{\min}-\mu_1}{\sigma_1}\right) \overset{(b)}{>} 0,\end{aligned} \tag{5.66}$$

where step (a) follows from that $\sigma_2 < \sigma_1$ and $d^{(\mathrm{W})}_{\min} \le \frac{1}{2}\sum_{u=k+1}^{n}\widetilde{\alpha}_u$. Step (b) then follows $\mu_2 > \mu_1$. For $0 \le u \le n$, it has been proved that $\mathrm{P}(u|\widetilde{\alpha}_u) < 1/2$ [18]. Thus, $\sigma_2 < \sigma_1$ and $\mu_2 > \mu_1$ holds.

Let $d^{(\mathrm{W})}_{\min,u}$ denote the recorded minimum WHD so far before the uth TEP $\mathbf{e}_i \in \mathcal{E}$ is being re-encoded and assume $d^{(\mathrm{W})}_{\min,u} \le \frac{1}{2}\sum_{u=k+1}^{n}\widetilde{\alpha}_u$; thus, there holds $\frac{1}{2}\sum_{u=k+1}^{n}\widetilde{\alpha}_u \ge d^{(\mathrm{W})}_{\min,1} \ge d^{(\mathrm{W})}_{\min,2} \ge \dots \ge d^{(\mathrm{W})}_{\min,N_{\max}}$. According to Proposition 5.2, it can be obtained that

$$\mathrm{P}^{\mathrm{pro}}_{\mathbf{e}_u} \ge \dots \ge \mathrm{P}^{\mathrm{pro}}_{\mathbf{e}_i} \ge \mathrm{P}^{\mathrm{pro}}_{t} \ge \dots \ge \mathrm{P}^{\mathrm{pro}}_{\mathbf{e}_{N_{\max}}}. \tag{5.67}$$

Equation (5.67) implies that the decoder does not need to calculate the promising probability for each TEP, but once one of them satisfies the SDR (5.57) and $d^{(\mathrm{W})}_{\min} \le \frac{1}{2}\sum_{u=k+1}^{n}\widetilde{\alpha}_u$, all following TEPs in $\mathcal{E}$ can be discarded. Therefore, the overhead of computing the promising probability can be reduced by applying $\mathcal{E}$.

5.3.3.3 Improved PB-OSD

The improved PB-OSD utilizes the optimal TEP sequence $\mathcal{E}$ introduced in Section 5.3.3.2. To determine the optimal TEP sequence on-the-fly, one can adopt the algorithms introduced in [28, 47, 57]. Tree structures are applied in [47, 57], and the approach [47] can generate a new TEP with the complexity of $\mathcal{O}(\log_2 k)$. In the approach of [28], the means of ordered reliabilities are determined and then the TEP list based on the determined means is re-arranged. Because the means of ordered reliabilities can be determined offline based on analytical methods, the approach [28] is particularly efficient for long block length codes.

Utilizing the monotonicity of the promising probability, if $d^{(\mathrm{W})}_{\min} \le \frac{1}{2}\sum_{u=k+1}^{n}\widetilde{\alpha}_u$, the improved PB-OSD only calculates the probability every q TEPs, where q is a positive integer. Thus, the number of SDR checks can be as low as about $\frac{N_a}{q}$, but the TEP-reduction capability of SDR will not be apparently degraded for $q \ll N_a$. We refer to this SDR implementation as "q-step SDR". For example, let us assume $q = 10$, so that the improved PB-OSD calculates the promising probability every 10 TEPs, and the overhead of SDR is significantly reduced.

The PB-OSD decoder combined with the optimal TEP sequence $\mathcal{E}$ is presented in Algorithm 3.

Algorithm 3 Improved PB-OSD [57]

Require:

Matrix $\mathbf{G}$, input signal $\mathbf{r}$, decoding order m, parameters P_t^{suc} and P_t^{pro}, SDR step q

Ensure:

The decoding output $\hat{\mathbf{c}}_{\text{opt}}$

//**Preprocessing**

1: Calculate reliability value $\alpha_i = |r_i|$

2: First permutation: $\boldsymbol{\alpha}' = \pi_1(\boldsymbol{\alpha})$, $\mathbf{r}' = \pi_1(\mathbf{r})$, $\mathbf{G}' = \pi_1(\mathbf{G})$

3: Gaussian elimination and second permutation: $\tilde{\boldsymbol{\alpha}} = \pi_2(\boldsymbol{\alpha}')$, $\tilde{\mathbf{r}} = \pi_2(\mathbf{r}')$, $\tilde{\mathbf{G}} = \pi_2(\mathbf{G}')$

4: Perform hard-decision: $\tilde{y}_i = \begin{cases} 1 & \text{for } \tilde{r}_i < 0 \\ 0 & \text{for } \tilde{r}_i \geq 0 \end{cases}$

5: Calculate $\text{Pe}(u|\tilde{\alpha}_u)$ for $1 \leq u \leq n$.

6: Initialize $d_{\text{min}}^{(\text{W})} = \infty$, $\tilde{\mathbf{c}}_{\text{opt}} = \tilde{\mathbf{y}}$

//**Reprocessing**

7: **for** $j = 1 : N_{\text{max}}$ **do**

8:     Take an unprocessed TEP $\mathbf{e}_j$ with the highest reliability by calling algorithms from [47] or [57].

9:     **if** $d_{\text{min}}^{(\text{W})} \leq \frac{1}{2} \sum_{u=k+1}^{n} \tilde{\alpha}_u$ and $j \mod q = 0$ **then**

10:         Calculate $P_{\mathbf{e}_j}^{\text{pro}}(d_{\text{min}}^{(\text{W})}|\tilde{\boldsymbol{\alpha}})$ according to (5.52).

11:         **if** $P_{\mathbf{e}_j}^{\text{pro}}(d_{\text{min}}^{(\text{W})}|\tilde{\boldsymbol{\alpha}}) < P_t^{\text{pro}}$ **then**

12:             **return** $\hat{\mathbf{c}}_{\text{opt}} = \pi_1^{-1}(\pi_2^{-1}(\tilde{\mathbf{c}}_{\text{opt}}))$

13:         **end if**

14:     **end if**

15:     Calculate $\tilde{\mathbf{c}}_{\mathbf{e}_j} = (\tilde{\mathbf{y}}_B \oplus \mathbf{e})\tilde{\mathbf{G}}$

16:     Calculate the WHD $d_{\mathbf{e}_j}^{(\text{W})}$

17:     **if** $d_{\mathbf{e}_j}^{(\text{W})} < d_{\text{min}}^{(\text{W})}$ **then**

18:         Declaim $\tilde{\mathbf{c}}_{\text{opt}} = \tilde{\mathbf{c}}_{\mathbf{e}_j}$ and $d_{\text{min}}^{(\text{W})} = d_{\mathbf{e}_j}^{(\text{W})}$

19:         Calculate $\tilde{P}_{\mathbf{e}_j}^{\text{suc}}(\mathbf{d}_{\mathbf{e}_j}|\tilde{\boldsymbol{\alpha}})$ according to (5.47).

20:         **if** $\tilde{P}_{\mathbf{e}_j}^{\text{suc}}(\mathbf{d}_{\mathbf{e}_j}|\tilde{\boldsymbol{\alpha}}) \geq P_t^{\text{suc}}$ **then**

21:             **return** $\hat{\mathbf{c}}_{\text{opt}} = \pi_1^{-1}(\pi_2^{-1}(\tilde{\mathbf{c}}_{\text{opt}}))$

22:         **end if**

23:     **end if**

24: **end for**

25: **return** $\hat{\mathbf{c}}_{\text{opt}} = \pi_1^{-1}(\pi_2^{-1}(\tilde{\mathbf{c}}_{\text{opt}}))$

5.3.3.4 Comparisons Between Latest OSD Approaches

This section presents simulation results to compare the performance of the PB-OSD algorithms and another latest modified OSD approach, i.e. the Segmentation-Discarding Decoding (SDD) algorithm [56]. The numbers of re-encoded TEPs are compared because they dominate the OSD algorithm's complexity and alternatives. The overall time complexity is further measured by comparing the average time consumption of one decoding using MATLAB implementation with a 3.0 GHz CPU. Let $\nu = \frac{1}{k}\sum_{u=1}^{k}\mathrm{Pe}(u|\widetilde{\alpha}_u)$ denote the average bit-wise error probabilities of MRB. The probability of m errors occurring in MRB can then be derived as $\epsilon(m) = \sum_{i=0}^{m}\binom{k}{i}\nu^i(1-\nu)^{k-i}$. In the implementation of order-m PB-OSD decoders, parameters are set to $\mathrm{P}_t^{\mathrm{suc}} = 0.99\epsilon(m)$ and $\mathrm{P}_t^{\mathrm{pro}} = 0.002\sqrt{\frac{1-\epsilon(m)}{N_{\max}}}$, respectively, where $N_{\max} = \sum_{i=0}^{m}\binom{k}{i}$ is the maximum number of TEPs allowed.

As can be seen in Figure 5.11, the SDD, the simple PB-OSD and improved PB-OSD exhibit nearly identical error rate performance compared to the original OSD in decoding (128, 64, 22) eBCH code, approaching the normal approximation bound [39]. However, the simple PB-OSD and improved PB-OSD both require TEPs less than about half of the SDD, and the PB-OSD is the best among its counterparts.

Under the same implementation environment, we further compared the decoding time of one codeword in Figure 5.12. As shown, the PB-OSD algorithms can significantly reduce the overall decoding time compared to the SDD and the original OSD, especially at high SNRs. It is worth noting that the PB-OSD algorithms only slightly outperform the SDD at low SNRs due to the overhead introduced by techniques applied; nevertheless, the advantages emerge as the SNR increases.

It is worth noting that the introduced PB-OSD algorithms only employ the stopping rule and discarding rule to reduce the decoding complexity by decreasing the number of required TEPs. One can further improve the decoder by combining techniques from literature [17, 26, 48, 53] with the PB-OSD. For example, the IISR technique can be applied [17], which further reduces the complexity by exploiting the information outside MRB.

5.4 Conclusion

This chapter has reviewed the most recent signs of progress in designing and implementing short block length channel codes for URLLC. Several candidate channel codes, including polar codes, turbo codes, LDPC codes, convolutional codes, and BCH codes, were considered and introduced. They were compared in terms of block error rate under optimal decoder and algorithmic complexity of decoding algorithms. Among these candidate codes, BCH codes provided the highest reliability under optimal decoding, since they have the highest minimum Hamming distance. Polar codes with SCL decoding provided reliability of $(1 - 10^{-4})$ with only 0.5 dB gap to the normal approximation with reasonable

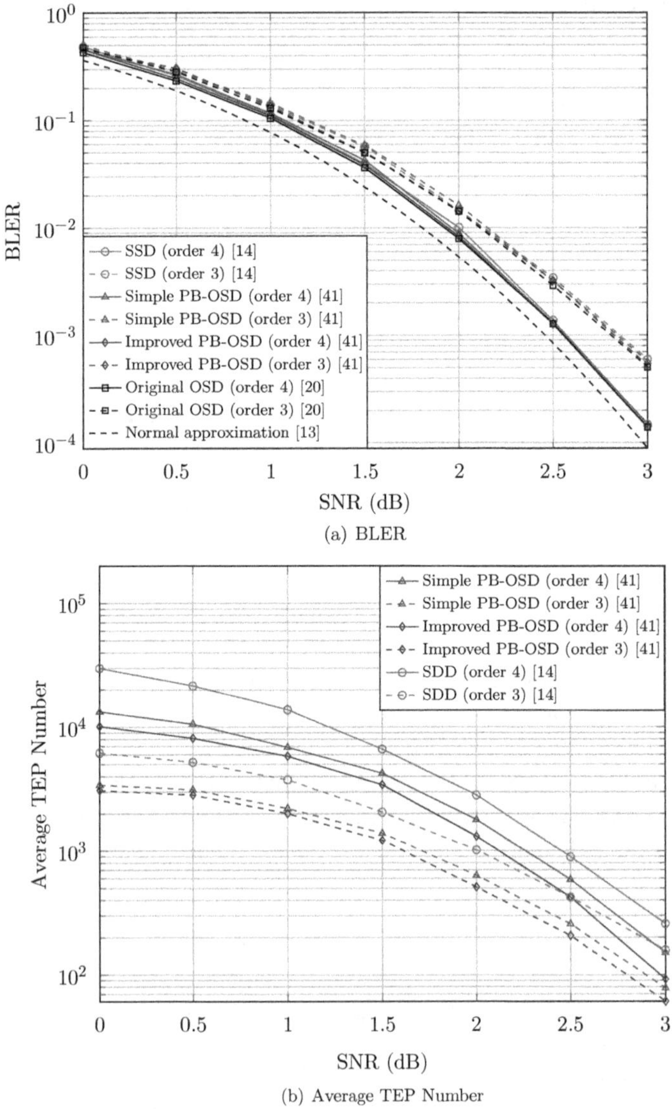

Figure 5.11 The comparisons of decoding (128,64,22) eBCH code. This figure is reused with permission from [57].

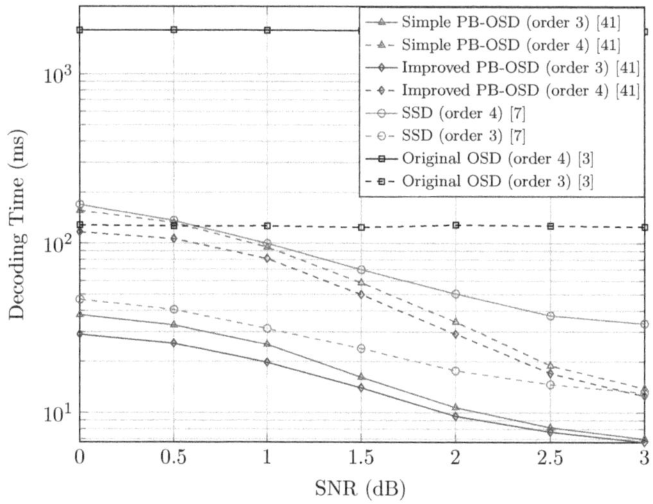

Figure 5.12 The average time consumption of decoding a single codeword. This figure is reused with permission from [57].

complexity, which offers a feasible trade-off between reliability and decoding latency. LDPC codes show a considerable gap to the normal approximation except for very low coding rate cases.

Furthermore, we promoted OSD as the possible universal near-optimal decoding algorithm for URLLC. It has been shown that OSD with appropriate orders can achieve near-ML performance of any linear block codes, with only knowledge of generator matrices. This feature benefits the flexible implementation of receiver devices and weakens the limitations in designing rate-compatible codes that fit specific decoders. For example, the design of rate-compatible polar codes must obey the positions of frozen bits for using SCL decoding, which degrades the optimality of codes. Although the decoding complexity of original OSD is impractical for applications in URLLC, several recent works have made efforts to reduce the complexity. Some of the notable results were introduced in this chapter and the latest OSD approaches were compared.

Bibliography

[1] R1-1611081 - Final Report of 3GPP TSG RAN WG1 #86bis v1.0.0. Final Minutes Report, 3rd Generation Partnership Project, Nov. 2016.

[2] R1-1608770- Flexibility evaluation of channel coding schemes for NR. 3GPP TSG RAN WG1 Meeting #86- Discussion and Decision, 3rd Generation Partnership Project, Oct. 2016.

[3] E. Arikan. Channel polarization: A method for constructing capacity-achieving codes for symmetric binary-input memoryless channels. *IEEE Transactions on Information Theory*, 55(7):3051–3073, July 2009. ISSN 0018-9448. doi: 10.1109/TIT.2009.2021379.

[4] Elwyn Berlekamp. On decoding binary bose-chadhuri-hocquenghem codes. *IEEE Transactions on Information Theory*, 11(4):577–579, 1965.

[5] Claude Berrou, Alain Glavieux, and Punya Thitimajshima. Near shannon limit error-correcting coding and decoding: Turbo-codes. 1. In *Proceedings of ICC'93-IEEE International Conference on Communications*, volume 2, pages 1064–1070. IEEE, 1993.

[6] Valerio Bioglio, Frederic Gabry, and Ingmar Land. Low-complexity puncturing and shortening of polar codes. In *2017 IEEE Wireless Communications and Networking Conference Workshops (WCNCW)*, pages 1–6. IEEE, 2017.

[7] Valerio Bioglio, Carlo Condo, and Ingmar Land. Design of polar codes in 5g new radio. *IEEE Communications Surveys & Tutorials*, 2020.

[8] Baptiste Cavarec, Hasan Basri Celebi, Mats Bengtsson, and Mikael Skoglund. A learning-based approach to address complexity-reliability tradeoff in os decoders. In *2020 54th Asilomar Conference on Signals, Systems, and Computers*, pages 689–692. IEEE, 2020.

[9] Hasan Basri Celebi, Antonios Pitarokoilis, and Mikael Skoglund. Latency and reliability trade-off with computational complexity constraints: Os decoders and generalizations. *IEEE Transactions on Communications*, 69(4):2080–2092, 2021.

[10] Kai Chen, Kai Niu, and Jiaru Lin. Improved successive cancellation decoding of polar codes. *IEEE Transactions on Communications*, 61(8):3100–3107, 2013.

[11] Tsung-Yi Chen, Kasra Vakilinia, Dariush Divsalar, and Richard D Wesel. Protograph-based raptor-like ldpc codes. *IEEE Transactions on Communications*, 63(5):1522–1532, 2015.

[12] Robert Chien. Cyclic decoding procedures for bose-chaudhuri-hocquenghem codes. *IEEE Transactions on information theory*, 10(4):357–363, 1964.

[13] Pawan Dhakal, Roberto Garello, Shree Krishna Sharma, Symeon Chatzinotas, and Björn Ottersten. On the error performance bound of ordered statistics decoding of linear block codes. In *Communications (ICC), 2016 IEEE International Conference on*, pages 1–6. IEEE, 2016.

[14] Mostafa El-Khamy, Jilei Hou, and Naga Bhushan. Design of rate-compatible structured ldpc codes for hybrid arq applications. *IEEE Journal on Selected Areas in Communications*, 27(6):965–973, 2009.

[15] Tomaso Erseghe. Coding in the finite-blocklength regime: Bounds based on laplace integrals and their asymptotic approximations. 62(12):6854–6883, 2016.

[16] Yi Fang, Guoan Bi, Yong Liang Guan, and Francis CM Lau. A survey on protograph ldpc codes and their applications. *IEEE Communications Surveys & Tutorials*, 17(4):1989–2016, 2015.

[17] M. P. C. Fossorier. Reliability-based soft-decision decoding with iterative information set reduction. *IEEE Transactions on Information Theory*, 48(12): 3101–3106, Dec 2002. ISSN 0018-9448. doi: 10.1109/TIT.2002.805089.

[18] M. P. C. Fossorier and Shu Lin. Soft-decision decoding of linear block codes based on ordered statistics. *IEEE Transactions on Information Theory*, 41(5): 1379–1396, Sep 1995. ISSN 0018-9448. doi: 10.1109/18.412683.

[19] Marc PC Fossorier and Shu Lin. First-order approximation of the ordered binary-symmetric channel. *IEEE Transactions on Information Theory*, 42(5): 1381–1387, 1996.

[20] Marc PC Fossorier and Shu Lin. Error performance analysis for reliability-based decoding algorithms. *IEEE Transactions on Information Theory*, 48(1):287–293, Jan 2002. doi: 10.1109/18.971758.

[21] Lorenzo Gaudio, Tudor Ninacs, Thomas Jerkovits, and Gianluigi Liva. On the performance of short tail-biting convolutional codes for ultra-reliable communications. In *SCC 2017; 11th International ITG Conference on Systems, Communications and Coding; Proceedings of*, pages 1–6. VDE, 2017.

[22] Jeongseok Ha, Jaehong Kim, and Steven W McLaughlin. Rate-compatible puncturing of low-density parity-check codes. *IEEE Transactions on information Theory*, 50(11):2824–2836, 2004.

[23] Michael Helmling, Stefan Scholl, Florian Gensheimer, Tobias Dietz, Kira Kraft, Stefan Ruzika, and Norbert Wehn. Database of Channel Codes and ML Simulation Results. www.uni-kl.de/channel-codes, 2017.

[24] Song-Nam Hong, Dennis Hui, and Ivana Marić. Capacity-achieving rate-compatible polar codes. *IEEE Transactions on Information Theory*, 63(12): 7620–7632, 2017.

[25] W. Jin and M. P. C. Fossorier. Reliability-based soft-decision decoding with multiple biases. *IEEE Transactions on Information Theory*, 53(1):105–120, Jan 2007. ISSN 0018-9448. doi: 10.1109/TIT.2006.887510.

[26] W. Jin and M. P. C. Fossorier. Reliability-based soft-decision decoding with multiple biases. *IEEE Transactions on Information Theory*, 53(1):105–120, Jan 2007. ISSN 0018-9448. doi: 10.1109/TIT.2006.887510.

[27] Wenyi Jin and Marc Fossorier. Probabilistic sufficient conditions on optimality for reliability based decoding of linear block codes. In *Information Theory, 2006 IEEE International Symposium on*, pages 2235–2239. IEEE, 2006.

[28] A. Kabat, F. Guilloud, and R. Pyndiah. New approach to order statistics decoding of long linear block codes. In *IEEE GLOBECOM 2007 - IEEE Global Telecommunications Conference*, pages 1467–1471, 2007.

[29] Bin Li, David Tse, Kai Chen, and Hui Shen. Capacity-achieving rateless polar codes. In *2016 IEEE International Symposium on Information Theory (ISIT)*, pages 46–50. IEEE, 2016.

[30] Jing Li and K Narayanan. Rate-compatible low density parity check codes for capacity-approaching arq scheme in packet data communications. In *Int. Conf. on Comm., Internet, and Info. Tech.(CIIT)*, pages 201–206. Citeseer, 2002.

[31] Shu Lin and Daniel J Costello. *Error control coding*. Pearson Education India, 2004.

[32] Gianluigi Liva, Lorenzo Gaudio, Tudor Ninacs, and Thomas Jerkovits. Code design for short blocks: A survey. *arXiv preprint arXiv:1610.00873*, 2016.

[33] Florence Jessie MacWilliams and Neil James Alexander Sloane. *The theory of error-correcting codes*. Elsevier, 1977.

[34] Kai Niu and Kai Chen. CRC-aided decoding of polar codes. *IEEE communications letters*, 16(10):1668–1671, 2012.

[35] Kai Niu, Kai Chen, and Jia-Ru Lin. Beyond turbo codes: Rate-compatible punctured polar codes. In *2013 IEEE International Conference on Communications (ICC)*, pages 3423–3427. IEEE, 2013.

[36] Toon Norp. 5g requirements and key performance indicators. *Journal of ICT Standardization*, 6(1):15–30, 2018.

[37] Athanasios Papoulis and S Unnikrishna Pillai. *Probability, random variables, and stochastic processes*. Tata McGraw-Hill Education, 2002.

[38] Wesley Peterson. Encoding and error-correction procedures for the bose-chaudhuri codes. *IRE Transactions on information theory*, 6(4):459–470, 1960.

[39] Yury Polyanskiy, H Vincent Poor, and Sergio Verdú. Channel coding rate in the finite blocklength regime. *IEEE Transactions on Information Theory*, 56(5): 2307–2359, 2010.

[40] Paul R Rider. Variance of the median of small samples from several special populations. *Journal of the American Statistical Association*, 55(289):148–150, 1960.

[41] C. E. Shannon. A mathematical theory of communication. *The Bell System Technical Journal*, 27(4):623–656, Oct 1948. ISSN 0005-8580. doi: 10.1002/j.1538-7305.1948.tb00917.x.

[42] Shuai Shao, Peter Hailes, Tsang-Yi Wang, Jwo-Yuh Wu, Robert G Maunder, Bashir M Al-Hashimi, and Lajos Hanzo. Survey of turbo, ldpc, and polar decoder asic implementations. *IEEE Communications Surveys & Tutorials*, 21 (3):2309–2333, 2019.

[43] Changyang She, Chenyang Yang, and Tony QS Quek. Radio resource management for ultra-reliable and low-latency communications. *IEEE Communications Magazine*, 55(6):72–78, 2017.

[44] M. Shirvanimoghaddam, M. S. Mohammadi, R. Abbas, A. Minja, C. Yue, B. Matuz, G. Han, Z. Lin, W. Liu, Y. Li, S. Johnson, and B. Vucetic. Short block-length codes for ultra-reliable low latency communications. *IEEE*

Communications Magazine, 57(2):130–137, February 2019. doi: 10.1109/MCOM.2018.1800181.

[45] Tao Tian and Christopher R Jones. Construction of rate-compatible ldpc codes utilizing information shortening and parity puncturing. *EURASIP Journal on wireless communications and networking*, 2005(5):1–7, 2005.

[46] Thibaud Tonnellier, Adam Cavatassi, and Warren J Gross. Length-compatible polar codes: a survey. In *2019 53rd Annual Conference on Information Sciences and Systems (CISS)*, pages 1–6. IEEE, 2019.

[47] A. Valembois and M. Fossorier. An improved method to compute lists of binary vectors that optimize a given weight function with application to soft-decision decoding. *IEEE Communications Letters*, 5(11):456–458, 2001.

[48] A. Valembois and M. Fossorier. Box and match techniques applied to soft-decision decoding. *IEEE Transactions on Information Theory*, 50(5): 796–810, May 2004. ISSN 0018-9448. doi: 10.1109/TIT.2004.826644.

[49] Antoine Valembois and Marc Fossorier. A comparison between "most-reliable-basis reprocessing" strategies. *IEICE TRANSACTIONS on Fundamentals of Electronics, Communications and Computer Sciences*, 85(7): 1727–1741, 2002.

[50] Johannes Van Wonterghem, Amira Alloum, Joseph Jean Boutros, and Marc Moeneclaey. On short-length error-correcting codes for 5G-NR. *Ad Hoc Networks*, 79:53–62, 2018.

[51] Runxin Wang and Rongke Liu. A novel puncturing scheme for polar codes. *IEEE Communications Letters*, 18(12):2081–2084, 2014.

[52] J. Van Wonterghem, A. Alloumf, J. J. Boutros, and M. Moeneclaey. Performance comparison of short-length error-correcting codes. In *2016 Symposium on Communications and Vehicular Technologies (SCVT)*, pages 1–6, Nov 2016. doi: 10.1109/SCVT.2016.7797660.

[53] Yingquan Wu and Christoforos N Hadjicostis. Soft-decision decoding using ordered recodings on the most reliable basis. *IEEE transactions on information theory*, 53(2):829–836, 2007.

[54] Yingquan Wu and Christoforos N Hadjicostis. Soft-decision decoding of linear block codes using preprocessing and diversification. *IEEE transactions on information theory*, 53(1):378–393, 2007.

[55] C. Yue, M. Shirvanimoghaddam, Y. Li, and B. Vucetic. Hamming distance distribution of the 0-reprocessing estimate of the ordered statistic decoder. In *2019 IEEE International Symposium on Information Theory (ISIT)*, pages 1337–1341, July 2019. doi: 10.1109/ISIT.2019.8849229.

[56] C. Yue, M. Shirvanimoghaddam, Y. Li, and B. Vucetic. Segmentation-discarding ordered-statistic decoding for linear block codes. In *2019 IEEE Global Communications Conference (GLOBECOM)*, pages 1–6, 2019.

[57] C. Yue, M. Shirvanimoghaddam, G. Park, O. S. Park, B. Vucetic, and Y. Li. Probability-based ordered-statistics decoding for short block codes. *IEEE Communications Letters*, pages 1–1, 2021. doi: 10.1109/LCOMM.2021.3058978.

[58] Chentao Yue, Mahyar Shirvanimoghaddam, Branka Vucetic, and Yonghui Li. A revisit to ordered statistics decoding: Distance distribution and decoding rules. *IEEE Transactions on Information Theory*, 2021.

6

Sparse Vector Coding for Ultra-reliable and Low-latency Communications

Byonghyo Shim

Institute of New Media and Communications and Department of Electrical and Computer Engineering, Seoul National University, Seoul, Korea

6.1 Introduction

Ultra Reliable and Low Latency Communication (URLLC) is a newly introduced use case in 5G to support mission-critical applications such as the Tactile Internet, autonomous driving, factory automation, cyber-physical systems, and remote robot surgery [1]. In order to support URLLC, the 3rd Generation Partnership Project (3GPP) sets an aggressive requirement that a packet should be delivered with 10^{-5} Block Error Rate (BLER) within a 1 ms period [2]. Basically, there are two types of information; the first is the image/video/audio/speech signal and the second is control, command, or sensing information. A large part of the URLLC information can be classified as the second type consisting of control (e.g. bandwidth, resource block, Modulation and Coding Scheme (MCS)), command type information (e.g. start/stop, move left/right, rotate/shift, and speed up/down) or sensing information (e.g. temperature, moisture, pressure, distance, speed, and gas density) so that the amount of information to be delivered in these cases is very small [3]. Since today's wireless transmission strategy, often called Shannon's principle, designed to achieve the channel capacity using a long codeblock might not an appropriate strategy in URLLC scenarios, we need to come up with a new transmission strategy optimized for the short packet transmission. While there have been some efforts to improve the connection density, medium access latency, and reliability of the re-transmission scheme for URLLC [4–8], not much work has been done for the short-sized packet transmission except for the channel coding scheme for the short packet [9].

Ultra-Reliable and Low-Latency Communications (URLLC) Theory and Practice: Advances in 5G and Beyond, First Edition. Edited by Trung Q. Duong, Saeed R. Khosravirad, Changyang She, Petar Popovski, Mehdi Bennis and Tony Q.S. Quek.
© 2023 John Wiley & Sons Ltd. Published 2023 by John Wiley & Sons Ltd.

In the 4G Long Term Evolution (LTE) and 5G New Radio (NR) systems, reliability of the data transmission is achieved mainly by the channel coding, diversity scheme, or a combination of these two [9, 11]. In the 5G NR system, for example, encoding at the Base Station (BS) is done by the polar encoding and the decoding at the mobile terminal is done by the Successive Cancellation List (SCL)-based polar decoder. While this approach is, to some extent, effective, it might not be easy in future URLLC scenarios since there is a stringent limitation on the packet length (and thus the parity size) yet the required reliability (target BLER $= 10^{-5}$) is very high [9].

The purpose of this chapter is to introduce a new type of short packet transmission for URLLC distinct from the conventional channel coding principle.[1] The key idea behind the proposed technique, henceforth referred to as Sparse Vector Coding (SVC), is to transmit the short-sized information after the sparse vector transformation[2]. To be specific, by mapping the transmit bit vector into a sparse vector and then transmitting it after the pseudo-random spreading, we obtain an underdetermined sparse system for which the principle of Compressed Sensing (CS) can be applied [12]. From the theory of CS, an accurate recovery of a sparse vector is guaranteed with a relatively small number of measurements when the system matrix (in the CS parlance, it is called a sensing matrix) is generated at random [13], which is achieved in our case via the random spreading. Since the sparsity of an input vector is guaranteed in the sparse vector transformation, SVC decoding can be done by the sparse signal recovery, an operation to identify the support (non-zero positions in the transmit sparse vector). Therefore, the SVC scheme is very simple to implement and can be used for a variety of future wireless applications in which the transmit packet size is sufficiently small.

There have been some previous efforts to use the support locations in the information encoding process [14–18]. For example, sparse mapping is conceptually related to the Position Modulation (PM) or the Index Modulation (IM) techniques [14–16] in which the indices of the building block of the communication systems, such as pulses in optical systems, transmit antennas at the BS, or subcarrier groups in Orthogonal Frequency Division Multiplexing (OFDM) systems, are used to convey additional information bits. Also, in the single/multiple PM techniques, information is transmitted via the time sparsity realized by the combinations of the positions of optical pulses. In the spatial modulation-based IM technique, additional information can be delivered by selectively choosing a part of the transmit antennas in the information transmission.

---

1 This chapter has been a part of [30, 31]

2 A vector $\mathbf{s}$ is called sparse if the number of non-zero entries is sufficiently smaller than the dimension of the vector.

SVC is distinct from these studies in that we fully utilize the physical resources in the data transmission and thus the loss, if any, caused by the underutilization of physical resources can be prevented. Also, in contrast to the IM technique where the receiver processing consists of two steps (index recovery and symbol detection), SVC decoding can be realized by the identification of the non-zero position (this step is called the support identification) [12]. Another distinctive point of SVC over the conventional techniques is that there is no random spreading mechanism in the conventional schemes so that the compression of the transmit vector is not possible for the conventional schemes.

The rest of this chapter is organized as follows. In Section 6.2, we briefly explain the short-sized packet transmission in 5G NR systems. In Section 6.3, we present the SVC scheme and explain the encoding and decoding operations. In Section 6.4, we analyze the success probability of SVC-encoded data transmission. Various implementation issues are discussed in Section 6.5. In Section 6.6, we present simulation results to verify the performance of the SVC scheme. In Sections 6.7 and 6.8, we introduce the advanced SVC schemes including the pilot-less SVC and the symbol-embedded SVC schemes and provide their simulation results. We conclude the chapter in Section 6.9.

6.2 Short-sized Packet in 5G NR Downlink

In this section, we briefly review the control-type data transmission to illustrate the short-sized packet transmission in the conventional 4G LTE and 5G NR systems. We note that an extension to data channel (e.g. physical downlink shared channel (PDSCH)) is straightforward.

In 4G and 5G, the Physical Downlink Control Channel (PDCCH) carries essential information for the mobile terminal called User Equipment (UE) when the UE tries to transmit or receive the data. Specifically, PDCCH carries small-sized control information (e.g. resource assignment, modulation order, code rate) needed for the decoding of the data channel. In PDCCH, a Cyclic Redundancy Check (CRC) is added to check the decoding error [19]. Since the CRC bit stream is scrambled with a user index called the Radio Network Temporary Identifier (RNTI), only the scheduled user can pass the CRC test.

After the channel coding and symbol mapping,[3] the modulated symbol vector $\mathbf{s} \in \mathbb{C}^{N\times 1}$ is transmitted. The corresponding received vector $\mathbf{y} \in \mathbb{C}^{m\times 1}$ is given by

$$\mathbf{y} = \mathbf{HRs} + \mathbf{v}, \tag{6.1}$$

---

3 For example polar coding with rate $\frac{1}{3}$ and Quadrature Phase Shift Keying (QPSK) modulation are employed.

where $\mathbf{H} \in \mathbb{C}^{m \times m}$ is the diagonal matrix whose diagonal entry h_{ii} is the channel component for each resource, $\mathbf{v} \sim \mathcal{CN}(\mathbf{0}, \sigma_v^2 \mathbf{I})$ is the additive Gaussian noise, and $\mathbf{R} \in \mathcal{C}^{m \times N}$ is the matrix describing the mapping between the symbol and resource element. For example, when one symbol is mapped to a single resource, $\mathbf{R}$ becomes the identity matrix ($\mathbf{R} = \mathbf{I}$). Whereas, if two resources are assigned to one symbol for the transmit diversity, then $\mathbf{R}$ would be $2N \times N$ matrix.

When one tries to improve the reliability by modifying PDCCH, one needs to consider three options. The first is to achieve the better coding gain by lowering the code rate (i.e. $r = \frac{b}{2N} < r_{pdcch} = \frac{1}{3}$). This option is easy and straightforward but when the coded symbol length N increases, the transmission and processing latency will also increase, resulting in the violation of the URLLC requirement. The second option is to use the multiple resources in time and/or frequency to achieve the diversity gain ($m > N$). By combining multiple copies of the same symbol at the receiver, the reliability of the symbol can be improved. A potential drawback of this approach is that it requires additional wireless resources to achieve the diversity gain, so the resource utilization efficiency can be degraded severely. The third option is to reduce the size of the control information b. By removing part of the scheduling parameters, one can save resources used for the control channel. While this option is somewhat doable, there is a limitation on the reduction of the packet size since one cannot entirely remove essential information (e.g. CRC and user index).

6.3 Sparse Vector Coding

6.3.1 SVC Encoding and Transmission

In a nutshell, SVC encoding is mapping the transmit information into the positions of a sparse vector $\mathbf{s}$. Conceptually, SVC encoding can be thought of as putting a few balls into empty boxes As illustrated in Figure 6.1, when one tries to pick two boxes out of nine to put a ball, there are $\binom{9}{2} = 36$ choices in total. In general, when we choose K out of N positions, $\lfloor \log_2 \binom{N}{K} \rfloor$ bits of information

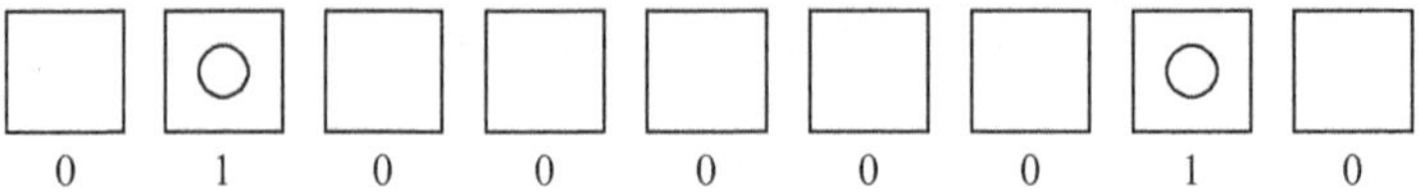

Figure 6.1 Metaphoric illustration of SVC encoding. Information is mapped into the position of a sparse vector.

can be encoded. An example of one-to-one mapping between the information bits **w** and transmit sparse vector **s** is (see Table 6.1 for the SVC mapping algorithm)

$$\underbrace{\begin{matrix} 0&0&0&0&0 \\ 0&0&0&0&1 \\ 0&0&0&1&1 \\ &&\vdots&& \\ 1&1&1&1&1 \end{matrix}}_{b\text{-bit information } \mathbf{w}\ (b=5)} \quad \begin{matrix} \longleftrightarrow \\ \longleftrightarrow \\ \longleftrightarrow \\ \vdots \\ \longleftrightarrow \end{matrix} \quad \underbrace{\begin{matrix} 0&0&0&0&0&0&0&1&1 \\ 0&0&0&0&0&0&1&0&1 \\ 0&0&0&0&0&1&0&0&1 \\ &&&&\vdots&&&& \\ 1&0&0&0&0&0&0&0&1 \end{matrix}}_{K\text{-sparse vector } \mathbf{s}\ (K=2)}$$

After the sparse mapping, each non-zero element in **s** is spread into $m(< N)$ resources using the codeword (spreading sequence) in the spreading codebook **C**. While it is possible to allocate resources either in time, frequency axis or a hybrid of these, here we assume that they are allocated in the frequency axis (see Figure 6.2(a)). We note that this choice will not affect the system model but minimizes the transmission latency. As a result of this spreading process, the resource mapping matrix **R** in (1) is replaced with the codebook matrix $\mathbf{C} = [\mathbf{c}_1\, \mathbf{c}_2\, \cdots\, \mathbf{c}_N]$ where $\mathbf{c}_i = [c_{1i}\, c_{2i}\, \cdots\, c_{mi}]^T$ is the spreading sequence. For example, if the second and the fourth element of **s** are non-zero, then the transmit vector after spreading is

$$\begin{aligned} \mathbf{x} &= \mathbf{C}\mathbf{s} \\ &= s_2\mathbf{c}_2 + s_4\mathbf{c}_4. \end{aligned} \tag{6.2}$$

Since the position of a non-zero element is chosen at random, the codebook matrix **C** should be designed such that the transmit vector **x** contains enough information to recover the sparse vector **s** irrespective of the selection of the non-zero positions. It has been shown that if entries of the codebook matrix **C** are generated at random (e.g. sampled from a Gaussian or Bernoulli distribution), then an accurate recovery of the sparse vector is possible as long as $m = \mathcal{O}(K \log N)$ [13]. Example of **C** for $m = 5$ and $N = 10$, when elements of $\mathbf{c}_i$ are chosen from the Bernoulli distribution, is given by

$$\mathbf{C} = \frac{1}{\alpha}\begin{bmatrix} 1 & 1 & 1 & 1 & -1 & 1 & -1 & 1 & -1 & -1 \\ 1 & -1 & 1 & -1 & 1 & -1 & 1 & -1 & -1 & 1 \\ 1 & 1 & -1 & -1 & 1 & 1 & -1 & -1 & 1 & 1 \\ 1 & -1 & -1 & 1 & 1 & -1 & 1 & 1 & -1 & -1 \\ -1 & 1 & 1 & 1 & -1 & -1 & -1 & -1 & 1 & 1 \end{bmatrix} \tag{6.3}$$

Table 6.1 Example of mapping between the information $\mathbf{w}$ and the sparse vector $\mathbf{s}$.

Input:

Size of sparse vector N, information vector $\mathbf{w}$

Output:

Sparse vector $\mathbf{s}$

```
a := 0
for i = 2 to N do
  for j = 1 to i − 1 do
    if a = (w)_(10)
      s := (2^i + 2^j)_(2)
    end if
    a := a + 1
  end for
end for
return ŝ^K
```

Note: $(\mathbf{w})_{(10)}$ is the decimal expression of the binary vector $\mathbf{w}$ and
$(w)_{(2)}$ is the binary expression of the integer w.

where α is the normalization factor depending on the modulated symbols. The corresponding received signal $\mathbf{y}$ is

$$\begin{aligned}\mathbf{y} &= \mathbf{Hx} + \mathbf{v}\\ &= \begin{bmatrix}\mathbf{Hc}_2 & \mathbf{Hc}_4\end{bmatrix}\begin{bmatrix}s_2\\ s_4\end{bmatrix} + \mathbf{v}.\end{aligned} \tag{6.4}$$

In general, the received vector $\mathbf{y}$ is given by

$$\begin{aligned}\mathbf{y} &= \mathbf{HCs} + \mathbf{v}\\ &= \begin{bmatrix}h_{11} & & \\ & \ddots & \\ & & h_{mm}\end{bmatrix}\begin{bmatrix}| & & |\\ \mathbf{c}_1 & \cdots & \mathbf{c}_N\\ | & & |\end{bmatrix}\begin{bmatrix}s_1\\ \vdots\\ s_N\end{bmatrix} + \begin{bmatrix}v_1\\ \vdots\\ v_m\end{bmatrix}.\end{aligned} \tag{6.5}$$

In the SVC decoding, an accurate recovery of the sparse vector $\mathbf{s}$ is unnecessary since the decoding process is done by the identification of non-zero positions, not

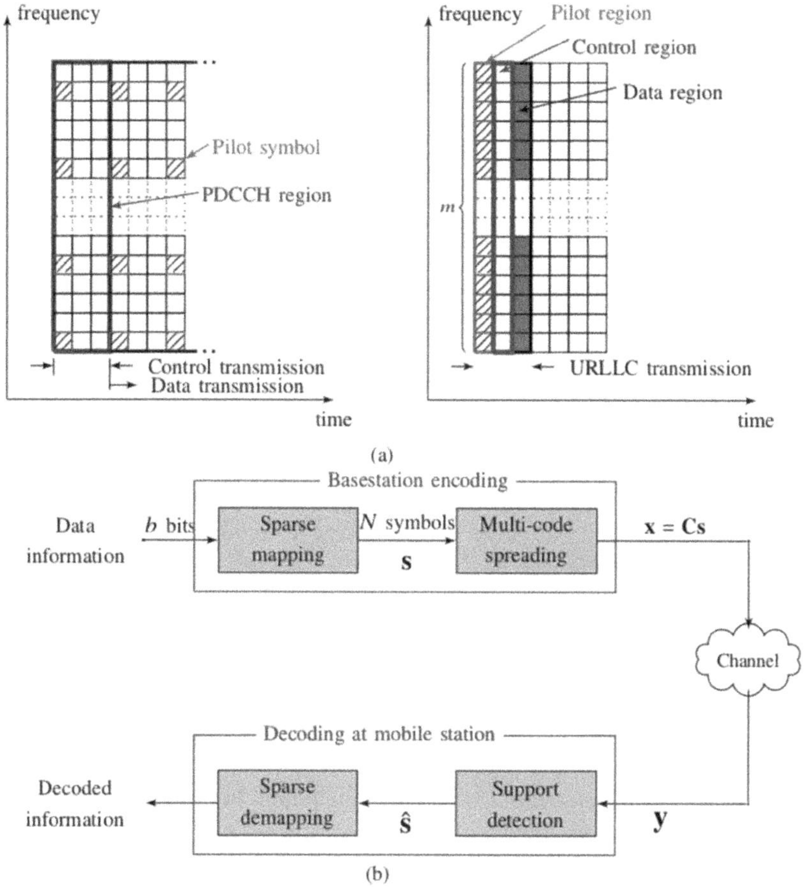

Figure 6.2 SVC-based packet transmission: (a) packet structure of 4G (left) and the URLLC packet (right) and (b) the overall block diagram. ©IEEE 2020. Reprinted with permission from [31].

the symbol values of the transmit vector[4] The fact that the decoding is done by the support identification greatly simplifies the decoding process[5] and also reduces the chance of decoding failure. The overall structure of the SVC is depicted in Figure 6.2(b).

---

4 In an advanced SVC algorithm, to improve the data rate, symbols can be encoded to non-zero positions.

5 Support is the set of non-zero elements. For example, if $\mathbf{s} = [0\,0\,1\,0\,0\,1]^T$, the support $\Omega_{\mathbf{s}} = \{3, 6\}$.

The advantages of SVC can be summarized as follows. First, the transmission power of the data channel is concentrated on the non-zero elements of an information vector. When compared to the conventional system in which the transmission power is uniformly distributed across all symbols, effective transmit power per symbol is higher. Second, the SVC decoding process achieved by the sparse recovery algorithm lends itself to the test of decoding success/failure so that the CRC operation is unnecessary. This directly implies that the code rate of SVC can be made smaller than the rate of PDCCH. Specifically, when the number of resources used for the data channel is m and the QPSK modulation is used, the code rate of SVC is $r_{svc} = \frac{b_i}{2m}$ (b_i is the number of information bits) and the code rate of PDCCH is $r_{pdcch} = \frac{(b_i+b_c)}{2m} \left(= \frac{1}{3}\right)$. If the number of CRC bits is $b_c = \beta b_i$ ($\beta > 0$), then $m = \frac{3}{2}(b_i + b_c) = \frac{3}{2}(b_i + \beta b_i)$. Thus, the code rate of SVC can be expressed in terms of β as $r_{svc} = \frac{b_i}{2m} = \frac{1}{3(1+\beta)} < \frac{1}{3} = r_{pdcch}$. Third, when m is sufficiently large, the BS can easily assign the distinct codebook $\mathbf{C}$ for each user. This is because the codebook matrices can be made nearly orthogonal by using a properly designed codebook generation mechanism.[6] For example, when $m = 64$ and the codebook is generated by the Bernoulli distribution, then there would be 2^{64} different spreading sequences $\mathbf{c}_i$. Thus, if $N = 128$, then the BS can support approximately $2^{57}(\approx \frac{2^{64}}{128})$ devices. The last but not least important benefit of SVC is that the implementation cost is small and the processing latency is low. Encoding is done via a simple one-to-one mapping and random spreading, which can be easily realized by the simple look-up table and addition/subtraction operations. Decoding is performed by the support identification and demapping. In particular, since the sparsity K is small and also known to the receiver, one can decode the SVC packet using a simple sparse signal recovery algorithm such as Orthogonal Matching Pursuit (OMP)[7] [12]. In Table 6.2, we compare PDCCH and SVC.

6.3.2 SVC Decoding

6.3.2.1 Support Identification

As mentioned, the SVC decoding is done by the identification of the support and basically any sparse recovery algorithm can be used to this end. In this

---

6 The correlation between two distinct columns of random matrix decreases exponentially as the dimension of a column increases [20].

7 Most CS algorithms discover the solution without prior knowledge of the sparsity K. However, when K is known in advance, one can recover the sparse vector more accurately by using the sparsity-aware recovery technique [29].

work, we use a greedy sparse recovery algorithm such as OMP in the decoding of the SVC-encoded packet. In the decoding process, one can directly use the system model in (6.5). Alternatively, one can use the modified system model after pre-multiplying the diagonal matrix constructed by the complex exponential $e^{-j\measuredangle h}$ to ensure that the observation vector is real. In case of the complex system, one can also stack real and imaginary parts and then perform the decoding.

Since $\mathbf{s}$ has K non-zero elements, the received vector $\mathbf{y} = \mathbf{HCs} + \mathbf{v}$ can be expressed as a linear combination of K columns of $\mathbf{\Phi} = \mathbf{HC}$ perturbed by the noise. This means that the main task of the SVC decoding is to identify the columns in $\mathbf{\Phi}$ participating in the received vector $\mathbf{y}$. In each iteration, the greedy sparse recovery algorithm identifies one column of $\mathbf{\Phi}$ at a time using a greedy strategy [21]. Specifically, a column of $\mathbf{\Phi}$ that is maximally correlated with the (modified) observation $\mathbf{r}^{j-1}$ is picked in each iteration. That is, an index of the non-zero column of $\mathbf{\Phi}$ chosen as the jth iteration is[8]

$$\omega_j = \arg\max_l |\langle \phi_l, \mathbf{r}^{j-1} \rangle|^2, \tag{6.6}$$

where $\mathbf{r}^{j-1} = \mathbf{y} - \mathbf{\Phi}_{\Omega_{\mathbf{s}}^{j-1}} \hat{\mathbf{s}}^{j-1}$ is the modified observation called the residual and $\hat{\mathbf{s}}^{j-1} = \mathbf{\Phi}^{\dagger}_{\Omega_{\mathbf{s}}^{j-1}} \mathbf{y}$ is the estimate of $\mathbf{s}$ at $(j-1)$th iteration[9].

Instead of the sparse recovery algorithm, one can use the Maximum Likelihood (ML) detection to achieve the better performance. Recalling that the sparsity

Table 6.2 PDCCH versus SVC technique.

	PDCCH	SVC technique
Coding (encoding/ decoding)	Polar code ($\frac{1}{3}$ rate)/ SCL decoding	Sparse encoding/sparse recovery algorithm
Transmission	Time/frequency mapping	Spreading in frequency/ time direction
User identification	CRC scrambled with user index	User codebook $\mathbf{C}$
Resource overhead (L repetitions, QPSK)	$L\frac{3b}{2}$	Lm where m is the size of spreading length

8 If $\Omega = \{1, 3\}$, then $\mathbf{\Phi}_{\Omega} = [\phi_1\ \phi_3]$.
9 $\mathbf{\Phi}^{\dagger} = (\mathbf{\Phi}^T\mathbf{\Phi})^{-1}\mathbf{\Phi}^T$ is the pseudo-inverse of $\mathbf{\Phi}$.

K is known to both transmitter and receiver, the ML detection problem can be expressed as

$$\mathbf{s}^* = \arg \max_{||\mathbf{s}||_0=K} \Pr(\mathbf{y}|\mathbf{s}, \mathbf{H}, \mathbf{C}), \tag{6.7}$$

where $||\mathbf{s}||_0$ is the ℓ_0-norm of $\mathbf{s}$ counting the number of non-zero elements in $\mathbf{s}$. Since our goal is to find the support of $\mathbf{s}$, we alternatively have

$$\Omega_{\mathbf{s}}^* = \arg \max_{|\Omega_{\mathbf{s}}|=K} \Pr(\mathbf{y}|\Omega_{\mathbf{s}}, \mathbf{H}, \mathbf{C}), \tag{6.8}$$

where $|\Omega_{\mathbf{s}}|$ is the cardinality of the set $\Omega_{\mathbf{s}}$.

To find the ML solution, we should enumerate all possible combinations of candidate supports with cardinality K. Obviously, this exhaustive search would not be feasible for most practical scenarios. As an alternative option, one can use the Multipath Matching Pursuit (MMP) algorithm [22], a near-ML sparse recovery algorithm. In a nutshell, MMP performs a deliberate tree search to find the near-ML solution. Unlike the single-path search algorithm, MMP selects multiple promising indices in each iteration since the candidate chosen in one iteration brings forth multiple new child candidates. After finishing K iterations, candidate $\mathbf{s}^*$ having the smallest cost function among all candidates is chosen as the final output (i.e. $\mathbf{s}^* = \arg\min_{\hat{\mathbf{s}}} J(\hat{\mathbf{s}}) = ||\mathbf{y} - \mathbf{\Phi}_{\Omega_{\hat{\mathbf{s}}}}\hat{\mathbf{s}}||_2$). Due to the fact that many candidates are redundant and hence counted only once, the actual number of candidates examined in MMP is quite moderate [22].

One advantage of MMP, in the perspective of SVC decoding, is that it deteriorates the quality of incorrect candidates yet does not impose any estimation error to the correct one. This is because the quality of incorrect candidates gets worse due to the error propagation while no such behavior occurs to the right one. On top of this, since the non-zero values of an original sparse vector $\mathbf{s}$ are known to the receiver[10], no estimation error will be introduced in the correct candidate.

We also note that the computational complexity of the SVC decoding is quite moderate since the computational complexity of the greedy sparse recovery algorithm is proportional to the sparsity K. In fact, in many sparse recovery algorithms, the number of iterations is set to the sparsity level K[11] Thus, the processing latency

10 Since the goal of SVC decoding is to find out the non-zero positions of a sparse vector, we can pre-define values of the non-zero elements in $\mathbf{s}$.
11 In each iteration, the greedy sparse recovery algorithm performs three operations: support identification, non-zero element estimation, and residual update. Since the non-zero values are known in advance, estimation of the non-zero elements is unnecessary.

of SVC decoding can also be made sufficiently small. This is in contrast to the conventional decoding algorithm (e.g. Viterbi decoding used for the PDCCH packet decoding) in which the computational complexity is proportional to the length of a codeblock [23].

6.3.2.2 Identification of False Alarm

In a nutshell, there are two types of false alarm events causing the decoding failure: (1) support detection when the BS transmits information to the different user and (2) support detection when there is no transmission at the BS. In order to minimize the detection error caused by these, we examine the residual magnitude in each iteration. Firstly, when a packet for a different user is received, the codebooks of the two distinct users would be different from each other so that the magnitude of the correlation μ_{ij} between two codewords, each being chosen from a distinct codebook would be small. In this case, clearly, one cannot expect a meaningful reduction in the residual magnitude. Secondly, when there is no transmission, the received vector contains the noise only (i.e. $\mathbf{y} = \mathbf{v}$) and thus some column in $\mathbf{\Phi}$, say ϕ_l, will be added to the residual in each iteration $\mathbf{r}^i = \mathbf{r}^{i-1} - \phi_l \hat{\mathbf{s}}_l$ (see Figure 6.3). Based on these observations, we declare a decoding failure when the residual magnitude is outside the confidence interval of the pure noise contribution. The selection of confidence interval will be discussed in the next section.

The MMP-based SVC decoding algorithm is summarized in Table 6.3.

6.4 SVC Decoding Success Probability Analysis

We now discuss the performance analysis of the SVC technique. Since the decoding of the SVC-encoded packet is successful when all support elements are chosen by the sparse recovery algorithm, we analyze the probability that the support is identified accurately. In our analysis, we assume that the greedy sparse recovery algorithm is used in the decoding process and analyze the lower bound of the success probability.

For analytic simplicity, we consider the $K = 2$ scenario initially and then extend to the general case. Without loss of generality, we assume that the pth and qth elements of $\mathbf{s}$ are non-zero (i.e. $\Omega_{\mathbf{s}} = \{p, q\}$). Further, by setting the information vector such that $s_p = 1$ and $s_q = j$, we can model the QPSK (see Section 6.5.2).

The following lemmas will be useful in our analysis.

Lemma 6.1: Consider the vector $\mathbf{a}_i (i = 1, \cdots, N)$ whose element is i.i.d. standard Gaussian. Then, $\frac{\mathbf{a}_i^T \mathbf{a}_j}{\|\mathbf{a}_i\|_2}$ is standard Gaussian. That is, $\frac{\mathbf{a}_i^T \mathbf{a}_j}{\|\mathbf{a}_i\|_2} \sim \mathcal{N}(0, 1)$.

Proof. See Appendix A.

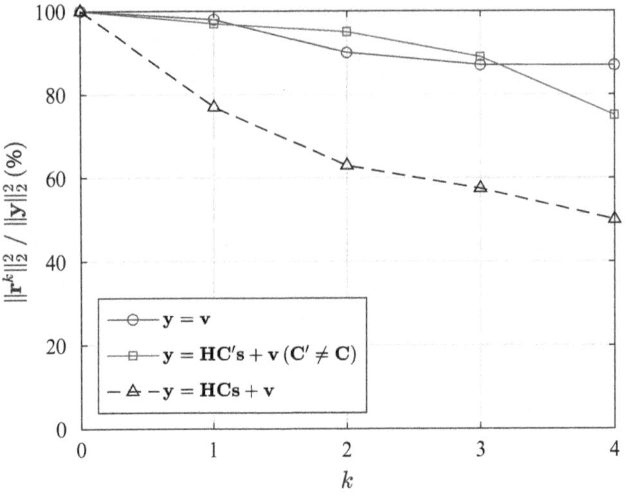

Figure 6.3 Snapshot of the ratio between residual magnitude $\|\mathbf{r}^k\|_2^2$ and $\|\mathbf{y}\|_2^2$ as a function of the number of iterations in the OMP algorithm. Signal-to-noise ratio (SNR) is set to 0 dB and the sparsity K is set to 4.

Lemma 6.2: Consider the vector $\tilde{\mathbf{h}} = [\tilde{h}_{11}\ \tilde{h}_{22}\ \cdots\ \tilde{h}_{mm}]^T$ where $\tilde{h}_{ii} = h_{ii}e^{j\measuredangle h_{ii}}$. The Probability Density Function (PDF) of the $\|\tilde{\mathbf{h}}\|_2^2$ is a Chi-squared distribution with

$$f_{\|\tilde{\mathbf{h}}\|_2^2}(x) = \frac{x^{m-1}\exp(-x)}{\Gamma(m)}, \tag{6.9}$$

where $\Gamma(m) = (m-1)!$ is the Gamma function and $\mathrm{E}[\|\tilde{\mathbf{h}}\|_2^2] = m$.

Proof. In the SVC decoding, $\|\tilde{\mathbf{h}}\|_2^2$ can be expressed as $\|\tilde{\mathbf{h}}\|_2^2 = \|\mathbf{h}\|_2^2 = \sum_{i=1}^m |h_{ii}|^2 = \sum_{i=1}^m (\Re(h_{ii})^2 + \Im(h_{ii})^2)$ where $\Re(c)$ and $\Im(c)$ are the real and imaginary part of c, respectively. Since $\Re(h_{ii}), \Im(h_{ii}) \sim \mathcal{N}(0, \frac{\sigma_v^2}{2})$, one can show after some manipulations that $2\|\tilde{\mathbf{h}}\|_2^2$ follows a Chi-squared distribution with $2m$ Degrees of Freedom (DoF) [28].

$$f_{2\|\tilde{\mathbf{h}}\|_2^2}(x) = \frac{x^{m-1}\exp(-\frac{x}{2})}{2^m\Gamma(m)}. \tag{6.10}$$

Since $f_Z(z) = 2f_{2Z}(2z)$, we have

$$f_{\|\tilde{\mathbf{h}}\|_2^2}(x) = \frac{x^{m-1}\exp(-x)}{\Gamma(m)}. \tag{6.11}$$

Table 6.3 The MMP-based SVC decoding algorithm.

Input:

Measurement $\mathbf{y}$, sensing matrix $\mathbf{\Phi} = \mathbf{HC}$, sparsity k, number of expansion L, max number of search candidate l_{max}, stop threshold ϵ, detection threshold η

Initialization:

$l := 0$ (candidate order), $\rho := \infty$ (minimum magnitude of residual)

While: $l < l_{max}$ and $\epsilon < \rho$ **do**

$l := l + 1$

$\mathbf{r}^0 := \mathbf{y}$

$[p_1, \cdots, p_K] :=$compute_$p_k(l, L)$ (compute layer order)

for $k = 1$ **to** $k = K$ **do** (investigate lth candidate)

$\tilde{\omega} :=$compute_$\omega(k, L)$ (choose L best indices)

$\Omega_l^k := \Omega_l^{k-1} \cup \{\tilde{\omega}_{p_k}\}$ (construct a path in kth layer)

$\mathbf{r}^k := \mathbf{y} - \mathbf{\Phi}_{\Omega_l^k} \mathbf{s}^k$ (update residual)

$\hat{\Omega}^k := \Omega_l^k$ (update support set)

end for

if $\|\mathbf{r}^K\|_2^2 < \rho$ **then** (update the smallest residual)

$\rho := \|\mathbf{r}^K\|_2^2$

if $\frac{\|\mathbf{r}^K\|_2^2}{\|\mathbf{y}\|_2^2} > 1 - \eta$ **then** (false-alarm identification)

$\hat{\Omega}^* := \mathbf{0}$

end if

$\hat{\Omega}^* := \hat{\Omega}^K$

end if

end while

return $\hat{\Omega}^*$

---

function compute_$p_k(l, L)$ $t := l - 1$ **for** $k = 1$ **to** K **do**

$p_k := \text{mod}(t, L) + 1$

$t := \text{floor}(t/L)$

end for return $[p_1, \cdots, p_K]$

end function

---

function compute_$\omega(k, L)$

if k =odd **then**

return $\arg\max_{|\pi|=L} \|(\Re\langle \frac{\phi^T}{\|\phi\|_2} \mathbf{r}^{k-1} \rangle)_\pi\|_2^2$

else

return $\arg\max_{|\pi|=L} \|(\Im\langle \frac{\phi^T}{\|\phi\|_2} \mathbf{r}^{k-1} \rangle)_\pi\|_2^2$

end if

end function

Let S^j be the success probability that the support element is chosen in the jth iteration. Since $K = 2$ and thus the required number of iterations is two, the probability that the SVC packet is successfully decoded can be expressed as

$$\begin{aligned} P_{succ} &= \mathrm{P}(\Omega_{\mathbf{s}}^* = \Omega_{\mathbf{s}}) \\ &= \mathrm{P}\left(S^1, S^2\right) \\ &= \mathrm{P}(S^2|S^1)\mathrm{P}(S^1). \end{aligned} \tag{6.12}$$

Our main result is provided in the following theorem.

Theorem 6.1: The probability that the SVC-encoded packet is decoded successfully satisfies

$$P_{succ} \geq \left(1 - \left(1 + \frac{(1-\mu^*)^2}{\sigma_{\mathbf{v}}^2}\right)^{-m} - \left(1 + \frac{1}{\sigma_{\mathbf{v}}^2}\right)^{-m}\right)^{2N}, \tag{6.13}$$

where m is the number of measurements (resources), N is the size of sparse vectors, $\sigma_{\mathbf{v}}^2$ is the noise variance, and $\mu^* = \max\limits_{i \neq j} |\mu_{ij}|$ is the maximum absolute value of correlation between two distinct columns of $\mathbf{\Phi}$.

When m is sufficiently large, we approximately have

$$P_{succ} \gtrsim \left(1 - \left(1 + \frac{(1-\mu^*)^2}{\sigma_{\mathbf{v}}^2}\right)^{-m}\right)^{2N}. \tag{6.14}$$

Also, since the block error rate is $\mathrm{BLER}_{svc} = 1 - P_{succ}$, the upper bound of BLER is

$$\mathrm{BLER}_{svc} \lesssim 1 - \left(1 - \left(1 + \frac{(1-\mu^*)^2}{\sigma_{\mathbf{v}}^2}\right)^{-m}\right)^{2N}. \tag{6.15}$$

In Figure 6.4, we plot the BLER performance of SVC as a function of SNR. In order to check the validity of Theorem 6.1, we perform the empirical simulation for $m = 42, N = 96$. From the empirical test, we obtain that $\mu^* \approx 0.7$. When we apply this value to the upper bound in (6.15), we see that the obtained bound is tight (see Figure 6.4(a)). To evaluate the performance of SVC, we plot the BLER as a function of μ^*, N, and m in Figure 6.4(b), 6.4(c), and 6.4(d). First, when the maximum correlation μ^* decreases, we see from Figure 6.4(b) that the BLER performance improves sharply. For example, if μ^* is reduced from 0.4 to 0.2, we can achieve 1.5 dB gain at the target reliability point (BLER$= 10^{-5}$). Next, we test the BLER performance for various sparse vector dimensions in Figure 6.4(c). While the BLER performance degrades with N, we see that the degradation is marginal. In contrast, as shown in Figure 6.4(d), the BLER performance is quite sensitive to the number of measurements m.

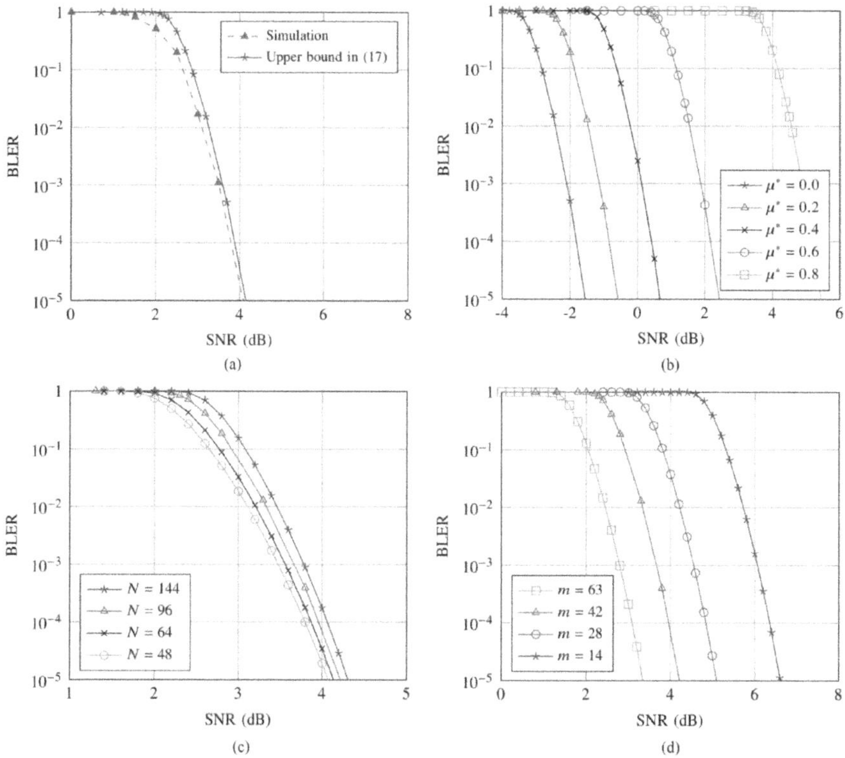

Figure 6.4 BLER performance of SVC-encoded packet from (6.15).

To prove Theorem 6.1, we first analyze the success probability $\mathrm{P}(S^1)$ for the first iteration.

Lemma 6.3: Consider the received signal $\tilde{\mathbf{y}} = \gamma\mathbf{\Phi}\mathbf{s} + \tilde{\mathbf{v}}$ where $\gamma = \frac{\sqrt{\text{SNR}}}{\alpha}$, $\mathbf{\Phi} = [\phi_1\ \phi_2\ \cdots\ \phi_N]$, and $\phi_i = [\tilde{h}_{11}c_{1i}\ \tilde{h}_{22}c_{2i}\ \cdots\ \tilde{h}_{mm}c_{mi}]^T$. The probability that the support element is chosen in the first iteration satisfies

$$\mathrm{P}(S^1) \geq \left(1 - \left(1 + \frac{(1-\mu^*)^2}{\sigma_{\mathbf{v}}^2}\right)^{-m} - \left(1 + \frac{1}{\sigma_{\mathbf{v}}^2}\right)^{-m}\right)^{N-1}. \tag{6.16}$$

Proof. As shown in Table 6.3, N decision statistics $\frac{\phi_l^T}{\|\phi_l\|_2}\mathbf{r}^{k-1}$ $(l = 1, \cdots, N)$ are computed in each iteration. For analytic simplicity, we take the real part of the decision statistic in the first iteration and the imaginary part in the second iteration.[12]

12 This choice is suboptimal but simplifies the analysis.

In order to identify the support element in the first iteration, we should have $\left|\Re\langle\frac{\phi_p}{\|\phi_p\|_2},\mathbf{r}^0\rangle\right| \geq \max_i \left|\Re\langle\frac{\phi_i}{\|\phi_i\|_2},\mathbf{r}^0\rangle\right|$ and thus the success probability for a given channel realization $\mathbf{h}$ is

$$\begin{aligned}
\mathrm{P}(S^1|\mathbf{h}) &= \mathrm{P}\left(\left|\Re\langle\frac{\phi_p}{\|\phi_p\|_2},\mathbf{r}^0\rangle\right| \geq \max_i \left|\Re\langle\frac{\phi_i}{\|\phi_i\|_2},\mathbf{r}^0\rangle\right|\right) \\
&= \prod_{i=1,i\neq p}^{N} \mathrm{P}\left(\left|\Re\langle\frac{\phi_p}{\|\phi_p\|_2},\mathbf{r}^0\rangle\right| \geq \left|\Re\langle\frac{\phi_i}{\|\phi_i\|_2},\mathbf{r}^0\rangle\right|\right),
\end{aligned} \tag{6.17}$$

where $\langle\mathbf{a},\mathbf{b}\rangle$ is the inner product between two vectors $\mathbf{a}$ and $\mathbf{b}$. First, noting that $s_p = 1$ and $s_q = j$, we have

$$\begin{aligned}
\langle\frac{\phi_p}{\|\phi_p\|_2},\mathbf{r}^0\rangle &= \langle\frac{\phi_p}{\|\phi_p\|_2},\phi_p s_p + \phi_q s_q + \tilde{\mathbf{v}}\rangle \\
&= \|\tilde{\mathbf{h}}\|_2 + j\|\tilde{\mathbf{h}}\|_2\mu_{qp} + \frac{\phi_l^T}{\|\phi_l\|_2}\tilde{\mathbf{v}},
\end{aligned} \tag{6.18}$$

where the equality follows from (see Appendix B)

$$\langle\frac{\phi_k}{\|\phi_k\|_2},\phi_l\rangle = \begin{cases} \|\tilde{\mathbf{h}}\|_2, & \text{for } k = l \\ \|\tilde{\mathbf{h}}\|_2\mu_{kl}, & \text{for } k \neq l. \end{cases} \tag{6.19}$$

Let $z_p = \Re\left(\frac{\phi_p^T}{\|\phi_p\|_2}\tilde{\mathbf{v}}\right)$, then

$$\Re\langle\frac{\phi_p}{\|\phi_p\|_2},\mathbf{r}^0\rangle = \|\tilde{\mathbf{h}}\|_2 + z_p. \tag{6.20}$$

In a similar way, we have

$$\Re\langle\frac{\phi_i}{\|\phi_i\|_2},\mathbf{r}^0\rangle = \|\tilde{\mathbf{h}}\|_2\mu_{ip} + z_i, \tag{6.21}$$

and hence

$$\begin{aligned}
&\mathrm{P}\left(\left|\Re\langle\frac{\phi_p}{\|\phi_p\|_2},\mathbf{r}^0\rangle\right| \geq \left|\Re\langle\frac{\phi_i}{\|\phi_i\|_2},\mathbf{r}^0\rangle\right|\right) \\
&= \mathrm{P}\left(\left|\|\tilde{\mathbf{h}}\|_2 + z_p\right| \geq \left|\|\tilde{\mathbf{h}}\|_2\mu_{ip} + z_i\right|\right) \\
&\overset{(a)}{=} \mathrm{P}(\|\tilde{\mathbf{h}}\|_2 + z_p > |\|\tilde{\mathbf{h}}\|_2\mu_{ip} + z_i|)\mathrm{P}(\|\tilde{\mathbf{h}}\|_2 + z_p > 0) \\
&\quad + \mathrm{P}(-\|\tilde{\mathbf{h}}\|_2 - z_p > |\|\tilde{\mathbf{h}}\|_2\mu_{ip} + z_i|)\mathrm{P}(\|\tilde{\mathbf{h}}\|_2 + z_p < 0) \\
&\geq \mathrm{P}(\|\tilde{\mathbf{h}}\|_2 + z_p > \|\tilde{\mathbf{h}}\|_2|\mu_{ip}| + |z_i|)\mathrm{P}(\|\tilde{\mathbf{h}}\|_2 + z_p > 0) \\
&\geq \mathrm{P}(\|\tilde{\mathbf{h}}\|_2 + z_p > \mu^*\|\tilde{\mathbf{h}}\|_2 + |z_i|)\mathrm{P}(\|\tilde{\mathbf{h}}\|_2 + z_p > 0),
\end{aligned} \tag{6.22}$$

where (a) follows from

$$\mathrm{P}(|A| \geq |B|) = \mathrm{P}(A > |B|)\mathrm{P}(A > 0) + \mathrm{P}(-A > |B|)\mathrm{P}(A < 0). \tag{6.23}$$

Since $z_i \sim \mathcal{N}(0, \frac{\sigma_{\mathbf{v}}^2}{2})$ from Lemma 6.1, the second term in (6.22) is lower bounded as

$$\begin{aligned}
\mathrm{P}(\|\tilde{\mathbf{h}}\|_2 + z_p > 0) &= \mathrm{P}(z_p > -\|\tilde{\mathbf{h}}\|_2) \\
&= 1 - \mathcal{Q}\left(-\frac{\|\tilde{\mathbf{h}}\|_2}{\frac{\sigma_{\mathbf{v}}}{\sqrt{2}}}\right) \\
&\geq 1 - \exp\left(-\frac{\|\tilde{\mathbf{h}}\|_2}{\sigma_{\mathbf{v}}^2}\right),
\end{aligned} \tag{6.24}$$

where the last inequality follows from $\mathcal{Q}(x) \leq \exp\left(-\frac{x^2}{2}\right)$. In a similar way, the first term in (6.22) is lower bounded as

$$\begin{aligned}
&\mathrm{P}(\|\tilde{\mathbf{h}}\|_2 + z_p > \mu^*\|\tilde{\mathbf{h}}\|_2 + |z_i|) \\
&\quad = 1 - \mathrm{P}(|z_i| - z_p \geq (1-\mu^*)\|\tilde{\mathbf{h}}\|_2) \\
&\quad = 1 - \mathrm{P}(z_i - z_p \geq (1-\mu^*)\|\tilde{\mathbf{h}}\|_2)\mathrm{P}(z_i > 0) \\
&\qquad - \mathrm{P}(-z_i - z_p \geq (1-\mu^*)\|\tilde{\mathbf{h}}\|_2)\mathrm{P}(z_i < 0) \\
&\quad \overset{(a)}{=} 1 - 2\mathrm{P}(z_i - z_p \geq (1-\mu^*)\|\tilde{\mathbf{h}}\|_2)\mathrm{P}(z_i > 0) \\
&\quad \overset{(b)}{\geq} 1 - \mathcal{Q}\left(-\frac{\|\tilde{\mathbf{h}}\|_2(1-\mu^*)}{\sigma_{\mathbf{v}}}\right) \\
&\quad \geq 1 - \exp\left(-\frac{\|\tilde{\mathbf{h}}\|_2^2(1-\mu^*)^2}{2\sigma_{\mathbf{v}}^2}\right),
\end{aligned} \tag{6.25}$$

where (a) is because $-z_i \sim \mathcal{N}(0, \frac{\sigma_{\mathbf{v}}^2}{2})$ and (b) is because $z_i - z_p \sim \mathcal{N}(0, \sigma_{\mathbf{v}}^2)$. By plugging (6.24) and (6.25) into (6.22), we have

$$\begin{aligned}
&\mathrm{P}\left(\left|\Re\langle\frac{\phi_p}{||\phi_p||_2}, \mathbf{r}^0\rangle\right| \geq \left|\Re\langle\frac{\phi_i}{||\phi_i||_2}, \mathbf{r}^0\rangle\right|\right) \\
&\geq \left(1 - \exp\left(-\frac{||\tilde{\mathbf{h}}||_2^2(1-\mu^*)^2}{2\sigma_{\mathbf{v}}^2}\right)\right)\left(1 - \exp\left(-\frac{||\tilde{\mathbf{h}}||_2}{\sigma_{\mathbf{v}}^2}\right)\right) \\
&\geq 1 - \exp\left(-\frac{||\tilde{\mathbf{h}}||_2^2(1-\mu^*)^2}{2\sigma_{\mathbf{v}}^2}\right) - \exp\left(-\frac{||\tilde{\mathbf{h}}||_2}{\sigma_{\mathbf{v}}^2}\right).
\end{aligned} \tag{6.26}$$

Note that $\mathrm{P}(S^1|\mathbf{h})$ in (6.17) is the success probability in the first iteration for a given channel realization $\mathbf{h}$. In order to obtain the unconditional probability, we need to take expectation with respect to the channel $\mathbf{h}$:

$$\mathrm{P}(S^1) = \int \mathrm{P}(S^1|\mathbf{h}) f_{\mathbf{h}}(x)dx = \mathrm{E}_{\mathbf{h}}\left[\mathrm{P}(S^1|\mathbf{h})\right]. \tag{6.27}$$

Thus,

$$\begin{aligned}
\mathrm{P}(S^1) &= \mathrm{E}_{\mathbf{h}}\left[\prod_{i=1,i\neq p}^{N} \mathrm{P}\left(\left|\Re\langle\frac{\phi_p}{||\phi_p||_2}, \mathbf{r}^0\rangle\right| \geq \left|\Re\langle\frac{\phi_i}{||\phi_i||_2}, \mathbf{r}^0\rangle\right|\right) \mid \mathbf{h}\right] \\
&= \prod_{i=1,i\neq p}^{N} \mathrm{E}_{\mathbf{h}}\left[\mathrm{P}\left(\left|\Re\langle\frac{\phi_p}{||\phi_p||_2}, \mathbf{r}^0\rangle\right| \geq \left|\Re\langle\frac{\phi_i}{||\phi_i||_2}, \mathbf{r}^0\rangle\right|\right) \mid \mathbf{h}\right] \\
&\geq \prod_{i=1,i\neq p}^{N} \mathrm{E}_{\mathbf{h}}\left[1 - \exp\left(-\frac{||\tilde{\mathbf{h}}||_2^2(1-\mu^*)^2}{2\sigma_{\mathbf{v}}^2}\right) - \exp\left(-\frac{||\tilde{\mathbf{h}}||_2^2}{\sigma_{\mathbf{v}}^2}\right) \mid \mathbf{h}\right] \\
&= \prod_{i=1,i\neq p}^{N} \left(1 - \mathrm{E}_{\mathbf{h}}\left[\exp\left(-\frac{||\tilde{\mathbf{h}}||_2^2(1-\mu^*)^2}{2\sigma_{\mathbf{v}}^2}\right) \mid \mathbf{h}\right] - \mathrm{E}_{\mathbf{h}}\left[\exp\left(-\frac{||\tilde{\mathbf{h}}||_2^2}{\sigma_{\mathbf{v}}^2}\right) \mid \mathbf{h}\right]\right).
\end{aligned} \tag{6.28}$$

Since $||\tilde{\mathbf{h}}||_2^2$ follows a Chi-squared distribution with $2m$ DoF (see Lemma 2), we have

$$\begin{aligned}
\mathrm{E}_{\mathbf{h}}\left[\exp\left(-\frac{||\tilde{\mathbf{h}}||_2^2}{\sigma_{\mathbf{v}}^2}\right) \mid \mathbf{h}\right] &= \int_0^\infty \exp\left(-\frac{x}{\sigma_{\mathbf{v}}^2}\right)\frac{x^{m-1}\exp(-x)}{(m-1)!}dx \\
&= \frac{1}{\left(\frac{1}{\sigma_{\mathbf{v}}^2} + 1\right)^m},
\end{aligned} \tag{6.29}$$

where the equality follows from $\int_0^\infty x^n \exp(-ax)dx = \frac{n!}{a^{n+1}}$ for $n = 0, 1, 2, \ldots,$ $a > 0$.

In a similar way, we have

$$\mathrm{E}_{\mathbf{h}}\left[\exp\left(-\frac{\|\tilde{\mathbf{h}}\|_2^2(1-\mu^*)^2}{2\sigma_{\mathbf{v}}^2}\right) \mid \mathbf{h}\right] = \left(1+\frac{(1-\mu^*)^2}{\sigma_{\mathbf{v}}^2}\right)^{-m}. \tag{6.30}$$

Finally, by plugging (6.29) and (6.30) into (6.28), we obtain the lower bound of $\mathrm{P}(S^1)$ as

$$\begin{aligned}
\mathrm{P}(S^1) &= \mathrm{E}_{\mathbf{h}}\left[\prod_{i=1,i\neq p}^{N} \mathrm{P}\left(\left|\Re\langle\frac{\phi_p}{\|\phi_p\|_2},\mathbf{r}^0\rangle\right| \geq \left|\Re\langle\frac{\phi_i}{\|\phi_i\|_2},\mathbf{r}^0\rangle\right|\right) \mid \mathbf{h}\right] \\
&= \prod_{i=1,i\neq p}^{N}\left(1-\left(1+\frac{(1-\mu^*)^2}{\sigma_{\mathbf{v}}^2}\right)^{-m} - \left(1+\frac{1}{\sigma_{\mathbf{v}}^2}\right)^{-m}\right) \\
&\geq \left(1-\left(1+\frac{(1-\mu^*)^2}{\sigma_{\mathbf{v}}^2}\right)^{-m} - \left(1+\frac{1}{\sigma_{\mathbf{v}}^2}\right)^{-m}\right)^{N-1}.
\end{aligned} \tag{6.31}$$

We now present the success probability for the second iteration under the condition that the first iteration is successful.

Lemma 6.4: The probability that the support element is chosen at the second iteration under the condition that the first iteration is successful satisfies

$$\mathrm{P}(S^2|S^1) \geq \left(1-\left(1+\frac{(1-\mu^*)^2}{\sigma_{\mathbf{v}}^2}\right)^{-m} - \left(1+\frac{1}{\sigma_{\mathbf{v}}^2}\right)^{-m}\right)^{N-2}. \tag{6.32}$$

Proof. When the first iteration is successful, the residual $\mathbf{r}^1$ can be expressed as

$$\begin{aligned}
\mathbf{r}^1 &= \mathbf{r}^0 - \mathbf{\Phi}_{\Omega_s^1}\hat{\mathbf{s}}^1 \\
&\overset{(a)}{=} \mathbf{r}^0 - \phi_p s_p \\
&= \phi_q s_q + \tilde{\mathbf{v}},
\end{aligned} \tag{6.33}$$

where (a) is because the transmit symbols are known in advance ($\hat{\mathbf{s}}^1 = s_p$). After taking similar steps to Lemma 6.3, one can show that $\mathrm{P}(S^2|S^1)$ satisfies (we skip the detailed steps for brevity)

$$\begin{aligned}
\mathrm{P}(S^2|S^1) &= \mathrm{P}\left(\left|\Im\langle\frac{\phi_q}{||\phi_q||_2},\mathbf{r}^1\rangle\right| \geq \max_i \left|\Im\langle\frac{\phi_i}{||\phi_i||_2},\mathbf{r}^1\rangle\right|\right) \\
&= \prod_{i=1,i\neq p,q}^{N} \mathrm{P}\left(\left|\Im\langle\frac{\phi_q}{||\phi_q||_2},\mathbf{r}^1\rangle\right| \geq \left|\Im\langle\frac{\phi_i}{||\phi_i||_2},\mathbf{r}^1\rangle\right|\right) \\
&\geq \left(1-\left(1+\frac{(1-\mu^*)^2}{\sigma_{\mathbf{v}}^2}\right)^{-m}-\left(1+\frac{1}{\sigma_{\mathbf{v}}^2}\right)^{-m}\right)^{N-2}.
\end{aligned} \tag{6.34}$$

It is worth mentioning that the lower bounds of $\mathrm{P}(S^1)$ and $\mathrm{P}(S^2|S^1)$ have the same form except for the exponent. We are now ready to prove the main theorem.

Proof of Theorem 1. By combining Lemmas 3 and 4, we can obtain the lower bound of the success probability P_{succ} as

$$\begin{aligned}
P_{succ} &= \mathrm{P}(S^2|S^1)\mathrm{P}(S^1) \\
&= \mathrm{P}\left(\left|\Im\langle\frac{\phi_q}{||\phi_q||_2},\mathbf{r}^1\rangle\right| \geq \max_i \left|\Im\langle\frac{\phi_i}{||\phi_i||_2},\mathbf{r}^1\rangle\right|\right) \\
&\quad \times \mathrm{P}\left(\left|\Re\langle\frac{\phi_p}{||\phi_p||_2},\mathbf{r}^0\rangle\right| \geq \max_i \left|\Re\langle\frac{\phi_i}{||\phi_i||_2},\mathbf{r}^0\rangle\right|\right) \\
&= \prod_{i=1,i\neq p,q}^{N} \mathrm{P}\left(\left|\Im\langle\frac{\phi_q}{||\phi_q||_2},\mathbf{r}^1\rangle\right| \geq \left|\Im\langle\frac{\phi_i}{||\phi_i||_2},\mathbf{r}^1\rangle\right|\right) \\
&\quad \prod_{i=1,i\neq p}^{N} \mathrm{P}\left(\left|\Re\langle\frac{\phi_p}{||\phi_p||_2},\mathbf{r}^0\rangle\right| \geq \left|\Re\langle\frac{\phi_i}{||\phi_i||_2},\mathbf{r}^0\rangle\right|\right) \\
&\geq \left(1-\left(1+\frac{(1-\mu^*)^2}{\sigma_{\mathbf{v}}^2}\right)^{-m}-\left(1+\frac{1}{\sigma_{\mathbf{v}}^2}\right)^{-m}\right)^{(N-2)+(N-1)} \\
&\geq \left(1-\left(1+\frac{(1-\mu^*)^2}{\sigma_{\mathbf{v}}^2}\right)^{-m}-\left(1+\frac{1}{\sigma_{\mathbf{v}}^2}\right)^{-m}\right)^{2N},
\end{aligned}$$

which completes the proof.

Finally, we present the decoding success probability bound for general sparsity K.

Theorem 6.2: The probability that the SVC-encoded packet can be successfully decoded for a given K satisfies

$$P_{succ} \geq \left(1 - \left(1 + \frac{(1-\mu^*)^2}{\sigma_{\mathbf{v}}^2}\right)^{-m} - \left(1 + \frac{1}{\sigma_{\mathbf{v}}^2}\right)^{-m}\right)^{KN}. \tag{6.35}$$

Proof. The success probability P_{succ} is expressed as

$$\begin{aligned} P(S^1, S^2, \cdots, S^K) &= P(S^K|S^{K-1}, \cdots, S^1) \cdots P(S^2|S^1)P(S^1) \\ &\geq \left(1 - \left(1 + \frac{(1-\mu^*)^2}{\sigma_{\mathbf{v}}^2}\right)^{-m} - \left(1 + \frac{1}{\sigma_{\mathbf{v}}^2}\right)^{-m}\right)^{(N-K)+\cdots+(N-1)} \\ &\geq \left(1 - \left(1 + \frac{(1-\mu^*)^2}{\sigma_{\mathbf{v}}^2}\right)^{-m} - \left(1 + \frac{1}{\sigma_{\mathbf{v}}^2}\right)^{-m}\right)^{KN}. \end{aligned} \tag{6.36}$$

Since the proof is similar to the proof of Theorem 1, we skip the details.

If $m \gg 1$, we approximately have

$$P_{succ} \gtrsim \left(1 - \left(1 + \frac{(1-\mu^*)^2}{\sigma_{\mathbf{v}}^2}\right)^{-m}\right)^{KN}. \tag{6.37}$$

It is clear from (6.37) that the decoding success probability improves (decreases) when the information vector is less sparse (i.e. K is large), and the number of observations is large (i.e. N is large), which matches with our expectation.

6.5 SVC Implementation

In this section, we discuss the implementation issues including codebook design, high-order modulation, diversity transmission, pilot-less transmission, and threshold selection to prevent the false alarm event.

6.5.1 Spreading Codebook

In order to improve the decoding success probability, a codebook whose codewords are less correlated is desired. As mentioned, as m increases, the correlation between two randomly generated codewords decreases, and thus we can use any kind of random sequence. For example, if we use the Bernoulli random matrix, then the maximum correlation satisfies $\mu^* \leq \sqrt{4m^{-1} \ln \frac{N}{\delta}}$ with probability exceeding $1 - \delta^2$ [24].

Instead of relying on the random sequence, we can also use the deterministic sequences. Well-known deterministic sequences include chirps, BCH, DFT, and Second-Order Reed–Muller (SORM) sequences [25]. For example, SORM is a sequence designed to generate low correlation sequences. SORM of length 2^m is defined as

$$\phi_{\mathbf{P},\mathbf{b}(\mathbf{a})} = \frac{(-1)^{w(\mathbf{b})}}{\sqrt{2^p}} \mathbf{i}^{(2\mathbf{b}+\mathbf{P}\mathbf{a})^T\mathbf{a}}, \tag{6.38}$$

where $\mathbf{P}$ is a $d \times d$ binary symmetric matrix, $\mathbf{a} = [a_0 \ a_1 \ \cdots \ a_{d-1}]^T$ and $\mathbf{b} = [b_0 \ b_1 \ \cdots \ b_{d-1}]^T$ are binary vectors in $\mathbb{Z}_2^d$, and $w(\mathbf{b})$ is the weight (number of ones) of $\mathbf{b}$. The corresponding SORM matrix can be expressed as

$$\boldsymbol{\Phi}_{rm} = \left[\mathbf{U}_{\mathbf{P}_1} \quad \mathbf{U}_{\mathbf{P}_2} \quad \cdots \quad \mathbf{U}_{\mathbf{P}_{2^{d(d-1)/2}}},\right] \tag{6.39}$$

where $\mathbf{U}_{\mathbf{P}_j}$ is the $2^d \times 2^d$ orthogonal matrix whose columns are the SORM sequences. The maximum correlation υ^* of the SORM sequence is

$$\upsilon^* = \begin{cases} \frac{1}{\sqrt{2^l}}, & l = \text{rank}(\mathbf{P}_i - \mathbf{P}_j) \\ \frac{1}{\sqrt{m}}, & l = d \end{cases} \tag{6.40}$$

For example, if $m = 64$ and $l = d$, then $\upsilon^* = 0.125$. The benefit of using SORM sequence is that the correlation between any two codewords is a constant and thus the performance variation can be reduced significantly.

6.5.2 Diversity Integration

To further improve the reliability, one can exploit the diversity schemes to SVC. One can easily think of the frequency diversity in which the SVC-encoded packet is repeated L times in L distinct frequency bands. Specifically, by applying the maximal-ratio combining at the receiver for the same symbol of the repeated packets, effective SNR can be increased and thus the BLER performance can be enhanced [26]. This scheme will be useful for wideband systems exploiting mmWave or THz bands. The benefit of the frequency diversity is that the diversity gain can be achieved without increasing the transmission latency. In Additive White Gaussian Noise (AWGN) environments, for example, when the SVC-encoded packet is repeated $L = 8$ times, due to the power gain of the combined symbol, the required SNR to achieve the desired URLLC performance (e.g. 10^{-5} BLER) can be reduced from 3 dB to $3 - 10\log_{10}(L) = -6$ dB. Other than the frequency diversity, diversity schemes such as time, antenna, and space diversity can also be added easily.

6.5.3 High-order Modulation

Since ensuring the reliability is the most important issue in URLLC, QPSK modulation would be the reasonable choice in practice. In order to use the QPSK modulation, we set one of the non-zero entries in $\mathbf{s}$ to 1 and the other to j. For example, if the non-zero positions are 4 and 6, then we set $\mathbf{s} = [0\,0\,0\,1\,0\,j\,0\,0\,0]^T$ and thus the transmit vector $\mathbf{x}$ can be expressed as

$$\mathbf{x} = 1\mathbf{c}_4 + j\mathbf{c}_6. \tag{6.41}$$

From (6.41), we see that elements of the transmit vector $\mathbf{x}$ are mapped to the QPSK symbol (i.e. $x_i \in \{1+j, 1-j, -1+j, -1-j\}$). Note, by distinguishing two possible choices (i.e. $[1, j]$ and $[j, 1]$) one additional bit can be encoded at the expense of a slight degradation in performance. When the higher sparsity ($K > 2$) is used, this mapping can be readily extended to the high order modulation (e.g. $K = 4$ for 16-QAM and $K = 6$ for 64-QAM). Specifically, if $K = 4$, we map the element in $\mathbf{x}$ to the 16-QAM symbol by setting two non-zero entries to 1,2 and the remaining non-zero entries to $j, 2j$. In a similar way, if $K = 6$, then we can transmit 64-QAM symbols by setting three non-zero entries to 1,2,3 and the remaining ones to j, $2j$, and $3j$. The normalization factor (α in (6.3)) corresponding M-QAM is $\alpha = \sqrt{\frac{2(M-1)}{3}}$.

6.5.4 Threshold to Prevent a False Alarm Event

To distinguish the false alarm event from the normal decoding process, we need to check whether the residual after the sparse recovery algorithm contains pure noise only. In fact, if the SVC decoding is finished successfully, the residual contains the noise contribution only ($\mathbf{r}^K = \mathbf{v}$). In this case, the residual power $\|\mathbf{r}^K\|_2^2$ can be modeled as a Chi-squared random variable with $2m$ degrees of freedom. Naturally, one can reject this hypothesis if the residual power is too large and lies outside the pre-defined confidence interval. In other words, if $\|\mathbf{r}^K\|_2^2 > F^{-1}_{\|\mathbf{v}\|_2^2}(1 - P_{th})$ where P_{th} is the pre-defined probability threshold (e.g. $P_{th} = 0.01$) and $F^{-1}_{\|\mathbf{v}\|_2^2}$ is the inverse cumulative distribution function of a Chi-squared random variable, then we declare the hypothesis is not true (i.e. decoding is not successful) and discard the decoded packet. To test the effectiveness of this thresholding approach, we simulate the probability of a false alarm as a function of SNR for the conventional 16-bit CRC transmission and the proposed residual-based thresholding. From Figure 6.5, we observe that the performance of residual-based thresholding is similar to the CRC-based error checking.

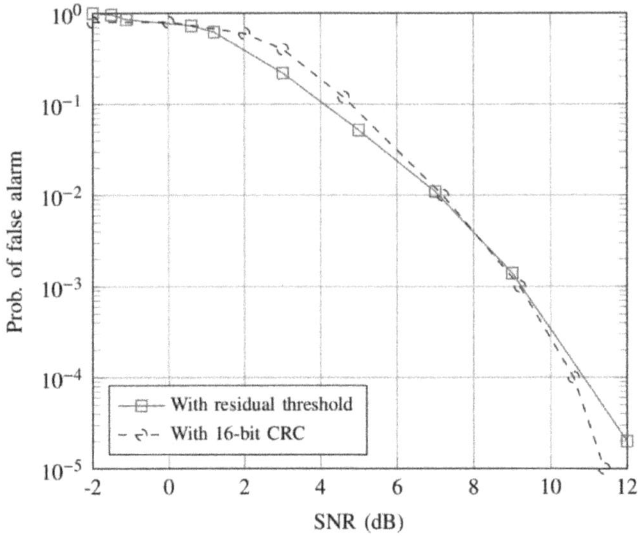

Figure 6.5 Decoding failure as a function of SNR ($P_{th} = 10^{-5}$).

6.6 SVC Performance

6.6.1 Simulation Setup

We here investigate the performance of the SVC technique. In our simulations, we consider the downlink OFDM system in the 3GPP LTE-Advanced [10, 11]. As a channel model, AWGN and realistic International Telecommunication Union (ITU) channel models including the Extended Typical Urban (ETU) and Extended Pedestrian-A (EPA) channel models are employed [11]. For comparison, we also investigate the performance of the conventional PDCCH of LTE-Advanced system using Viterbi decoding, polar code-based PDCCH in 5G NR [27] using the Successive Cancellation List (SCL) decoding. Also, we added the lower bound obtained from the AWGN model. The transmission of b bit information consists of information bit b_i and CRC bit b_c. In the conventional PDCCH method, the convolution code with rate $\frac{1}{3}$ with the 16-bit CRC is employed. Since the block size of the polar code is not flexible, we set the rate $\frac{1}{4}$ to test similar conditions ($b = 2^4$ and $m = 32$). In the SVC, we use the random binary spreading codebook with $N = 96$ and $K = 2$. For a fair comparison, we use the same number of resources ($m = 42$ with $L = 8$ repetitions) in the control packet transmission.

In Figure 6.6(a), we investigate the BLER performance of SVC and competing schemes under AWGN channel conditions. One can see that SVC outperforms the conventional PDCCH and polar code-based scheme, achieving more than a 4 dB

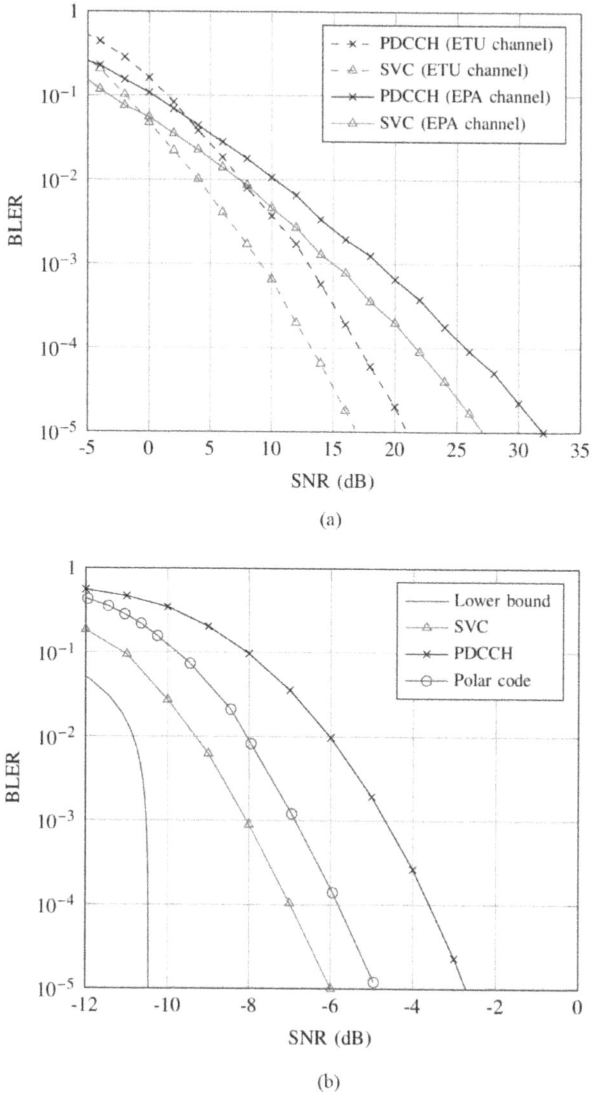

Figure 6.6 BLER performance as a function of SNR ($b_i = 12$, $b_c = 16$, $m = 42$, $L = 8$, and $N = 96$) for (a) AWGN channel and (b) ETU and EPA channel. ©IEEE 2020. Reprinted with permission from [31].

gain over the conventional PDCCH and about a 1.1 dB gain over the polar code-based scheme at 10^{-5} BLER point. Even in realistic scenarios such as EPA and EVA channels in LTE-Advanced, we observe that the performance gain of SVC against the competing schemes is maintained (see Figure 6.6(b)).

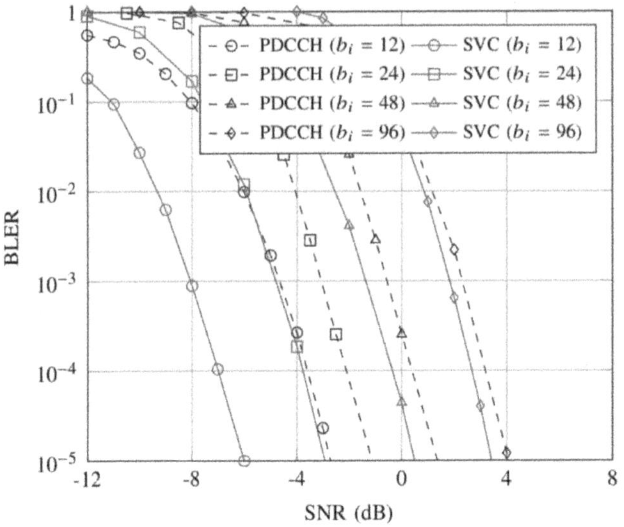

Figure 6.7 BLER performance for various sizes of control information ($L = 8$). ©IEEE 2020. Reprinted with permission from [31].

In Figure 6.7, we evaluate the BLER performance of PDCCH and SVC as a function of SNR for various information bit sizes ($b_i = 12, 24, 48,$ and 96). These results demonstrate that SVC can transmit more information bits than the conventional PDCCH. For example, SVC can deliver twice as much information than PDCCH in the low SNR region (e.g. $b_i = 12$ of PDCCH and $b_i = 24$ of SVC in Figure 6.7). To confirm this behavior, we plot the minimum SNR to achieve the target BLER as a function of the information bit size in Figure 6.8. For example, to achieve 10^{-5} BLER with $b = 10$, it requires -2.9 dB for PDCCH and -6.2 dB SNR for SVC, resulting in a 3.3 dB gain in performance. It is worth mentioning that the coding gain of the conventional PDCCH improves with the codeblock size so that the gap between the SVC and PDCCH diminishes gradually as the number of information bits increases.

We next evaluate the latency performance of the SVC and PDCCH. In this experiment, we plot the distribution of transmission latency to achieve 10^{-5} BLER when an n-repetition scheme is employed. Transmission latency is defined as the time from the initial transmission to the time that the packet is successfully decoded at the mobile terminal.[13] From Figure 6.9, we see that most of the SVC packets satisfy the URLLC requirement (1 ms latency).

13 In our experiment, we ignored the decoding latency.

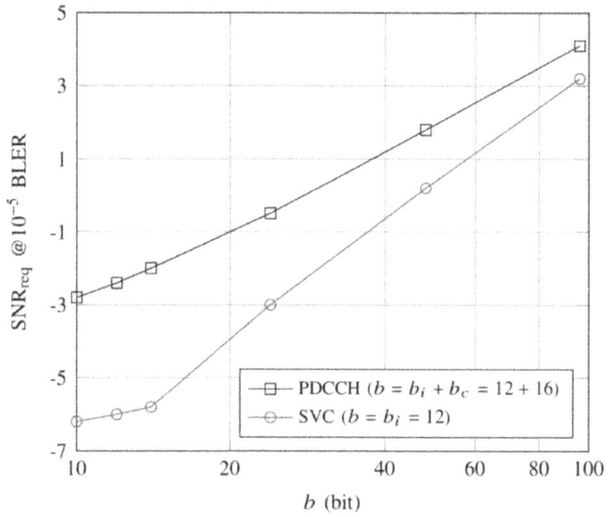

Figure 6.8 Required SNR for achieving 10^{-5} BLER ($m = 42$, $L = 8$, and $N = 96$). This figure is reused with permission from IEEE.

Figure 6.9 Probability of transmission latency for achieving 10^{-5} BLER ($b_i = 12, m = 42, L = 8, N = 96$, and SNR = -12 dB). ©IEEE 2020. Reprinted with permission from [31].

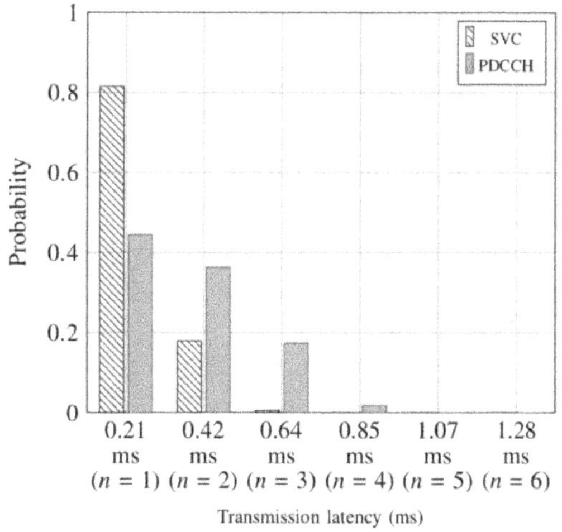

In Figure 6.10, we investigate the performance of SVC in the small cell scenarios (e.g. Ultra Dense Network (UDN) in urban areas) where the received signal contains a considerable amount of interference from adjacent BSs. Note that densely deployed small cell (pico, femto, and micro) environments are important to enhance the cell throughput in 5G and 6G, so the interference management is

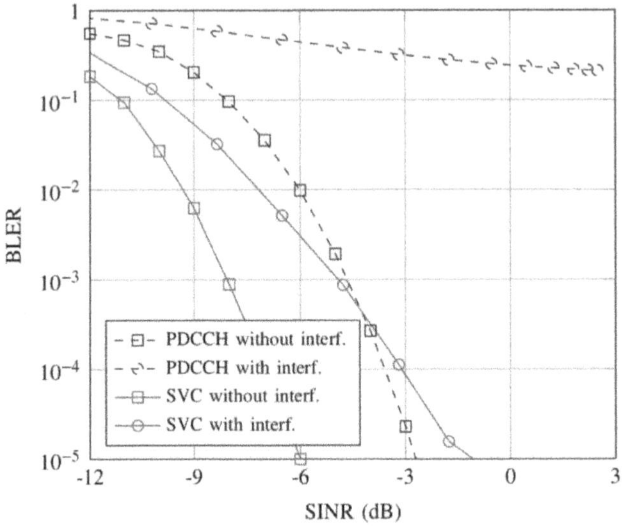

Figure 6.10 BLER performance as a function of SINR ($b_i = 12, m = 42, L = 8, N = 96$, and interference power is half of the signal power). ©IEEE 2020. Reprinted with permission from [31].

crucial for the success of UDN networks. In our simulations, we set the power level of interference to half of the desired cell signal. Since the SVC transmission is based on the pseudo-random spreading and also the effective transmit power per symbol is large, SVC can control the interference effectively. Since the conventional PDCCH has no such interference control mechanism, error correction capability of PDCCH is degraded significantly and thus the PDCCH does not perform well as shown in Figure 6.10.

6.7 Pilot-less Sparse Vector Coding

6.7.1 Pilot-less SVC Processing

When the channel is a constant or channel variation is very small (i.e. $\mathbf{h} \approx$const), which is true for static or slowly varying environments, SVC decoding can be further simplified and performed even without channel information, resulting in savings of pilot resources, transmission power, receiver processing time, and cost. This scenario can also occur when the short packet is transmitted in a narrowband channel. In fact, when the packet size is short, the packet transmission time nT_s (n is the number of symbols in a packet and T_s is the symbol duration) is in general much smaller than the channel coherence time T_c for the moderate mobility ν ($T_c \gg nT_s$). For example, when the carrier frequency $f_c = 3.5$ GHz and the mobile speed is $\nu = 15$ km h^{-1}, $T_c = \frac{9c}{16\pi\nu f_c} = 1.52$ ms and thus T_c is much larger than $T_s = 0.07$ ms [26].[14] Thus the channel remains unchanged or at least channel variation is negligible for the short packet (small n). In these scenarios, one can readily assume that the diagonal elements of $\mathbf{H}$ are the same (i.e. $\mathbf{H} = \text{diag}(h, \cdots, h)$). The key idea of this is to design the system model such that the sparse vector contains the channel component as well as the sparse information vectors and thus the system matrix is equal to the codebook $\mathbf{C}$. To be specific,

$$
\begin{aligned}
\mathbf{y} &= \mathbf{Hx} + \mathbf{v} \\
&= \begin{bmatrix} h & & \\ & \ddots & \\ & & h \end{bmatrix} \mathbf{x} + \mathbf{v} \\
&= \sum_i (hs_i)\mathbf{c}_i + \mathbf{v}.
\end{aligned}
\tag{6.42}
$$

Let $\hat{\mathbf{s}} = \begin{bmatrix} hs_1 & \cdots & hs_N \end{bmatrix}^T$, then $\mathbf{y} = \mathbf{C}\hat{\mathbf{s}} + \mathbf{v}$. Recalling that the goal of the Pilot-less-SVC (PL-SVC) decoding is to identify the non-zero positions of $\mathbf{s}$ vector (equivalently, $\hat{\mathbf{s}}$ vector), we can perform the decoding without the channel knowledge, meaning that the pilot transmission is unnecessary.

In the PL-SVC decoding, any sparse signal recovery algorithm can be employed. In this work, we employ an OMP [12], well-known greedy sparse recovery algorithm, as a baseline. In a nutshell, the role of a decoding algorithm is to identify the support Ω (locations of non-zero elements in a vector). For example, if $\hat{\mathbf{s}} = [0\ hs_2\ 0\ 0\ hs_5\ 0]^T$, then $\Omega = \{2, 5\}$. When columns of $\mathbf{C}$ are approximately orthogonal, $\tilde{\mathbf{s}} = \mathbf{C}_\Omega^H \mathbf{y} = \mathbf{C}_\Omega^H(\mathbf{C}\hat{\mathbf{s}} + \mathbf{v})$ would be an approximate estimate of $\mathbf{s}$. Based on this observation, OMP finds that the support element is one at each iteration. To be specific, let $\mathbf{C}_{\Omega^j}$ be the submatrix of $\mathbf{C}$ that only contains columns indexed by the element of Ω^j (support of jth iteration), then an index ω_{j+1} chosen at the $(j+1)$th iteration of the OMP algorithm is

$$
\omega_{j+1} = \arg\max_l |\mathbf{c}_l^H \bar{\mathbf{r}}^j|^2,
\tag{6.43}
$$

where $\bar{\mathbf{r}}^j = \mathbf{y} - \mathbf{C}_{\Omega^j}\tilde{\mathbf{s}}^j$ is the residual vector and $\tilde{\mathbf{s}}^j = \mathbf{C}_{\Omega^j}^\dagger \mathbf{y}$.

14 We use LTE symbol length, i.e. 14 symbols in 1 subframe (1 ms) [11].

6.7.2 PL-SVC Decoding in Multiple BS Antennas

As discussed, an element of $\hat{\mathbf{s}}^j$ is hs_i (composite of channel h and symbol s_i) so that the detection performance is affected by the channel magnitude $||h||$. Clearly, when the channel is under deep fading (i.e. $||h|| \ll 1$), the performance degradation would be severe. When multiple antennas are available at BS, one can obtain the multi-antenna diversity gain and therefore improve the reliability and also mitigate the performance degradation caused by the deep fading. When multiple, say L, antennas are receiving the PL-SVC packet, a received vector $\tilde{\mathbf{y}}$ can be expressed as the sum of a linear combination of LN columns. That is,

$$\begin{aligned}\tilde{\mathbf{y}} &= \begin{bmatrix}\mathbf{y}^1 & \mathbf{y}^2 & \cdots & \mathbf{y}^L\end{bmatrix}^T \\ &= \begin{bmatrix}\mathbf{C} & & & \\ & \mathbf{C} & & \\ & & \ddots & \\ & & & \mathbf{C}\end{bmatrix}\begin{bmatrix}h^1\mathbf{s} \\ h^2\mathbf{s} \\ \vdots \\ h^L\mathbf{s}\end{bmatrix} + \begin{bmatrix}\mathbf{v}^1 \\ \mathbf{v}^2 \\ \vdots \\ \mathbf{v}^L\end{bmatrix} \\ &= \tilde{\mathbf{C}}\tilde{\mathbf{s}} + \tilde{\mathbf{v}}\end{aligned} \tag{6.44}$$

where $\mathbf{y}^l$ is the received signal at the lth antennas, h^l is the channel response at the lth antenna, and $\mathbf{v}^l$ is the corresponding noise vector. In order to exploit the block sparsity of the aggregated sparse vector $\tilde{\mathbf{s}}$, we rearrange the $\tilde{\mathbf{s}}$ vector as $[h^1\mathbf{s} \cdots h^L\mathbf{s}]^T \longrightarrow [h^1 s_1 \cdots h^L s_1 \ \cdots \ h^1 s_L \cdots h^L s_L]^T$ and thus,

$$\tilde{\mathbf{y}} = \begin{bmatrix}\bar{\mathbf{C}}^{(1)} & & & \\ & \bar{\mathbf{C}}^{(2)} & & \\ & & \ddots & \\ & & & \bar{\mathbf{C}}^{(N)}\end{bmatrix}\begin{bmatrix}\bar{\mathbf{h}}s^1 \\ \bar{\mathbf{h}}s^2 \\ \vdots \\ \bar{\mathbf{h}}s^N\end{bmatrix} + \begin{bmatrix}\bar{\mathbf{v}}^1 \\ \bar{\mathbf{v}}^2 \\ \vdots \\ \bar{\mathbf{v}}^N\end{bmatrix} \tag{6.45}$$

where $\bar{\mathbf{C}}^{(i)} = \mathrm{diag}(\mathbf{c}_i, \mathbf{c}_i, \cdots, \mathbf{c}_i)$ is the block diagonalization of the ith spreading sequence, $\bar{\mathbf{h}} = [h^1 \ h^2 \ \cdots \ h^L]^T$ is the stacked channel response, and $\bar{\mathbf{v}}^i = [\mathbf{v}^1(i) \ \mathbf{v}^2(i) \ \cdots \ \mathbf{v}^N(i)]^T$ is the rearranged noise vector. One can use the block sparse recovery algorithm (e.g. block OMP algorithm [12]) to find a non-zero block. The non-zero block selection rule of block OMP at $(j+1)$th iteration is

$$\omega_{j+1} = \arg\max_l ||\bar{\mathbf{C}}^{(l)}\bar{\mathbf{r}}^j||_2^2 \tag{6.46}$$

where $\bar{\mathbf{r}}^j = \tilde{\mathbf{y}} - \bar{\mathbf{C}}^{(\Omega^j)}\hat{\mathbf{s}}^j$ is the residual vector.

Note that the computational burden to process the single-input and multiple-output (SIMO) channel decoding is fairly small because the decoding operation is simply finished in the Kth iteration regardless of the number of antennas. In fact, since the sparsity K is small (e.g. $K = 2$ or 3) and known to the receiver, the computational overhead of the PL-SVC decoding is negligible. The PL-SVC decoding algorithm based on OMP is summarized in Table 6.4.

Table 6.4 The PL-SVC Decoding Algorithm.

Input:

Measurement $\tilde{\mathbf{y}}$, sensing matrix $\tilde{\mathbf{C}}$, sparsity K

Output:

Support set $\hat{\mathcal{S}}$

Initialization:

$\rho := \infty$ (minimum magnitude of residual)

$\bar{\mathbf{r}}^0 := \tilde{\mathbf{y}}$

for $j = 0$ **to** $K - 1$ **do**

$\omega_{j+1} := \arg\max_{l} \|\bar{\mathbf{C}}^{(l)}\bar{\mathbf{r}}^j\|_2^2$ (choose best index)

$\mathcal{S}^{j+1} := \mathcal{S}^j \cup \{\omega_{j+1}\}$ (update support set)

$\hat{\mathbf{s}^{j+1}} := \bar{\mathbf{C}}^{(\mathcal{S}^{j+1})^{\dagger}}\tilde{\mathbf{y}}$ (estimate non-zero element)

$\bar{\mathbf{r}}^{j+1} := \tilde{\mathbf{y}} - \bar{\mathbf{C}}^{(\mathcal{S}^{j+1})}\hat{\mathbf{s}}^{j+1}$ (update residual)

$\hat{\mathcal{S}}^{j+1} := \mathcal{S}^{j+1}$

end for

return $\hat{\mathcal{S}}^K$

6.7.3 Simulation Results

We here examine the performance of the PL-SVC technique in a 5G uplink scenario. Our simulation setup is based on the OFDM systems in the 5G NR. For comparison, we investigate the performance of the conventional SVC with pilot transmission and the Physical Uplink Share Channel (PUSCH) transmission in 4G LTE. In the SVC processing, we set $N = 96$ with $K = 2$ and $r = 84$ resources are used in total. For the conventional SVC with pilot transmission, 20% and 50% of resources are used for the pilot symbols and the rest are used for the data symbols. As a channel model, we use the frequency-flat Rayleigh fading model and consider 0, 7, and 70 Hz for the mobility of a mobile device in 2 GHz center frequency. In the PL-SVC transmission scheme, we set $m = r = 84$ for comparison. As a performance measure, we use the Packet Error Rate (PER).

In Figure 6.11, we evaluate the PER performance of PL-SVC, original SVC, and PUSCH with and without the pilot transmission as a function of SNR ($b = 12$). First, we compare the performance of the conventional SVC and PUSCH (dotted lines in Figure 6.11) with 50% pilot overhead. We observe that SVC outperforms the conventional PUSCH, achieving more than 3 dB gain at PER $= 10^{-4}$. We next compare the PER performance of PL-SVC and PUSCH without the pilot transmission. We observe that the performance of PUSCH-based transmission is poor while PL-SVC performs slightly worse than the conventional SVC at PER $= 10^{-4}$. Finally, we compare the performance of the conventional SVC and PL-SVC. When the pilot overhead is 20% and 50%, the PER performance of PL-SVC is worse than SVC in a low SNR regime but the gap decreases with SNR, achieving 0.3 dB and 1.8 dB gain at PER = 10^{-5}, respectively.

In Figure 6.12, we evaluate the PER performance of PL-SVC and PUSCH under the mobility scenario. Since PL-SVC does not exploit the channel information at the receiver, its performance is degraded when the mobility is introduced.

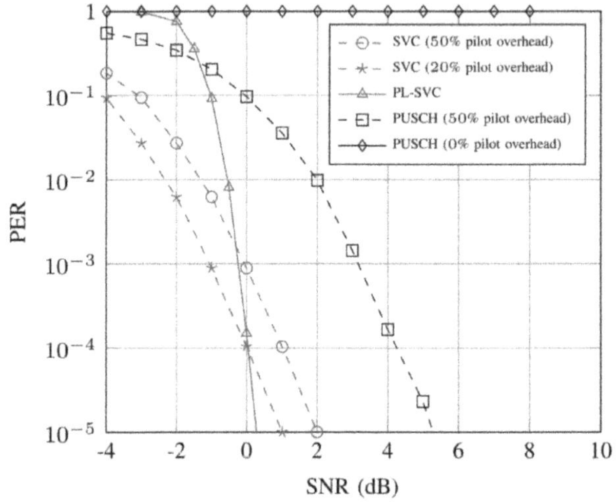

Figure 6.11 PER performance of PL-SVC and the conventional schemes ($b = 12, m = 84$).

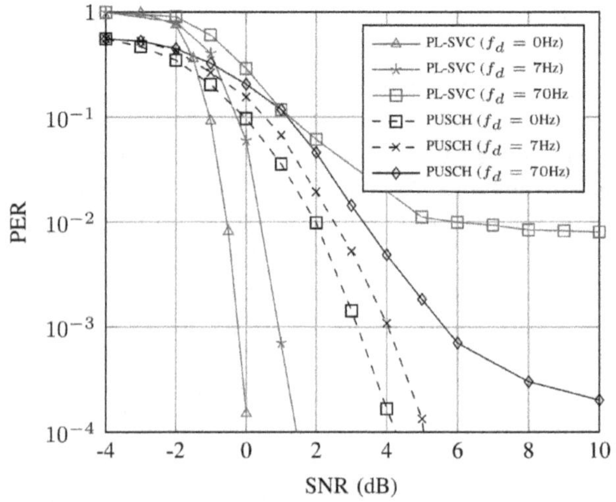

Figure 6.12 PER performance of PL-SVC and the conventional schemes under mobility. ©IEEE 2019. Reprinted with permission from [30].

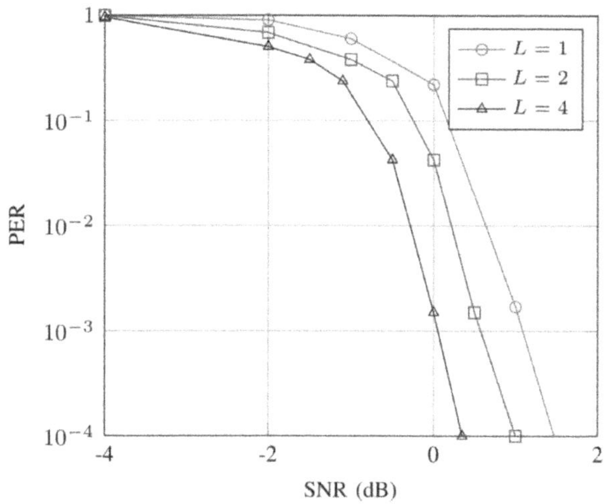

Figure 6.13 PER performance of PL-SVC for the SIMO channels as a function of SNR ($b = 12, m = r = 42$). ©IEEE 2019. Reprinted with permission from [30].

Indeed, in the low mobility condition ($f_d = 7$ Hz), PL-SVC outperforms PUSCH but the performance degradation is considerable when the mobility increases ($f_d = 70$ Hz).

In order to test the performance of SVC in the SIMO channel scenarios, we consider three distinct cases (i.e. number of received antennas is $L = 1, 2$, and 4). In Figure 6.13, we observe that the performance of PL-SVC improves with L. For example, if $L = 4$, we observe 1.5 dB gain over the case with $L = 1$. Although PL-SVC cannot achieve the full diversity gain since the PL-SVC receiver blindly decodes the packet without the channel information, it achieves the multi-antenna gain being proportional to the number of antennas (0.3~0.5 dB gain per antenna).

6.8 Symbol-embedded Sparse Vector Coding

6.8.1 Description of Symbol-embedded SVC

Thus far, we have only considered the scenario where the location of the sparse vector is used for the encoding. In this section, we discuss the extended SVC scheme considering both the location and symbol. Figure 6.14 depicts the block diagram of the symbol-embedded SVC (SE-SVC) scheme. In the SE-SVC scheme, the information is mapped to the support and symbols in the non-zero positions of $\mathbf{s}$. In this scheme, total number of bits b_t is divided into b_i and b_m bits and then b_i bit of information is encoded into the positions of sparse vector and b_m bit of

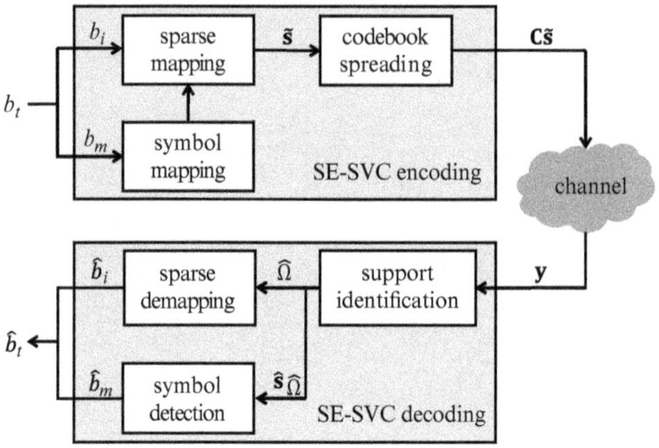

Figure 6.14 Transceiver structure of the SE-SVC. ©IEEE 2020. Reprinted with permission from [31].

information is encoded in the form of a symbol in the support ($b_m = kb_s$ where b_s is bits per symbol). For example, when $b_i = 5$, $b_m = 4$, $k = 2$, and $\Omega = \{3, 7\}$, the SE-SVC-encoded sparse vector is $\tilde{\mathbf{s}} = [0\ 0\ \tilde{s}_3\ 0\ 0\ 0\ \tilde{s}_7\ 0\ 0]^T$ where $\tilde{s}_3$ and $\tilde{s}_7$ are the QPSK modulated symbols. After the sparse vector mapping, the spreading codebook $\mathbf{C}$ is applied to $\tilde{\mathbf{s}}$ and then transmitted. The corresponding received signal $\mathbf{y}$ is

$$\mathbf{y} = \mathbf{HC}\tilde{\mathbf{s}} + \mathbf{v} = \mathbf{\Phi}\tilde{\mathbf{s}} + \mathbf{v}, \tag{6.47}$$

where $\mathbf{\Phi} = \mathbf{HC}$ is the sensing matrix.

The key factor bringing the performance gain of SE-SVC over the original SVC technique is as follows. For convenience, we denote the dimension of sparse vectors for SVC and SE-SVC as N_1 and N_2 respectively. Given the total number of bits b_t and sparsity k, N_1 is determined by the relationship $b_t = \left\lfloor \log_2 \binom{N_1}{k} \right\rfloor$. In a similar way, N_2 is obtained from $b_t - kb_s = \left\lfloor \log_2 \binom{N_2}{k} \right\rfloor$. Since kb_s is a positive integer, it is clear that N_2 is smaller than N_1 ($N_2 < N_1$). As mentioned, using the principle of CS, one can recover the sparse vector accurately as long as $m \approx ck \log N$ where c is the properly chosen constant [12]. Under the same set of conditions, therefore, the required number of measurements of SE-SVC is smaller than that of SVC. In a different perspective, if the number of measurements is the same, the underdetermined ratio $\frac{N_2}{m}$ of the SE-SVC-encoded vector will be smaller than that of SVC-encoded vector (i.e. $\frac{N_1}{m} > \frac{N_2}{m}$), and therefore the mutual coherence[15] of $\mathbf{\Phi}$ will be reduced.

15 The mutual coherence μ^* is defined as the largest magnitude of column correlation in the sensing matrix (i.e. $\mu^* = \max_{i \neq j} |\langle \phi_i, \phi_j \rangle|$)

Since the information is encoded in both support and symbol, the decoding process of SE-SVC process is divided into two steps: (1) support identification and (2) symbol detection. In the first step, we find k non-zero positions of $\tilde{\mathbf{s}}$. To this end, any sparse recovery algorithms can be employed [24]. Once the support is identified, we can convert the underdetermined system into the overdetermined system. For example, if $\Omega_{\mathbf{s}} = \{2, 5\}$, then the system model in (6.47) is simplified to $\mathbf{y} = [\phi_2 \quad \phi_5] \begin{bmatrix} \tilde{s}_2 \\ \tilde{s}_5 \end{bmatrix} + \mathbf{v}$. To detect the symbols, conventional technique such as the Maximum Likelihood (ML) detection or Minimum Mean Square Error (MMSE) detector followed by the symbol slices can be used.

6.8.2 SE-SVC performance analysis

In this section, we present the BLER analysis of the SE-SVC technique. For fair comparison with the conventional SVC, we set $k = 2$ and $\Omega_{\tilde{\mathbf{s}}} = \{p, q\}$ (i.e. $\tilde{\mathbf{s}}_\Omega = [\tilde{s}_p, \tilde{s}_q]^T$ where $\tilde{s}_p = \mathfrak{R}_p + j\mathfrak{I}_p$ and $\tilde{s}_q = \mathfrak{R}_q + j\mathfrak{I}_q$[16]). In essence, block error occurs in the following two cases:

1. The sparse recovery algorithm fails to detect the correct support (i.e. $\Omega^*_{\tilde{\mathbf{s}}} \neq \Omega_{\tilde{\mathbf{s}}}$).
2. The ML detector fails to find the symbols correctly after the successful support identification (i.e. $\mathbf{s}^*_\Omega \neq \tilde{\mathbf{s}}_\Omega$).

Using these, the BLER of the SE-SVC-encoded packet can be expressed as

$$\begin{aligned} \text{BLER} &= \text{P}\left(\{\Omega^*_{\tilde{\mathbf{s}}} \neq \Omega_{\tilde{\mathbf{s}}}\} \cup \{\mathbf{s}^*_\Omega \neq \tilde{\mathbf{s}}_\Omega | \Omega^*_{\tilde{\mathbf{s}}} = \Omega_{\tilde{\mathbf{s}}}\}\right) \\ &\overset{(a)}{\leq} \text{P}\left(\Omega^*_{\tilde{\mathbf{s}}} \neq \Omega_{\tilde{\mathbf{s}}}\right) + \text{P}\left(\mathbf{s}^*_\Omega \neq \tilde{\mathbf{s}}_\Omega | \Omega^*_{\tilde{\mathbf{s}}} = \Omega_{\tilde{\mathbf{s}}}\right) \end{aligned} \tag{6.48}$$

where (a) follows from the union bound of probability theory. Our main result is as follows.

Theorem 6.3: The block error rate of the SE-SVC packet can be upper bounded as,

$$\begin{aligned} \text{BLER} \leq & \left[1 - \left(1 - \left(1 + \frac{\alpha_1}{2\sigma_v^2}\right)^{-m} - \left(1 + \frac{\beta_1}{\sigma_v^2}\right)^{-m}\right)^{N-1}\right. \\ & \left.\left(1 - \left(1 + \frac{\alpha_2}{2\sigma_v^2}\right)^{-m} - \left(1 + \frac{\beta_2}{\sigma_v^2}\right)^{-m}\right)^{N-2}\right] + k\left(1 + \frac{\mu^* d_{min}^2}{4\sigma_v^2}\right)^{-m} \end{aligned} \tag{6.49}$$

where m is the number of measurements (resources), k is the sparsity, σ_v^2 is the noise variance, $\mu^* = \max_{i \neq j} |\mu_{ij}|$ is the absolute value of the maximum correlation

16 $\mathfrak{R}_i$ and $\mathfrak{I}_i$ are the real and imaginary parts of the ith symbol, respectively.

between two columns of $\mathbf{\Phi}$, d_{min} is the minimum euclidean distance between the symbols, and $\alpha_1 = \{\Re_p + \Re_q\mu^* - (|\Re_p| + |\Re_q|)\mu^*\}^2$, $\beta_1 = \{\Re_p + \Re_q\mu^*\}^2$, $\alpha_2 = \{\Im_q - |\Im_q|\mu^*\}^2$, $\beta_2 = \Im_q^2$.

The following two lemmas are used to prove Theorem 6.3.

Lemma 6.5: The probability that the support elements are chosen correctly satisfies

$$P_{succ} \geq \left(1-\left(1+\frac{\alpha_1}{2\sigma_v^2}\right)^{-m}-\left(1+\frac{\beta_1}{\sigma_v^2}\right)^{-m}\right)^{N-1} \left(1-\left(1+\frac{\alpha_2}{2\sigma_v^2}\right)^{-m}-\left(1+\frac{\beta_2}{\sigma_v^2}\right)^{-m}\right)^{N-2}. \tag{6.50}$$

Proof: Let $\mathcal{S}^j$ be the success probability of the chosen support element in the jth iteration, then the probability that the support elements are correctly chosen can be expressed as

$$P_{succ} = \mathrm{P}(\Omega_{\tilde{\mathbf{s}}}^* = \Omega_{\tilde{\mathbf{s}}}) = \mathrm{P}\left(\mathcal{S}^1, \mathcal{S}^2\right) = \mathrm{P}\left(\mathcal{S}^2|\mathcal{S}^1\right)\mathrm{P}\left(\mathcal{S}^1\right).$$

For analytic simplicity, we take real part of the decision statistics in the first iteration and then the imaginary part in the second iteration. For a given channel realization $\mathbf{h} = [h_1, \cdots, h_m]^T$, to identify the support element in the first iteration, we should have $\left|\Re\langle\frac{\boldsymbol{\phi}_p}{\|\boldsymbol{\phi}_p\|_2}, \mathbf{r}^0\rangle\right| \geq \max_i \left|\Re\langle\frac{\boldsymbol{\phi}_i}{\|\boldsymbol{\phi}_i\|_2}, \mathbf{r}^0\rangle\right|$. That is,

$$\begin{aligned}
\mathrm{P}(\mathcal{S}^1|\mathbf{h}) &= \mathrm{P}\left(\left|\Re\langle\frac{\boldsymbol{\phi}_p}{\|\boldsymbol{\phi}_p\|_2}, \mathbf{r}^0\rangle\right| \geq \max_i \left|\Re\langle\frac{\boldsymbol{\phi}_i}{\|\boldsymbol{\phi}_i\|_2}, \mathbf{r}^0\rangle\right|\right) \\
&= \prod_{i=1, i\neq p}^{N} \mathrm{P}\left(\left|\Re\langle\frac{\boldsymbol{\phi}_p}{\|\boldsymbol{\phi}_p\|_2}, \mathbf{r}^0\rangle\right| \geq \left|\Re\langle\frac{\boldsymbol{\phi}_i}{\|\boldsymbol{\phi}_i\|_2}, \mathbf{r}^0\rangle\right|\right)
\end{aligned} \tag{6.51}$$

Noting that $\tilde{s}_p = \Re_p + j\Im_p$ and $\tilde{s}_q = \Re_q + j\Im_q$ we have,

$$\Re\langle\frac{\boldsymbol{\phi}_p}{\|\boldsymbol{\phi}_p\|_2}, \mathbf{r}^0\rangle = \Re\langle\frac{\boldsymbol{\phi}_p}{\|\boldsymbol{\phi}_p\|_2}, \boldsymbol{\phi}_p\tilde{s}_p + \boldsymbol{\phi}_q\tilde{s}_q + \mathbf{v}\rangle = \Re_p\|\mathbf{h}\|_2 + \Re_q\mu_{pq}\|\mathbf{h}\|_2 + z_r, \tag{6.52}$$

where μ_{ij} is the correlation between $\mathbf{c}_i$ and $\mathbf{c}_j$ and $z_r = \Re\left(\frac{\boldsymbol{\phi}_p^T}{\|\boldsymbol{\phi}_p\|_2}\mathbf{v}\right)$. Similarly, we have

$$\Re\langle\frac{\boldsymbol{\phi}_i}{\|\boldsymbol{\phi}_i\|_2}, \mathbf{r}^0\rangle = \Re_p\mu_{ip}\|\mathbf{h}\|_2 + \Re_q\mu_{iq}\|\mathbf{h}\|_2 + z_i, \tag{6.53}$$

where $z_i = \Re\left(\frac{\phi_i^T}{\|\phi_i\|_2}\mathbf{v}\right)$. Using (6.52), (6.53), and $P(|A| \geq |B|) = P(A > |B|) P(A > 0) + P(-A > |B|) P(A < 0)$, we have

$$\begin{aligned} &P\left(\left|\Re\langle\frac{\phi_p}{\|\phi_p\|_2}, \mathbf{r}^0\rangle\right| \geq \left|\Re\langle\frac{\phi_i}{\|\phi_i\|_2}, \mathbf{r}^0\rangle\right|\right) \\ &= P\left(\left|(\Re_p + \Re_q\mu_{pq})\|\mathbf{h}\|_2 + z_r\right| \geq \left|(\Re_p\mu_{ip} + \Re_q\mu_{iq})\|\mathbf{h}\|_2 + z_i\right|\right) \quad (6.54) \\ &\overset{(a)}{\geq} 1 - \exp\left(-\frac{\|\mathbf{h}\|_2^2\alpha_1}{2\sigma_v{}^2}\right) - \exp\left(-\frac{\|\mathbf{h}\|_2^2\beta_1}{\sigma_v^2}\right) \quad (6.55) \end{aligned}$$

where (a) is because $P(z_p < -\|\mathbf{h}\|_2) = 1 - Q\left(-\frac{\|\mathbf{h}\|_2}{\sigma_v/\sqrt{2}}\right) \geq 1 - \exp\left(-\frac{\|\mathbf{h}\|_2^2}{\sigma_v^2}\right)$. Then, taking the expectation with respect to the channel realization $\mathbf{h}$,[17] we have

$$\begin{aligned} P(\mathcal{S}^1) &= \int P(\mathcal{S}^1|\mathbf{h}) f_{\mathbf{h}}(x)dx = E_{\mathbf{h}}\left[P(\mathcal{S}^1|\mathbf{h})\right] \\ &\geq \prod_{i=1,i\neq p}^{N}\left(1 - E_{\mathbf{h}}\left[\exp\left(-\frac{\|\mathbf{h}\|_2^2\alpha_1}{2\sigma_v{}^2}\right)\middle|\mathbf{h}\right] - E_{\mathbf{h}}\left[\exp\left(-\frac{\|\mathbf{h}\|_2^2\beta_1}{\sigma_v^2}\right)\middle|\mathbf{h}\right]\right) \quad (6.56) \\ &\overset{(a)}{\geq} \left(1 - \left(1 + \frac{\alpha_1}{2\sigma_v^2}\right)^{-m} - \left(1 + \frac{\beta_1}{\sigma_v^2}\right)^{-m}\right)^{N-1} \quad (6.57) \end{aligned}$$

where (a) is because $E_{\mathbf{h}}\left[\exp\left(-\frac{\|\mathbf{h}\|_2^2}{\sigma_v^2}\right)\middle|\mathbf{h}\right] = \left(1 + \frac{1}{\sigma_v^2}\right)^{-m}$.

In the second iteration, the success probability is given by

$$\begin{aligned} P\left(\mathcal{S}^2|\mathcal{S}^1\right) &= P\left(\left|\Im\langle\frac{\phi_q}{\|\phi_q\|_2}, \mathbf{r}^1\rangle\right| \geq \max_i \left|\Im\langle\frac{\phi_i}{\|\phi_i\|_2}, \mathbf{r}^1\rangle\right|\right) \\ &= \prod_{i=1,i\neq p,q}^{N} P\left(\left|\Im\langle\frac{\phi_q}{\|\phi_q\|_2}, \mathbf{r}^1\rangle\right| \geq \left|\Im\langle\frac{\phi_i}{\|\phi_i\|_2}, \mathbf{r}^1\rangle\right|\right) \quad (6.58) \end{aligned}$$

where $\mathbf{r}^1 = \phi_q s_q + \mathbf{v}$. In a similar way, one can show that

$$P\left(\mathcal{S}^2|\mathcal{S}^1\right) \geq \left(1 - \left(1 + \frac{\alpha_2}{2\sigma_v^2}\right)^{-m} - \left(1 + \frac{\beta_2}{\sigma_v^2}\right)^{-m}\right)^{N-2}. \quad (6.59)$$

In the following Lemma, we obtain the Symbol Error Rate (SER) of SE-SVC.

17 $\|\mathbf{h}\|_2^2$ is a Chi-squared distribution [28].

Lemma 6.6: The SER of SE-SVC satisfies

$$\mathrm{SER}_{\text{SE-SVC}} \leq k\left(1 + \frac{\mu^* d_{\min}^2}{4\sigma_v^2}\right)^{-m} \tag{6.60}$$

where $d_{\min}$ is the minimum Euclidean distance between the symbols (i.e. $d_{\min} = \min_{i,j\in\Omega, i\neq j} \|\tilde{s}_i - \tilde{s}_j\|_2^2$).

Proof : Let $\tilde{s}_p$ be the transmit symbol and $\tilde{s}_{\hat{p}}$ be the incorrectly detected symbol for $\tilde{s}_p$. The pairwise error probability of choosing $\tilde{s}_{\hat{p}}$ incorrectly can be expressed as

$$\begin{aligned}
\Pr(\tilde{s}_p \to \tilde{s}_{\hat{p}} | \, \mathbf{y}, \mathbf{h}) &= \mathrm{Q}\left(\sqrt{\frac{\|\Phi_i(\tilde{s}^i - \tilde{s}^j)\|^2}{2\sigma_v^2}}\right) \\
&\overset{(a)}{\leq} \exp\left(\frac{-\mu^* \|\mathbf{h}\|^2 d_{\min}^2}{4\sigma_v^2}\right),
\end{aligned} \tag{6.61}$$

where $\mathrm{Q}(x) = \int_x^\infty \mathcal{N}(0,1)$ is the Gaussian Q-function and (a) is because $\mathrm{Q}(x) \leq \exp\left(-\frac{x^2}{2}\right)$. The unconditional Packet Error Probability (PEP) can be obtained by taking the expectation with respect to the channel realization $\mathbf{h}$ as $\Pr(\tilde{s}^i \to \tilde{s}^j) = \mathrm{E}_{\mathbf{h}}\left[\Pr(\tilde{s}^i \to \tilde{s}^j | \, \mathbf{y}, \mathbf{h})\right] = \mathrm{E}_{\mathbf{h}}\left[\exp\left(-\frac{\|\mathbf{h}\|^2 \Delta_s^T \Upsilon_\Omega \Delta_s}{4\sigma_v^2}\right)\right] = \left(1 + \frac{\mu^* d_{\min}^2}{4\sigma_v^2}\right)^{-m}$. Using the union bound, the SER of SE-SVC can be upper bounded as

$$\mathrm{SER}_{\text{SE-SVC}} = \Pr\left(\tilde{\mathbf{s}}_\Omega^i \to \tilde{\mathbf{s}}_\Omega^j\right) \leq k\left(1 + \frac{\mu^* d_{\min}^2}{4\sigma_v^2}\right)^{-m}. \tag{6.62}$$

By combining Lemma 6.5 and Lemma 6.6, we finally obtain the upper bound of BLER in (6.49). It is clear from (6.49) that the BLER increases when the number of measurement m is small and information vector $\tilde{\mathbf{s}}$ is less sparse (i.e. k is large), which matches well with our expectation.

6.8.3 Simulation Results

In this section, we compare the BLER performance of SE-SVC and SVC. To ensure a fair comparison, we set $b_t = 16$, $k = 2$, $b_i = 12$ and $b_m = 4$. In the codebook generation, we use a randomly generated Bernoulli sequence. As a channel model, we use AWGN and realistic ITU models (EPA channel and EVA channel) [11].

In Figure 6.15, we evaluate the BLER performance of the SE-SVC and conventional SVC scheme in the AWGN scenario. We observe that SE-SVC outperforms SVC. For example, the SE-SVC scheme achieves 1.5 dB gain over the SVC at

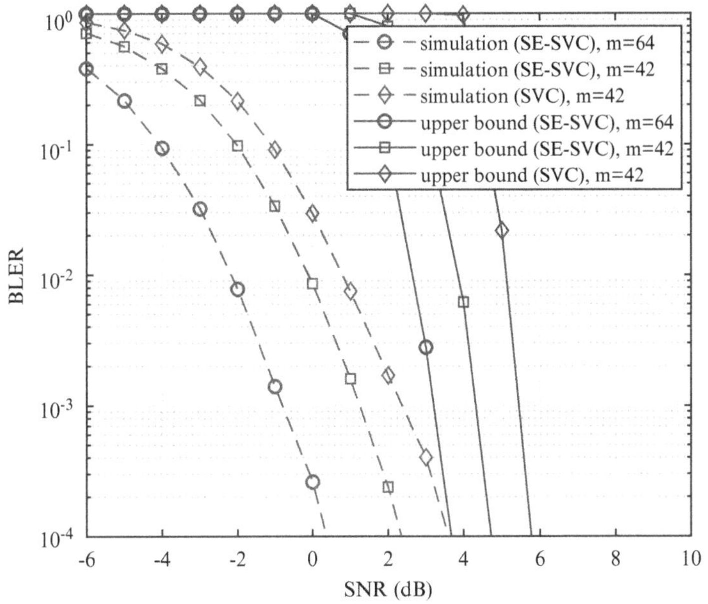

Figure 6.15 Empirical simulation and upper bound of BLER. ©IEEE 2020. Reprinted with permission from [31].

BLER= 10^{-4}. Also, we see that the BLER decreases sharply when the number of measurements m increases. For instance, if m is changed from 42 to 64, we obtain around 2 dB gain at BLER= 10^{-4}.

In Figure 6.16, we plot the BLER performance of the SE-SVC scheme and competing schemes under the EVA and EPA channels. We observe that the SE-SVC outperforms the conventional PDCCH and SVC by a large margin. For example, the SE-SVC scheme achieves 6.5 dB gain over the conventional PDCCH at BLER= 10^{-3} under EVA channel and around 2 dB gain over SVC at BLER= 10^{-3} under EPA channel.

Next, we plot the BLER performance of SE-SVC using various b_s ($b_s = 1$ for BPSK, $b_s = 2$ for QPSK, and $b_s = 3$ for 8PSK) in Figure 6.17. We can clearly see that the performance of SE-SVC depends highly on b_s because there is a trade-off between the support identification performance and the symbol detection performance. Specifically, when b_s increases, N can be reduced and thus the support identification performance will be improved at the expense of the performance degradation in the symbol detection. In fact, the symbol detection performance of the Low Order Modulation (LOM) is better than that of the High Order Modulation (HOM) but the support identification performance is poor due to the increase in the sparse vector dimension N.

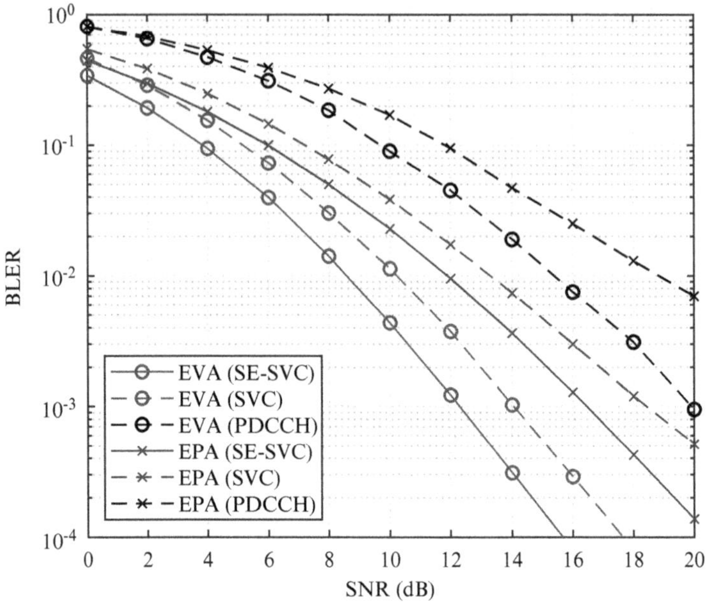

Figure 6.16 BLER performance as a function of SNR under EPA and EVA channels. ©IEEE 2020. Reprinted with permission from [31].

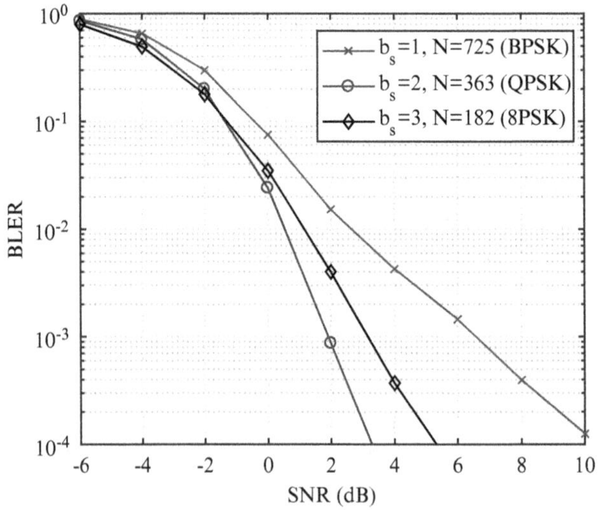

Figure 6.17 BLER performance as a function of SNR for various b_s ($b_t = 20$, $m = 50$, and $k = 2$). ©IEEE 2020. Reprinted with permission from [31].

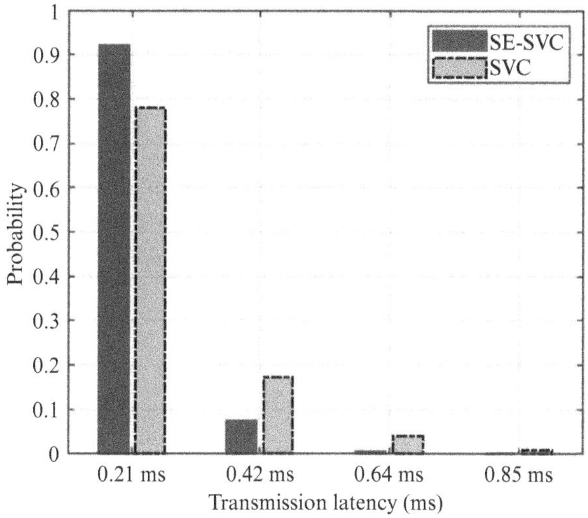

Figure 6.18 Probability of transmission latency to complete the packet transmission ($b_t = 16$, $m = 42$ and SNR = -2 dB). ©IEEE 2020. Reprinted with permission from [31].

In order to verify the latency gain of SE-SVC over SVC, we plot the distribution of transmission latency which is defined as the time from the initial transmission to the successful packet decoding at the mobile terminal (see Figure 6.18). From the result, we observe that the transmission latency of SE-SVC is much smaller (15% on average) than the SVC transmission latency.

6.9 Conclusion

URLLC is one of the key services in 5G communications having a variety of applications including automated controls, Tactile Internet, remote operations, and intelligent transportation systems. Since URLLC needs to satisfy two conflicting requirements (ultra-high reliability and low latency), special attention should be paid to the design of URLLC transmission schemes. In this chapter, we introduced a new type of short packet transmission framework named the Sparse Vector Coding (SVC) technique. The key idea behind SVC is to transform an information vector into the sparse vector in the transmitter and to exploit the sparse recovery algorithm in the receiver. Metaphorically, SVC can be thought as a marking dots on an empty table. As long as the number of dots is small enough and the measurements contain enough information to figure out the marked cell positions, accurate decoding of the SVC packet can be guaranteed. We demonstrated from the numerical evaluations that SVC is very effective in the short packet transmission for URLLC scenarios. Also, we introduced the advanced SVC schemes including PL-SVC and SE-SVC. In PL-SVC, the input information is mapped as a composite of the sparse vector and the channel vector so that the decoding process can be performed without the channel information. In SE-SVC, by conveying the information bits using both the non-zero positions and the symbols, we achieved the decoding performance gain and the transmission latency gain.

Appendix A

6.10 Proof of Lemma 1

Let $\mathbf{u}_i = \frac{\mathbf{a}_i}{\|\mathbf{a}_i\|_2}$, then it is clear that $\mathbf{u}_i$ is a random vector with zero mean and unit variance. Also, let $X = \frac{\mathbf{a}_i^T \mathbf{a}_j}{\|\mathbf{a}_i\|_2}$, then $X = \mathbf{u}_i^T \mathbf{a}_j$. One can easily show that X conditioned on any realization of $\mathbf{u}_i = u$ is a standard Gaussian. This is because $\mathrm{E}\left[X|\mathbf{u}_i = u\right] = \mathrm{E}\left[u^T \mathbf{a}_j \mathbf{a}_j^T\right] = u^T u = 1$. Further,

$$\begin{aligned} f_X(x) &= \int_u f_{X|\mathbf{u}_i}(x|u) f_{\mathbf{u}_i}(u) du \\ &= \frac{1}{\sqrt{2\pi}} \exp\left(-\frac{x^2}{2}\right) \int_u f_{\mathbf{u}_i}(u) du \\ &= \frac{1}{\sqrt{2\pi}} \exp\left(-\frac{x^2}{2}\right), \end{aligned} \tag{6.63}$$

which is the desired result.

Appendix B

6.11 Derivation of (6.19)

Noting that $\phi_i = [\tilde{h}_{11} c_{1i} \ \tilde{h}_{22} c_{2i} \ \cdots \ \tilde{h}_{mm} c_{mi}]^T$ and $\mu_{ij} = \frac{\phi_i^T \phi_j}{\|\tilde{\mathbf{h}}\|_2^2}$, we have

$$\langle \frac{\phi_i}{\|\phi_i\|_2}, \phi_j \rangle = \frac{\phi_i^T \phi_j}{\|\phi_i\|_2}. \tag{6.64}$$

Since $||\phi_i||_2 = \sqrt{|\tilde{h}_{11}c_{1i}|^2 + \cdots + |\tilde{h}_{mm}c_{mi}|^2} = ||\tilde{\mathbf{h}}||_2$, we have

$$\begin{aligned}\langle \frac{\phi_i}{||\phi_i||_2}, \phi_j \rangle &= ||\tilde{\mathbf{h}}||_2 \frac{\phi_i^T \phi_j}{||\tilde{\mathbf{h}}||_2} \\ &= ||\tilde{\mathbf{h}}||_2 \mu_{ij}. \end{aligned} \tag{6.65}$$

In particular, $i = j$, $\mu_{ij} = 1$ and thus

$$\langle \frac{\phi_i}{||\phi_i||_2}, \phi_i \rangle = ||\tilde{\mathbf{h}}||_2. \tag{6.66}$$

From (6.65) and (6.66), we have

$$\langle \frac{\phi_i}{||\phi_i||_2}, \phi_j \rangle = \begin{cases} ||\tilde{\mathbf{h}}||_2, & \text{for } i = j \\ ||\tilde{\mathbf{h}}||_2 \mu_{ij}, & \text{for } i \neq j. \end{cases} \tag{6.67}$$

Bibliography

[1] Rec. ITU-R M.2083-0, "Sparse Vector Coding for Ultra Short Packet Transmission," Sep. 2015.

[2] P. Schulz, M. Matthe, H. Klessig, M. Simsek, G. Fettweis, J. Ansari, S. A. Ashraf, B. Almeroth, J. Voigt, I. Riedel, A. Puschmann, A. Mitschele-Thiel, M. Muller, T. Elste, and M. Windisch, "Latency Critical IoT Applications in 5G: Perspective on the Design of Radio Interface and Network Architecture," *IEEE Commun. Mag.*, vol. 55, no. 2, pp. 70-78, 2017.

[3] 3GPP Technical Report 38.802, "Study on New Radio Access Technology Physical Layer Aspects (Release 14)," v14.1.0, 2017.

[4] C. Bockelmann, N. Pratas, H. Nikopour, K. Au, T. Svensson, C. Stefanovic, and A. Dekorsy, "Massive Machine-type Communications in 5G: Physical and MAC-layer solutions," *IEEE Commun., Mag.*, vol. 54, no. 9, pp. 59-65, 2016

[5] H. Ji, S. Park, J. Yeo, Y. Kim, J. Lee, and B. Shim, "Ultra Reliable and Low Latency Communications in 5G:Physical Layer Aspects," *IEEE Wireless Commun.*, vol. 25, no. 3, pp. 124-130, Jul. 2018.

[6] B. Lee, S. Park, D. Love, H. Ji, and B. Shim, "Packet Structure and Receiver Design for Low Latency Wireless Communications with Ultra Short Packets," *IEEE Trans. on Commun.*, vol. 66, no. 2, pp. 796-807, Feb. 2018.

[7] O. Yilmaz, Y. Wang, N. A. Johansson, N. Brahmi, S. A. Ashraf, and Joachim Sachs, "Analysis Ultra-Reliable and Low-Latency 5G Communication for a Factory Automation Use Case," *in Proc. IEEE Int. Conf. on Comm. (ICC) Workshop*, pp. 1190-1195, 2015.

[8] N. A. Johansson, Y.Wang, E. Eriksson, and M. Hessler, “Radio Access for Ultra-Reliable and Low-Latency 5G Communications,” *In Proc. IEEE Int. Conf. on Comm. (ICC) Workshop,* pp. 1185-1189, 2015.

[9] 3GPP Technical Report 38.913, “Study on Scenarios and Requirements for Next Generation Access Technologies (Release 14),” v14.2.0, 2017.

[10] C. Lim, T. Yoo, B. Clerckx, B. Lee, and B. Shim, “Recent trend of multiuser MIMO in LTE-Advanced,” *IEEE Comm Mag.,* vol. 51, no. 3, pp. 127-135, March 2013.

[11] S. Sesia, M. Baker, and I. Toufik, *LTE - the UMTS Long Term Evolution: From Theory to Practice,* John Wiley & Sons., 2011.

[12] J. Choi, B. Shim, Y. Ding, B. Rao and D. Kim, “Compressed sensing for wireless communications: useful tips and tricks,” *IEEE Commun. Surveys & Tutorials,* vol. 19, no. 3, pp. 1527-1550, Feb. 2017.

[13] D. L. Donoho, “Compressed Sensing,” *IEEE Trans. on Inform. Theory,* vol.52, no. 4, pp. 1289-1306, 2006.

[14] H. Sugiyama, K. Nosu “MPPM: A Method for Improving the Bandutilization Efficiency in Optical PPM,” *Journal of Lightwave Technology,* vol.7, no.3, pp.465-472, 1989.

[15] E. Basar, M. Wen, R. Mesleh, M. D. Renzo, Y. Xiao, and H. Haas, “Index Modulation Techniques for Next-Generation Wireless Networks,” *IEEE Access,* vol. 5, pp. 16693-16746, Aug. 2017.

[16] G. Kaddoum, Y. Nijsure, and H. Tran, “Generalized code index modulation technique for high-data-rate communication systems,” *IEEE Trans. Veh. Technol.,* vol. 65, no. 9, pp. 7000-7009, Sep. 2016.

[17] X. Chen, T. Chen, and D. Guo, “Capacity of Gaussian Many-Access Channels,” *IEEE Trans. Inform. Theory,* vol. 63. no. 6, pp. 3516-3539, 2017.

[18] W. Zhang and L. Huang, “On OR many-access channels,” *In Proc. IEEE Int. Symp. on Inform. Theory (ISIT),* Aachen, Germany, June 2017.

[19] G. Castagnoli, S. Brauer, and M. Herrmann, “Optimization of Cyclic Redundancy-Check Codes with 24 and 32 Parity Bits,” *IEEE Trans. on Commun.,* vol. 41, no. 6, pp. 883-892, 1993.

[20] T. T. Cai and J. Tiefeng, “Phase Transition in Limiting Distributions of Coherence of High-dimensional Random Matrices,” *Journal of Multivariate Analysis,* vol. 107, pp. 24-39, 2012.

[21] J. Wang, S. Kwon, and B. Shim, “Generalized Orthogonal Matching Pursuit,” *IEEE Trans. on Sig. proc.,* vol. 60, no. 12, pp.6202-6216, 2012.

[22] S. Kwon, J. Wang, and B. Shim, “Multipath matching pursuit,” *IEEE Trans. Sig. Process.,* vol. 60, no. 12, pp. 6202-6216, Dec. 2012.

[23] A. J. Viterbi, and K. O. Omura. *Principles of digital communication and coding,* Courier Corporation, 2013.

[24] J. A. Tropp and A. C. Gilbert, "Signal recovery from random measurement via orthogonal matching pursuit," *IEEE Trans. Inf. Theory,* vol. 53, no. 12, pp.4655-4666, Dec. 2007.

[25] S. D. Howard, A. R. Calderbank, and S. J. Searle, "A Fast Reconstruction Algorithm for Deterministic Compressive Sensing using Second Order Reed-Muller Codes," *In Proc. of IEEE Conf. on Information Sciences and Systems,* pp. 11-15, 2008.

[26] D. Tse and P. Viswanath, *Fundamentals of Wireless Communication,* Cambridge University Press, 2005.

[27] H. Vangala, Y. Hong, and E. Viterbo, "Efficient Algorithms for Systematic Polar Encoding," *IEEE Commun. Lett.,* vol. 20, no. 1, pp. 17-20, Jan. 2016.

[28] K. Venkatarama, *Probability and Random Processes,* John Wiley & Sons., 2006.

[29] R. Xie, H. Yin, X. Chen, and Z. Wang, "Many Access for Small Packets Based on Precoding and Sparsity-Aware Recovery," *IEEE Trans. Commun.,* vol. 64. no. 11, pp. 4680-4694, 2016.

[30] Ji, Hyoungju and Kim, Wonjun and Shim, Byonghyo, "Pilot-less sparse vector coding for short packet transmission," *IEEE Wireless Commun. Lett.,* vol. 8. no. 4, pp. 1036-1039, 2019.

[31] Kim, Wonjun and Bandari, Shravan Kumar and Shim, Byonghyo, "Enhanced sparse vector coding for ultra-reliable and low latency communications," *IEEE Trans. Veh. Technol.,* vol. 69. no. 5, pp. 5698-5702, 2020.

7

Network Slicing for URLLC

Peng Yang[1,], Xing Xi[2], Tony Q. S. Quek[1], Jingxuan Chen[2], Xianbin Cao[2], and Dapeng Wu[3]*

[1] *Information Systems Technology and Design, Singapore University of Technology and Design, 487372, Singapore*
[2] *School of Electronic and Information Engineering, Beihang University, 100083, Beijing, China*
[3] *Department of Electrical and Computer Engineering, University of Florida, 32611, Gainesville, USA*
[*] *Corresponding Author*

7.1 Introduction

Future wireless networks are desired to provide diverse service requirements concerning throughput, latency, reliability, availability, as well as operational requirements, e.g. energy efficiency and cost efficiency [17, 25]. These service requirements are made by mobile networks and some novel application areas such as Industry 4.0, airborne communication, vehicular communication, and smart grid.

The International Telecommunication Union (ITU) has categorized these services into three primary use cases: Enhanced Mobile Broadband (eMBB), Massive Machine-Type Communications (mMTC), and Ultra-Reliable and Low Latency Communications (URLLC) [27]. In order to provide cost-efficient solutions, it is agreed by some telecommunication organizations including the Third Generation Partnership Project (3GPP) and the Next Generation Mobile Network Alliance (NGMA), on the convergence of each use case onto a shared physical infrastructure instead of deploying individual network solutions for each use case [3].

To satisfy the requirement of reducing cost efficiency, the concept of network slicing has been proposed. The fundamental idea of network slicing is to logically isolate network resources and functions customized for specific requirements on a common physical infrastructure [25]. A network slice as a virtual End-to-End (E2E) network for efficiently implementing resource isolation and

Ultra-Reliable and Low-Latency Communications (URLLC) Theory and Practice: Advances in 5G and Beyond, First Edition. Edited by Trung Q. Duong, Saeed R. Khosravirad, Changyang She, Petar Popovski, Mehdi Bennis and Tony Q.S. Quek.
© 2023 John Wiley & Sons Ltd. Published 2023 by John Wiley & Sons Ltd.

increasing statistical multiplexing is self-contained with its virtual network resources, topology, traffic flow, and provisioning rules [11, 25]. Due to the significant role in constructing flexible and scalable future wireless networks, network slicing for mMTC, eMBB, and URLLC service (multiplexing) has received much attention from academia [2, 4, 22].

However, most of the current works do not study the impact of a time-varying channel on slice creation and benefits of exploiting advanced Radio Access Techniques (RATs) in network slicing systems. For example, the actual channel may vary in short timescales (e.g. milliseconds) while slice creation may be conducted in relatively long timescales (e.g. minutes or hours). Therefore, network slicing needs to mitigate a multi-timescale issue. Additionally, the utilization of advance RATs (e.g. coordinated multipoint, CoMP) has been considered as a promising way of satisfying spectrum challenges and improving system throughput [12, 21].

To tackle the multi-timescale issue of Radio Access Network (RAN) slicing, recent works [18, 31] developed a software-defined networking (SDN)-based radio resource allocation framework. This framework could facilitate spectrum exploitation among different network slices in different time scales. In addition, the [28] a CoMP-based RAN slicing framework was developed for eMBB and URLLC service multiplexing and it was proposed to tackle the multi-timescale issue of RAN slicing via an alternating direction method of multipliers (ADMM). This work[1], however, assumed that URLLC traffic was uninterruptedly generated and ignored the significant bursty characteristic of URLLC traffic [7]. The bursty URLLC traffic (e.g. remote surgery and remote robot control traffic) will further exacerbate the difficulty of slicing the RAN for URLLC involved service multiplexing from the following two aspects:

- *Resource efficiency*: one of the efficient proposals in future wireless communication networks to handle the uncertainty (including bursty) is to reserve network resources, which may waste a large number of valuable network resources. Therefore, it is important to develop resource orchestration schemes with high utilization for future networks, especially for some resource-constrained networks.
- *Immediate resource orchestration*: bursty URLLC packets need to be immediately scheduled if there are available resources and the system utility can be maximized. Therefore, under the premise of improving resource efficiency, immediate resource orchestration schemes related to the number of flashing URLLC packets should be developed.

1 This chapter has been a part of [29, 30]

The difficulty motivates us to address the CoMP-enabled RAN slicing for bursty URLLC and eMBB service provision. The CoMP transmission technique is explored because it can significantly improve the service quality of URLLC and eMBB via spatial diversity [15] and is easy and cost-efficient for the RAN slicing implementation [30].

7.2 CoMP-enabled RAN Slicing System Prototype

In this section, we present a prototype of the CoMP-enabled RAN slicing system to describe how to implement the convergence of eMBB and bursty URLLC services onto the common CoMP-enabled RAN through network slicing. In Figure 7.1, we consider a CoMP-enabled RAN slicing system prototype with four major parties: *End User Equipment (UE):* includes eMBB UEs and bursty URLLC UEs; UEs run their services on slices managed by the virtualized network slice management; *Software-Defined RAN Coordinator (SDRAN-C):* calculates and updates accepted network slices to accommodate service requirements of end UEs; *Network Slice Management (NSM):* a virtualized function aiming to control and manage network slices; *Core network control function (C^2F):* configures the core network based on the slice requirements.

We consider two types of end UE, i.e. multi-cast eMBB UEs and uni-cast bursty URLLC UEs. Each type of UE possesses several specific features regarding their Quality of Service (QoS) requirements. End UEs will send slice requests to SDRAN-C. Upon receiving slice requests, SDRAN-C analyzes slice requirements and decides to accept or decline requests based on the optimal policy.

The SDRAN-C block mainly consists of five components: Radio Resource Manage and Control (RRMC), Distributed Unit (DU) pool, Centralized Unit (CU), User Plane-Anchor (UP-Anchor), and the algorithm. The slice creation and

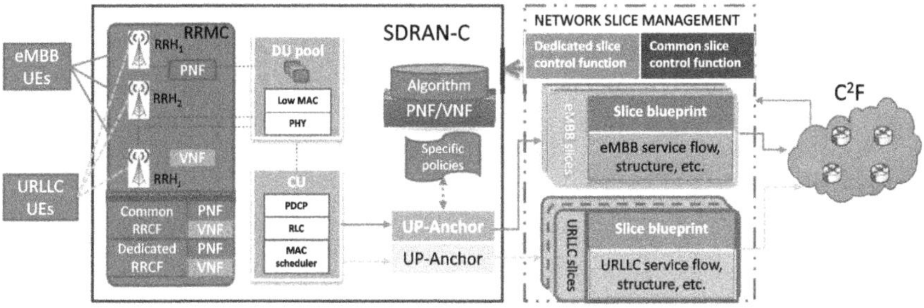

Figure 7.1 A RAN slicing system prototype. ©IEEE 2021. Reprinted with permission from [30]

configuration processes may take up multiple minislots. If some slices are created, then RRMC is responsible for configuring RAN protocol stacks and QoS according to slice requirements. For example, for slices with high throughput requirements, radio bearers should be configured to support CoMP transmission or multi-connectivity. For slices with high reliable and low-latency requirements, lower frame error rates, Reduced Round Trip time (RTT), shortened Transmission Time Interval (TTI), and/or multipoint diversity schemes are desired to be utilized after configuring the CoMP structure. Besides, the Radio Resource Control Function (RRCF) in RRMC, which can be further split into dedicated and common RRCF related functions, will be activated for UE-specific radio resource control on the basis of Virtualized Network Function (VNF) and/or Physical Network Function (PNF) [25]. The DU pool is exploited to realize the Physical Layer (PHY) collaboration facilitating the deployment of CoMP. The PHY and low MAC functions are split to the DU pool to execute some operations such as fast Fourier transform, modulation and precoding. The MAC scheduler, Radio Link Control (RLC) and Packet Data Convergence Protocol (PDCP) functions are split to the CU [1]. The MAC scheduler will schedule traffic based on the network condition to alleviate network congestion. There are many MAC scheduling schemes such as random access, back-off, access class barring. Depending on underlying services, CU can configure and tailor UP protocol stacks. For example, for slices supporting low-latency services, Internet Protocol (IP) and related header compression may not be used, and RLC function may be configured in the transparent mode [25]. Additionally, owing to the high transmission capacity requirements between the DUs and the CU, optical fiber communication can be considered as the ideal fronthaul or non-constrained fronthaul. Several typical protocols such as the Common Public Radio Interface (CPRI) and the Open Base Station Architecture Initiative (OBSAI) [10] aiming at unifying the interface of fronthaul should also be enabled. The UP-Anchor is responsible for distributing the traffic according to the configured slice policy, and for encryption with slice-specific security keys. For example, for slices requested from industry, security, resilience, and reliability of services are of higher priority. Then, policy requirements, e.g. security, resilience, and reliability, should be specified for this type of slice. Based on slice requests and available resources, the algorithm will periodically calculate, update, and reconfigure accepted slices to achieve the maximum utility. It is also responsible for RAN and CN mapping that can be implemented via some configuration protocols [23].

The virtualized NSM, which operates on the top of a physical and/or virtualized infrastructure, is responsible for creating, activating, maintaining, configuring, and releasing slices during their life cycle. Via the dedicated/common slice control function, NSM will generate a network slice blueprint (i.e. a template) for each accepted network slice, which describes the structure, configuration, control signals, and service flows for instantiating and controlling the network slice instance of a type of service during its life cycle. The slice instance includes a set of network functions and resources to meet the E2E service requirements.

For the C^2F, it will interpret the slice blueprint when the slice request arrives. Accordingly, the C^2F will arrange the network configuration according to the interpreted blueprint and find the optimal servers and paths to place VNFs to meet the required E2E service of the slice. Besides, C^2F possesses multiple data centers for network slicing services. Each data center contains a set of servers (e.g. home subscribe server) with diverse resources, e.g. computing, storage, which are used to support VNFs' services such as identity, independent subscription, session for each network slice. Data centers are connected via backhaul links and can provide services jointly or separately.

7.3 System Model

We consider a CoMP-enabled RAN slicing system for URLLC and eMBB multiplexing service provision. In this system, the time is discretized and partitioned into time slots and minislots, and a time slot includes T minislots. The URLLC deadlines are within a single minislot. There are N^e ground eMBB UEs, N^u ground URLLC UEs and J Remote Radio Heads (RRHs), which connect to a Baseband Unit (BBU). The eMBB UE set and the URLLC UE set are denoted as $\mathcal{I}^e = \{1, \dots, N^e\}$, $\mathcal{I}^u = \{1, \dots, N^u\}$, respectively. We assume that eMBB and URLLC UEs are randomly distributed in a considered communication area, and RRHs are regularly deployed. Besides, each RRH is assumed to be equipped with K antennas, and each UE is equipped with an antenna. All RRHs cooperate to transmit signals to a UE such that the Signal-to-Noise Ratio (SNR) of it can be significantly enhanced[2]. Meanwhile, a flexible Frequency Division Multiple Access (FDMA) technique is exploited to achieve the inter-slice and intra-slice interference isolation.

7.3.1 eMBB Slice Model

According to the above mentioned concept of a network slice (especially from the perspective of the QoS requirement of a slice), we can define an eMBB network slice request as the following.

2 This chapter exploits the optimization of transmit beamformers, and the issues of beam alignment and beam selection are out of the scope of this chapter.

Definition 7.1 (Multicast eMBB slice request): For any multicast eMBB slice $s \in \mathcal{S}^e$, its network slice request is composed of two components [28]: *The number of eMBB UEs:* the symbol I_s^e is utilized to represent the number of eMBB UEs grouped into the slice s; *QoS requirements of eMBB UEs:* the minimum required data rate of eMBB UEs in s, denoted by C_s^{th}, is characterized as their QoS requirements. To this end, the tuple $\{I_s^e, C_s^{th}\}$ is used to represent an eMBB slice request of s.

In this definition, eMBB UEs are partitioned into $|\mathcal{S}^e|$ groups according to the data rate requirement of a UE. UEs in the same slice have the same data rate requirement. The slice request of each group of eMBB UEs will always be admitted by the SDRAN-C in this chapter, and coordinated beamformers and Physical Resource Blocks (PRB) will be effectively configured to accommodate data rate requirements of all eMBB UEs by way of multicast transmission.

7.3.2 Bursty URLLC Slice Model

Different from eMBB UEs in terms of QoS requirements, URLLC packets need to be successfully transmitted and decoded with extremely low latency (1 ms) and extremely high reliability (99.999%). Thus, the definition of a bursty URLLC slice request can be described as follows.

Definition 7.2 (Unicast URLLC slice request): For any bursty URLLC slice $s \in \mathcal{S}^u$, its network slice request is composed of four components. *The number of unicast URLLC UEs:* the symbol I_s^u represents the number of unicast URLLC UEs classified into slice s. *QoS requirements of URLLC UEs:* URLLC UEs require low-latency end-to-end transmissions, which is significantly different from eMBB UEs. Thus, the communication latency D_s is leveraged to characterize the QoS requirements of URLLC UEs. *Packet blocking probability:* α, which should be lower than a threshold, is utilized to denote the URLLC packet blocking probability. The packet blocking probability is defined as the probability that a URLLC packet is blocked. *Codeword error decoding probability:* β that should not be greater than a threshold is used to represent the error probability of decoding a URLLC codeword[3]. In this way, a four-tuple $\{I_s^u, D_s, \alpha, \beta\}$ can be involved to represent a bursty URLLC slice request of s.

In this definition, unicast URLLC UEs are classified into $|\mathcal{S}^u|$ clusters according to the latency requirement of each UE. Owing to the ultra-low latency requirement URLLC traffic should be immediately scheduled upon arrival; thus, unicast

3 A URLLC packet will usually be coded before transmission, and the generated codeword will be transmitted in the air interface such that the transmission reliability can be improved.

URLLC slice requests will always be accepted by the SDRAN-C in this chapter[4]. Then, coordinated beamformers will be correspondingly generated to cover UEs by way of the unicast transmission at the beginning of each minislot.

7.4 Problem Formulation

On the basis of the above system model, this section aims to formulate the problem of RAN slicing for URLLC and eMBB multiplexing service provision.

7.4.1 QoS Requirements

7.4.1.1 Data Rate Requirements of eMBB UEs

The generated transmit beamformers for UEs of slice s ($s \in \mathcal{S}^e$) on RRH j at minislot t is denoted by $\boldsymbol{v}_{j,s}(t) \in \mathbb{C}^K$. The channel coefficient between RRH j and eMBB UE i of s at minislot t is denoted by $\boldsymbol{h}_{ij,s}(t) \in \mathbb{C}^K$, which does not greatly change in each minislot. Suppose that the instantaneous channel coefficient $\boldsymbol{h}_{ij,s}(t)$ can be effectively estimated by exploiting some machine learning methods [33] at the beginning of minislot t and the channel fading process is ergodic over a time slot for each (i, j) pair. The SNR received at UE i of slice s at t can then be written as

$$SNR_{i,s}^e(t) = \frac{|\sum_{j\in\mathcal{J}} \boldsymbol{h}_{ij,s}^{\mathrm{H}}(t)\boldsymbol{v}_{j,s}(t)|^2}{\sigma_{i,s}^2}, \text{for all } s \in \mathcal{S}^e, i \in \mathcal{I}_s^e, \tag{7.1}$$

where $\sigma_{i,s}^2$ denotes the noise power, $\mathcal{I}_s^e = \{1, \dots, I_s^e\}$ is the set of eMBB UEs of s. Since the multicast transmission and flexible FDMA mechanism are exploited the interference is not involved.

According to Shannon's formula, the achievable data rate $\gamma_{i,s}^e(t)$ of UE i of slice s at t can be expressed as

$$\gamma_{i,s}^e(t) = \omega_s^e(t)\log_2(1 + SNR_{i,s}^e(t)), \text{for all } s \in \mathcal{S}^e, i \in \mathcal{I}_s^e, \tag{7.2}$$

where $\omega_s^e(t)$ denotes the bandwidth allocated to s at t.

The following rate-related condition should be satisfied to admit the eMBB slice request

$$\gamma_{i,s}^e(t) \geq C_s^{th}, \text{for all } s \in \mathcal{S}^e, i \in \mathcal{I}_s^e. \tag{7.3}$$

---

4 We consider that the SDRAN-C can always accept eMBB and URLLC service requests here. However, whether these service requests can always be accommodated may be determined by both the QoS requirements of eMBB and URLLC UEs and the system service capability in practice.

7.4.1.2 QoS Requirements of URLLC UEs

As is known, it is challenging to design a RAN slicing system to support the transmission of URLLC traffic owing to the stringent QoS requirements of URLLC UEs. What makes the issue more difficult is that URLLC traffic may be bursty. Bursty URLLC traffic, which may cause severe packet blocking, may significantly degrade the system performance of RAN slicing when URLLC slices are not well configured. To understand the characteristics of bursty URLLC traffic and mitigate the effect of bursty URLLC traffic on RAN slicing, we will address the following two questions.

- *How can we model bursty URLLC traffic?*
- *What schemes can be developed for the RAN slicing system such that the URLLC packet blocking probability can be significantly reduced?*

During a time slot, bursty URLLC data packets destined to UEs of each URLLC slice and aggregated at the SDRAN-C are modelled as a compound Poisson process [8], where arrivals happen in bursts (or batches, i.e. several arrivals can happen at the same instant) and the inter-batch duration is independent and exponentially distributed. The vector of URLLC packet arrival rates is denoted by $\boldsymbol{\lambda} = \{\lambda_1, \dots, \lambda_s, \dots, \lambda_{|\mathcal{S}^u|}\}$, where $\lambda_s = \lambda_{b,s}/\lambda_{a,s}$ is a constant and represents the average arrival rate of packets destined to UEs of slice s during a unit of time, $\lambda_{a,s}$ is the average inter-batch time interval, and $\lambda_{b,s}$ is the average number of arrivals in a batch.

On the basis of the URLLC traffic model, we next discuss how to reduce the URLLC packet blocking probability via re-cutting PRBs. To satisfy QoS requirements of URLLC UEs, a portion of PRBs should be allocated to them. In the RAN slicing system, a URLLC UE i of s will be allocated a block of network bandwidth of size $\omega^u_{i,s}(t)$ for a period of time d_s at minislot t. Since URLLC packets in s have the deadline of D_s seconds for E2E transmission latency, we shall always choose $d_s \leq D_s$. Besides, a packet destined to the UE i will be coded before sending out to improve the reliability[5]; and the transmission of a codeword needs $r^u_{i,s}(t)$ channel uses that measure the speed and capacity of a specific information channel. The channel use, bandwidth, and transmission latency are related by $r^u_{i,s}(t) = \kappa \omega^u_{i,s}(t) d_s$ [5], where κ is a constant representing the number of channel uses per unit time per unit bandwidth of the FDMA frame structure and numerology. We denote the channel use set of URLLC UEs as $\boldsymbol{r(t)} = \{\boldsymbol{r}_1(t), \dots, \boldsymbol{r}_s(t), \dots, \boldsymbol{r}_{|\mathcal{S}^u|}(t)\}$ with $\boldsymbol{r}_s(t) = \{r^u_{1,s}(t), \dots, r^u_{I^u_s,s}(t)\}$.

Let us model the aggregation and departure of URLLC packets in the SDRAN-C as an $M/M/W^u$ queueing system, which has higher key performance indicators than some other queueing systems like $M/M/1$, with finite bandwidth W^u

---

5 *Packet* and *codeword* are different terminologies. *Packet* is a term used in the network layer (measuring in bits), and *codeword* is the one adopted in the physical layer (measuring in symbols).

and arrival data rate $\boldsymbol{\lambda}$. However, the $M/M/W^u$ queueing system itself may not capture deadline-constrained jobs. To tackle this issue, we introduce a queueing probability concept into the queueing system. Once entering a queue, URLLC packets should be immediately served. Nevertheless, due to stochastic variations in the packet arrival process, occasionally there may not be enough spare bandwidth to serve new arrivals , and new arrivals should wait in the queue and may be blocked. Then, we should take actions to reduce the blocking probability of URLLC packets. Effectively re-cutting the PRBs is a way of significantly reducing the URLLC packet blocking probability. Denote $p_b(\boldsymbol{\omega}^u(t), \boldsymbol{\lambda}, \boldsymbol{d}, W^u(t))$ as the blocking probability experienced by arrival URLLC packets at minislot t where $\boldsymbol{\omega}^u(t) = \{\omega^u_{1,1}(t), \dots, \omega^u_{i,s}(t), \dots, \omega^u_{I^u_{|\mathcal{S}^u|}, |\mathcal{S}^u|}(t)\}$ and $\boldsymbol{d} = \{d_1, \dots, d_s \dots, d_{|\mathcal{S}^u|}\}$. The following theorem provides us with a clue of re-cutting PRBs for URLLC packet transmission.

Theorem 7.1: At any minislot t, for the given $\boldsymbol{\omega}^u(t)$, $\boldsymbol{d}$, and a positive integer q, define $\tilde{\boldsymbol{\omega}}^u(t) = \{\omega^u_{1,1}(t), \dots, \omega^u_{i,s}(t)/q, \dots, \omega^u_{I^u_{|\mathcal{S}^u|}, |\mathcal{S}^u|}(t)\}$ and $\tilde{\boldsymbol{d}} = \{d_1, \dots, qd_s, \dots, d_{|\mathcal{S}^u|}\}$. If $\lambda_s d_s < 1$, then there exists a value $\tilde{W}^u(t)$ such that for $W^u(t) > \tilde{W}^u(t)$ we have $p_b(\boldsymbol{\omega}^u(t), \boldsymbol{\lambda}, \boldsymbol{d}, W^u(t)) \geq p_b(\tilde{\boldsymbol{\omega}}^u(t), \boldsymbol{\lambda}, \tilde{\boldsymbol{d}}, W^u(t))$ [30].

This theorem tells us that if we shorten the packet latency, then fewer resource blocks will be available in the frequency plane, which will definitely cause more severe queueing effects and will significantly increase the packet blocking probability. If we narrow the resource blocks in the frequency plane, then more concurrent transmissions are available, which is beneficial for decreasing the packet blocking probability.

Therefore, we should scale up d_s and select d_s and $\omega^u_{i,s}(t)$ for any URLLC slice s at any minislot according to the following equation

$$d_s = D_s \text{ and } \omega^u_{i,s}(t) = \frac{r^u_{i,s}(t)}{\kappa D_s}, \text{for all} i \in \mathcal{I}^u_s, s \in \mathcal{S}^u. \tag{7.4}$$

With (7.4), a square-root staffing rule [14] can be exploited to derive the minimum upper bound of bandwidth allocated to URLLC slices such that the QoS requirements URLLC UEs can be satisfied. The following lemma presents the bound.

Lemma 7.1: At any minislot t, for a given $M/M/W^u$ queueing system with packet arrival rates $\boldsymbol{\lambda}$ and packet transmit speeds $\{\kappa/r^u_{i,s}(t)\}$, let $W^u(\boldsymbol{r}(t))$ denote the minimum upper bound of bandwidth allocated to all URLLC slices to ensure that $P_Q^{M/M/W^u} \leq \varsigma$ and $p_b(\boldsymbol{\omega}^u(t), \boldsymbol{\lambda}, \boldsymbol{D}, W^u(\boldsymbol{r}(t)))$ is of the order of α, where $P_Q^{M/M/W^u}$ represents the queueing probability. If $\varsigma > \alpha$, then [30]

$$W^u(\boldsymbol{r}(t)) \geq A(\boldsymbol{r}(t)) + c(\varsigma, \alpha)\sqrt{B(\boldsymbol{r}(t))}, \tag{7.5}$$

where $A(\mathbf{r}(t)) = \sum_{s\in\mathcal{S}^u}\sum_{i\in\mathcal{I}_s^u} \lambda_s \frac{r_{i,s}^u(t)}{\kappa}$, $B(\mathbf{r}(t)) = \sum_{s\in\mathcal{S}^u}\sum_{i\in\mathcal{I}_s^u} \lambda_s \frac{r_{i,s}^u(t)^2}{\kappa^2 D_s}$, $\mathbf{D} = \{D_1, \dots, D_s\}$, and

$$c(\varsigma, \alpha) = \frac{\alpha - \varsigma\alpha}{\varsigma - \alpha}\sqrt{\frac{\sum_{s\in\mathcal{S}^u} I_s^u \lambda_s^2 D_s^2}{\min_{s\in\mathcal{S}^u}\{\lambda_s D_s\}}}. \tag{7.6}$$

In (7.5), the first summation item denotes the mean value of the bandwidth allocated to URLLC slices, and the second summation item can be regarded as the redundant bandwidth allocated to mitigate the impact of stochastic variations in the arrival process.

We next discuss the URLLC capacity and channel uses. For URLLC slice $s \in \mathcal{S}^u$, let $\boldsymbol{g}_{ij,s}(t) \in \mathbb{C}^K$ be the transmit beamformer to UE i from RRH j at t, $\boldsymbol{h}_{ij,s}(t)$ is the corresponding channel coefficient, the corresponding SNR received at UE i can then be expressed as

$$SNR_{i,s}^u(t) = \frac{|\sum_{j\in\mathcal{J}} \boldsymbol{h}_{ij,s}^{\mathrm{H}}(t)\boldsymbol{g}_{ij,s}(t)|^2}{\phi\sigma_{i,s}^2}, \text{for all } i \in \mathcal{I}_s^u, s \in \mathcal{S}^u. \tag{7.7}$$

The perception of Channel State Information (CSI) or channel fading distribution may require the signal exchange before transmission, which entails extra transmit latency and potential reliability loss as well. Therefore it may be impossible to obtain perfect CSI for URLLC service provision, and a constant $\phi > 1$ is involved in (7.7) to model the SNR loss for URLLC traffic transmission [19]. Meanwhile, interference signals are not included in (7.7) as a flexible FDMA mechanism is exploited.

On the other hand, owing to the stringent low latency requirements, URLLC packets typically have very short blocklength. We therefore utilize the capacity result for the finite blocklength regime in [24, 26] to calculate the URLLC capacity rather than Shannon's formula that cannot effectively capture the reliability of packet transmission. Particularly, for each UE i in $s \in \mathcal{S}^u$, the number of information bits $L_{i,s}^u(t)$ of a URLLC packet that is transmitted at t with a codeword decoding error probability of the order of β in $r_{i,s}^u(t)$ channel uses can be calculated by [24, 26]

$$L_{i,s}^u(t) \approx r_{i,s}^u(t) C(SNR_{i,s}^u(t)) - Q^{-1}(\beta)\sqrt{r_{i,s}^u(t) V(SNR_{i,s}^u(t))}, \tag{7.8}$$

where $C(SNR_{i,s}^u(t)) = \log_2(1 + SNR_{i,s}^u(t))$ is the AWGN channel capacity per Hz, $V(SNR_{i,s}^u(t)) = \ln^2 2\left(1 - \frac{1}{(1+SNR_{i,s}^u(t))^2}\right)$ is the channel dispersion.

The expression of (7.8) is complicated; yet, the following lemma gives the approximate expression of channel uses in terms of codeword decoding error probability β and SNR.

Lemma 7.2: For any UE i in $s \in \mathcal{S}^u$, the required channel uses $r_{i,s}^u(t)$ of transmitting a URLLC packet of size of $L_{i,s}^u(t)$ to UE i can be approximated as [30]

$$r_{i,s}^u(t) \leq \frac{L_{i,s}^u(t)}{C(SNR_{i,s}^u(t))} + \frac{(Q^{-1}(\beta))^2}{2(C(SNR_{i,s}^u(t)))^2} + \frac{(Q^{-1}(\beta))^2}{2(C(SNR_{i,s}^u(t)))^2}\sqrt{1 + \frac{4L_{i,s}^u(t)C(SNR_{i,s}^u(t))}{(Q^{-1}(\beta))^2}}. \tag{7.9}$$

7.4.2 Physical Resource Constraints

As each RRH has a limitation on the maximum transmit power E_j ($j \in \mathcal{J}$), we can obtain the following power constraint

$$\sum_{s \in \mathcal{S}^e} \boldsymbol{v}_{j,s}^{\mathrm{H}}(t)\boldsymbol{v}_{j,s}(t) + \sum_{s \in \mathcal{S}^u} \sum_{i \in \mathcal{I}_s^u} \boldsymbol{g}_{ij,s}^{\mathrm{H}}(t)\boldsymbol{g}_{ij,s}(t) \leq E_j, \tag{7.10}$$

where the first term on the Left-Hand-Side (LHS) of (7.10) is the total transmit power consumption of RRHs for serving eMBB UEs and the second item for serving URLLC UEs.

Besides, since the multicast eMBB and the unicast URLLC service provisions are considered, and network bandwidth resources allocated to eMBB and URLLC slices are separated in the frequency plane, the network bandwidth constraint can be written as

$$\sum_{s \in \mathcal{S}^e} \omega_s^e(\bar{t}) + W^u(\boldsymbol{r}(t)) \leq W, \tag{7.11}$$

where $\omega_s^e(\bar{t})$ represents the bandwidth allocated to eMBB slice $s \in \mathcal{S}^e$ over a time slot, W denotes the maximum network bandwidth.

7.4.3 Long-term Utility Function Design

We next discuss the design of the objective function of service multiplexing. To achieve the maximum utility of service multiplexing, utilities of eMBB and URLLC service provisions should be maximized simultaneously. In this chapter, we leverage a key performance indicator, i.e. energy efficiency, which is popularly exploited in resource allocation problems, to model the utility.

On the one hand, as network states of any two adjacent slots can be seen as independent in the time-discrete RAN slicing system, we focus on the problem formulation in a time slot of a duration of T. On the other hand, during a time slot, channel coefficients followed by the beamforming may change over minislots; as a result, time-varying utility functions in terms of channel coefficients and beamforming should be designed. Besides, as the CoMP technique can significantly improve the SNR of UEs, we define the following two utility functions.

Definition 7.3 (eMBB long-term utility function): Over a time slot, the eMBB long-term utility is defined as the time average energy efficiency of serving all eMBB UEs, which is calculated as

$$\begin{aligned}\bar{U}^e &= \frac{1}{T}\sum_{t=1}^{T} U^e(t) = \frac{1}{T}\sum_{t=1}^{T}\sum_{s\in\mathcal{S}^e} u_s^e(\boldsymbol{v}_{j,s}(t)) \\ &= \frac{1}{T}\sum_{t=1}^{T}\sum_{s\in\mathcal{S}^e}\left(\sum_{i\in\mathcal{I}_s^e} SNR_{i,s}^e(t) - \eta\sum_{j\in\mathcal{J}} \boldsymbol{v}_{j,s}^{\mathrm{H}}(t)\boldsymbol{v}_{j,s}(t)\right),\end{aligned} \tag{7.12}$$

where η is an energy efficiency coefficient reflecting the trade-off between the total transmit power consumption of RRHs and the benefit quantified by the total SNR of eMBB UEs.

Definition 7.4 (URLLC long-term utility function): Over a time slot, the URLLC long-term utility is defined as the time average energy efficiency of serving all URLLC UEs, which can be calculated as

$$\begin{aligned}\bar{U}^u &= \frac{1}{T}\sum_{t=1}^{T} U^u(t) = \frac{1}{T}\sum_{t=1}^{T}\sum_{s\in\mathcal{S}^u} u_s^u(\boldsymbol{g}_{ij,s}(t)) \\ &= \frac{1}{T}\sum_{t=1}^{T}\sum_{s\in\mathcal{S}^u}\Big(\sum_{i\in\mathcal{I}_s^u} SNR_{i,s}^u(t) - \eta\sum_{j\in\mathcal{J}}\sum_{i\in\mathcal{I}_s^u} \boldsymbol{g}_{ij,s}^{\mathrm{H}}(t)\boldsymbol{g}_{ij,s}(t)\Big).\end{aligned} \tag{7.13}$$

With the above description, we can formulate the problem of RAN slicing for bursty URLLC and eMBB service multiplexing as follows

$$\underset{\{\omega_s^e(\bar{t}),\boldsymbol{v}_s(t),\boldsymbol{g}_{i,s}(t)\}}{\text{maximize}} \quad \bar{U} = \bar{U}^e + \hat{\rho}\bar{U}^u \tag{7.14a}$$

$$\text{s.t:} \quad \text{constraints (7.3),(7.10),(7.11) are satisfied,} \tag{7.14b}$$

$$\omega_s^e(\bar{t}) \geq 0, \forall s \in \mathcal{S}^e. \tag{7.14c}$$

where $\hat{\rho}$ is a slice priority coefficient representing the priority of serving inter-slices, $\bar{U}$ denotes the long-term total slice utility, beamformers $\boldsymbol{v}_s(t) = [\boldsymbol{v}_{1,s}(t); \dots ; \boldsymbol{v}_{J,s}(t)] \in \mathbb{C}^{JK\times 1}$, and $\boldsymbol{g}_{i,s}(t) = [\boldsymbol{g}_{i1,s}(t); \dots ; \boldsymbol{g}_{iJ,s}(t)] \in \mathbb{C}^{JK\times 1}$.

The mitigation of (7.14) is highly challenging mainly because: (1) future channel information is needed: the optimization should be conducted at the beginning of the time slot; yet the objective function needs to be exactly computed according to channel information during the time slot; (2) timescale issue: the bandwidth $\{\omega_s^e(\bar{t})\}$ and the beamformers $\{\boldsymbol{v}_s(t)\}$ and $\{\boldsymbol{g}_{i,s}(t)\}$ should be optimized at two different time scales. $\{\omega_s^e(\bar{t})\}$ needs be optimized at the beginning of the time slot. $\{\boldsymbol{v}_s(t)\}$ and $\{\boldsymbol{g}_{i,s}(t)\}$ should be optimized at the beginning of each minislot.

In the following sections, we discuss *how to address the challenging problem effectively.*

7.5 Problem Solution with System Generated Channel Coefficients

In this section, we resort to a Sample Average Approximate (SAA) technique [16] and a distributed optimization method to tackle the above issues.

7.5.1 Sample Average Approximation

Owing to the ergodicity of the channel fading process over the time slot, the objective function can be approximated as

$$\frac{1}{T}\sum_{t=1}^{T} U^e(t) + \frac{1}{T}\sum_{t=1}^{T} \rho U^u(t) \approx E_{\hat{\boldsymbol{h}}}\left[\hat{U}^e + \rho\hat{U}^u\right], \tag{7.15}$$

where $\hat{\boldsymbol{h}}$ denotes a set of all channel coefficient samples collected at the beginning of the time slot, $\hat{U}^e$ and $\hat{U}^u$ are functions of $\hat{\boldsymbol{h}}$.

For SAA, its fundamental idea is to approximate the expectation of a random variable by its sample average. The following Corollary shows that if the number of samples M is reasonably large, then for all $m \in \mathcal{M} = \{1, \dots, M\}$, $\{\bar{U}_m\}$ converges to $\bar{U}$ uniformly on the feasible region constructed by constraints (7.14b) and (7.14c).

Corollary 7.1: Let Θ be a non-empty compact set formed by constraints (7.14b) and (7.14c), $Y(\boldsymbol{x}, \hat{\boldsymbol{h}}) = \hat{U}^e + \rho\hat{U}^u$ and $\boldsymbol{x} = \{\omega_s^e(\bar{t}), \boldsymbol{v}_s(t), \boldsymbol{g}_{i,s}(t)\}$. For any fixed $\boldsymbol{x} \in \Theta$, suppose that there exists $\varepsilon > 0$ such that the family of random variables $\{Y(\boldsymbol{y}, \hat{\boldsymbol{h}}) : \boldsymbol{y} \in B(\boldsymbol{x}, \varepsilon)\}$ is uniformly integrable, where $B(\boldsymbol{x}, \varepsilon) = \{\boldsymbol{y} : ||\boldsymbol{y} - \boldsymbol{x}||_2 \le \varepsilon\}$ denotes the closed ball of radius ε around $\boldsymbol{x}$. Then $\{\bar{U}_m\}$ converges to $\bar{U}$ uniformly on Θ almost surely as $M \to \infty$ [16].

Based on the conclusion in Corollary 1, given a set of samples of channel coefficients $\{\boldsymbol{h}_m\}$ with $\boldsymbol{h}_m = [\boldsymbol{h}_{11,1m}; \dots; \boldsymbol{h}_{1J,sm}; \dots; \boldsymbol{h}_{(N^e+N^u)J,(|\mathcal{S}^e|+|\mathcal{S}^u|)m}]$ that are assumed to be independent and identically distributed (i.i.d), the original problem (7.14) can be approximated as a single timescale one

$$\underset{\{\omega_{sm}^e, \omega_s^e, \boldsymbol{v}_{sm}, \boldsymbol{g}_{i,sm}\}}{\text{maximize}} \quad \{\bar{U}_m\} = \frac{1}{M}\sum\nolimits_{m=1}^{M} U_m^e + \frac{\rho}{M}\sum\nolimits_{m=1}^{M} U_m^u \tag{7.16a}$$

$$\text{s.t:} \quad \omega_{sm}^e = \omega_s^e, \forall s \in \mathcal{S}^e, \forall m \in \mathcal{M}, \tag{7.16b}$$

$$\sum\nolimits_{s\in\mathcal{S}^e} \boldsymbol{v}_{j,sm}^{\mathrm{H}} \boldsymbol{v}_{j,sm} + \sum\nolimits_{s\in\mathcal{S}^u}\sum\nolimits_{i\in\mathcal{I}_s^u} \boldsymbol{g}_{ij,sm}^{\mathrm{H}} \boldsymbol{g}_{ij,sm}$$

$$\le E_j, j \in \mathcal{J}, m \in \mathcal{M}, \tag{7.16c}$$

$$\sum\nolimits_{s\in\mathcal{S}^e} \omega_{sm}^e + W^u(\boldsymbol{r}_m) \le W, m \in \mathcal{M}, \tag{7.16d}$$

$$\gamma_{i,sm}^e \ge C_s^{th}, \forall i \in \mathcal{I}_s^e, s \in \mathcal{S}^e, m \in \mathcal{M}, \tag{7.16e}$$

$$\omega_{sm}^{e} \geq 0, s \in \mathcal{S}^e, m \in \mathcal{M}, \tag{7.16f}$$

where $\{\cdot\}_m$ denotes a variable corresponding to the mth coefficient sample $\boldsymbol{h}_m$. The constraint (7.16b) is imposed to explicitly describe the two timescale issue of the original problem (7.14). Note that, the optimization of (7.16) is not conducted at each minislot t; thus, the minislot index t is not involved in (7.16).

We consider $\{\omega_{sm}^e\}$ as a family of local variables and $\{\omega_s^e\}$ as a family of global variables in (7.16). In this case, (7.16) can be effectively mitigated by distributed optimization routines, such as ADMM [9]; that is, distributedly optimizing the local variables and then forcing them to the global variables at convergence.

7.5.2 Distributed Optimization

According to the fundamental principle of ADMM, the ADMM for (7.16) can be derived from the following augmented partial Lagrange problem

$$\underset{\substack{\{\omega_{sm}^e, \omega_s^e, \\ \boldsymbol{v}_{sm}, \boldsymbol{g}_{i,sm}\}}}{\text{minimize}} \sum_{m=1}^{M} \left\{ -\frac{U_m^e}{M} - \frac{\hat{\rho} U_m^u}{M} + \sum_{s \in \mathcal{S}^e} \left[\psi_{sm} \left(\omega_{sm}^e - \omega_s^e \right) + \frac{\mu \| \omega_{sm}^e - \omega_s^e \|_2^2}{2} \right] \right\} \tag{7.17a}$$

$$\text{s.t:} \quad \text{constraints (7.16c)} - (7.16f)\text{are satisfied.} \tag{7.17b}$$

where, ψ_{sm} is a Lagrangian multiplier, μ is a penalty coefficient.

For all channel samples, the distributed framework of mitigating (7.17) can then be summarized to alternatively calculate equations from (7.18) to (7.20).

$$\left\{ \begin{array}{l} \omega_{sm}^{e(k+1)}, \\ \boldsymbol{v}_{sm}^{(k+1)}, \boldsymbol{g}_{i,sm}^{(k+1)} \end{array} \right\} = \underset{\left\{ \begin{array}{l} \omega_{sm}^e, \\ \boldsymbol{v}_{sm}, \boldsymbol{g}_{i,sm} \end{array} \right\}}{\text{argmin}} \bar{\mathcal{L}}(\omega_{sm}^e, \boldsymbol{v}_{sm}, \boldsymbol{g}_{i,sm}) \tag{7.18a}$$

$$\text{s.t:} \quad \text{for a sample } m, \text{ (7.16c)} - (7.16f)\text{are satisfied.} \tag{7.18b}$$

$$\omega_s^{e(k+1)} = \frac{1}{M} \sum_{m=1}^{M} \left(\omega_{sm}^{e(k+1)} + \frac{1}{\mu} \psi_{sm}^{(k)} \right), \ \forall s \in \mathcal{S}^e, \tag{7.19}$$

$$\psi_{sm}^{(k+1)} = \psi_{sm}^{(k)} + \mu \left(\omega_{sm}^{e(k+1)} - \omega_s^{e(k+1)} \right), \ \forall s \in \mathcal{S}^e, \tag{7.20}$$

where,

$$\bar{\mathcal{L}}(\omega_{sm}^e, \boldsymbol{v}_{sm}, \boldsymbol{g}_{i,sm}) = -\frac{U_m^e}{M} - \frac{\hat{\rho} U_m^u}{M} + \sum_{s \in \mathcal{S}^e} \left[\psi_{sm}^{(k)} \left(\omega_{sm}^e - \omega_s^{e(k)} \right) + \frac{\mu}{2} \left\| \omega_{sm}^e - \omega_s^{e(k)} \right\|_2^2 \right]. \tag{7.21}$$

In our RAN slicing system, the SDRAN-C is responsible for executing the distributed framework, and M Virtual Machines (VM) are activated to conduct (7.18) and (7.20). An Aggregation VM (AVM) is utilized to aggregate the local variables. Additionally, in this framework, local dual variables $\{\psi_{sm}\}$ are updated to drive local variables $\{\omega_{sm}^{e}\}$ into consensus, and quadratic items in (7.18) help pull $\{\omega_{sm}^{e}\}$ towards their average value.

Unfortunately, the mitigation of (7.18) on each VM is difficult due to the existence of non-convex constraints. We next attempt to tackle the non-convexity of (7.18).

7.5.3 Semidefinite Relaxation Scheme

Let $\boldsymbol{V}_{sm} = \boldsymbol{v}_{sm}\boldsymbol{v}_{sm}^{\mathrm{H}} \in \mathbb{R}^{JK\times JK}$ for all $s \in \mathcal{S}^{e}$, $m \in \mathcal{M}$, and $\boldsymbol{G}_{i,sm} = \boldsymbol{g}_{i,sm}\boldsymbol{g}_{i,sm}^{\mathrm{H}} \in \mathbb{R}^{JK\times JK}$ for all $i \in \mathcal{I}_{s}^{u}$, $s \in \mathcal{S}^{u}$, $m \in \mathcal{M}$. Next, if we recall the properties: $\boldsymbol{V}_{s} = \boldsymbol{v}_{s}\boldsymbol{v}_{s}^{H} \Leftrightarrow \boldsymbol{V}_{s} \succeq 0$, $\mathrm{rank}(\boldsymbol{V}_{s}) \le 1$, and $\boldsymbol{G}_{i,sm} = \boldsymbol{g}_{i,sm}\boldsymbol{g}_{i,sm}^{H} \Leftrightarrow \boldsymbol{G}_{i,sm} \succeq 0$, $\mathrm{rank}(\boldsymbol{G}_{i,sm}) \le 1$, (7.18) can then be reformulated as

$$\left\{ \begin{array}{l} \omega_{sm}^{e(k+1)}, \\ \boldsymbol{V}_{sm}^{(k+1)}, \boldsymbol{G}_{i,sm}^{(k+1)} \end{array} \right\} = \underset{\left\{ \substack{\omega_{sm}^{e}, \\ \boldsymbol{V}_{sm}, \boldsymbol{G}_{i,sm}} \right\}}{\mathrm{argmin}} \ \bar{\mathcal{L}}(\omega_{sm}^{e}, \boldsymbol{V}_{sm}, \boldsymbol{G}_{i,sm}) \tag{7.22a}$$

$$\text{s.t:} \quad \omega_{sm}^{e}\log_2\left(1 + \frac{\mathrm{tr}(\boldsymbol{H}_{i,sm}\boldsymbol{V}_{sm})}{\sigma_{i,s}^{2}}\right) \ge C_{s}^{th}, \forall s \in \mathcal{S}^{e}, i \in \mathcal{I}_{s}^{e}, \tag{7.22b}$$

$$\sum_{s\in\mathcal{S}^{e}} \mathrm{tr}(\boldsymbol{Z}_{j}\boldsymbol{V}_{sm}) + \sum_{s\in\mathcal{S}^{u}}\sum_{i\in\mathcal{I}_{s}^{u}} \mathrm{tr}(\boldsymbol{Z}_{j}\boldsymbol{G}_{i,sm}) \le E_{j}, \forall j \in \mathcal{J}, \tag{7.22c}$$

$$\boldsymbol{V}_{sm} \succeq 0, \forall s \in \mathcal{S}^{e}, \tag{7.22d}$$

$$\boldsymbol{G}_{i,sm} \succeq 0, \forall i \in \mathcal{I}_{s}^{u}, s \in \mathcal{S}^{u}, \tag{7.22e}$$

$$\mathrm{rank}(\boldsymbol{V}_{sm}) \le 1, \forall s \in \mathcal{S}^{e}, \tag{7.22f}$$

$$\mathrm{rank}(\boldsymbol{G}_{i,sm}) \le 1, \forall i \in \mathcal{I}_{s}^{u}, s \in \mathcal{S}^{u}, \tag{7.22g}$$

$$\text{constraints (7.16d), (7.16f) are satisfied,} \tag{7.22h}$$

where $\boldsymbol{H}_{i,sm} = \boldsymbol{h}_{i,sm}\boldsymbol{h}_{i,sm}^{\mathrm{H}} \in \mathbb{R}^{JK\times JK}$, $\boldsymbol{h}_{i,sm} = [\boldsymbol{h}_{i1,sm}; \dots; \boldsymbol{h}_{iJ,sm}] \in \mathbb{C}^{JK\times 1}$, $\boldsymbol{Z}_{j} \in \mathbb{R}^{JK\times JK}$ is a square matrix with $J \times J$ blocks, and each block in $\boldsymbol{Z}_{j}$ is a $K \times K$ matrix. Besides, in $\boldsymbol{Z}_{j}$, the block in the jth row and jth column is a $K \times K$ identity matrix, and all other blocks are zero matrices.

As power matrices $\boldsymbol{V}_{sm}$ ($s \in \mathcal{S}^{e}$, $m \in \mathcal{M}$) and $\boldsymbol{G}_{i,sm}$ ($i \in \mathcal{I}_{s}^{u}$, $s \in \mathcal{S}^{u}$, $m \in \mathcal{M}$) are positive semidefinite, we then resort to the Semidefinite Relaxation (SDR) scheme to handle the low-rank non-convex constraints (7.22f) and (7.22g). That is, directly drop the constraints (7.22f) and (7.22g). However, owing to the relaxation, power matrices $\boldsymbol{V}_{sm}$ and $\boldsymbol{G}_{i,sm}$ obtained by mitigating the problem (7.22) without low-rank constraints will not satisfy the low-rank constraint in general. This is due to the fact that the (convex) feasible set of the relaxed (7.22) is a superset of the (non-convex) feasible set of (7.22). If they satisfy, then the relaxation is tight; if not, then some manipulation, e.g. *a randomization/scale method* [20], should be performed on them to obtain their approximate solutions.

Although non-convex constraints are removed, constraints related to $W^u(\boldsymbol{r}_m)$ are complicated, which hinders the optimization of the relaxed (7.22). Therefore, we next discuss *how to equivalently transform the complicated constraints via a variable slack scheme.*

7.5.4 Variable Slack Scheme

From (7.5), we observe that $W^u(\boldsymbol{r}_m)$ is a quadratic function with respect to (w.r.t) $\boldsymbol{r}_m$. Therefore, via introducing a family of slack variables $\boldsymbol{f}_m = \{f^u_{i,sm}\}$, $i \in \mathcal{I}^u_s$, $s \in \mathcal{S}^u$, $m \in \mathcal{M}$, the following lemma shows the equivalent expressions of (7.16d).

Lemma 7.3: Given the family of slack variables $\boldsymbol{f}_m = \{f^u_{i,sm}\}$, (7.16d) is equivalent to the following inequalities [30],

$$\sum_{s \in \mathcal{S}^e} \omega^e_{sm} + A(\boldsymbol{f}_m) + c(\varsigma, \alpha)\sqrt{B(\boldsymbol{f}_m)} \leq W, \tag{7.23}$$

and

$$f^u_{i,sm} \geq r^u_{i,sm}, \tag{7.24}$$

for all $i \in \mathcal{I}^u_s$, $s \in \mathcal{S}^u$, $m \in \mathcal{M}$.

Besides, we can know that the objective function (7.22a) is convex. This is because it is linear w.r.t variables $\boldsymbol{V}_{sm}$ and $\boldsymbol{G}_{i,sm}$ with an addition of affine terms and non-negative quadratic terms w.r.t ω^e_{sm}. (7.22c) is an affine constraint. Other constraints are non-linear. Based on the above equivalent transformation, we show that (7.22) can be further transformed into a standard convex problem in the following lemma.

Lemma 7.4: By introducing a family of slack variables, the problem (7.22) without low-rank constraints can be equivalently transformed into the following Semidefinite Programming (SDP) problem [30].

$$\left\{\begin{matrix} \omega_{sm}^{e(k+1)}, \boldsymbol{V}_{sm}^{(k+1)} \\ \boldsymbol{G}_{i,sm}^{(k+1)}, \dots, \tau_{i,sm}^{u(k+1)} \end{matrix}\right\} = \underset{\left\{\begin{matrix} \omega_{sm}^{e}, \boldsymbol{V}_{sm}, \\ \boldsymbol{G}_{i,sm}, \dots, \tau_{i,sm}^{u} \end{matrix}\right\}}{\text{argmin}} \quad \bar{\mathcal{L}}(\omega_{sm}^{e}, \boldsymbol{V}_{sm}, \boldsymbol{G}_{i,sm}) \tag{7.25a}$$

$$\text{s.t:} \quad \text{affine constraints (7.22c),(7.22e),(7.22f)}, \tag{7.25b}$$

$$\text{other conic constraints of (25) in [Yang et al., 2019].} \tag{7.25c}$$

Then, some standard optimization tools such as CVX [13] and MOSEK [6] can be used to mitigate (7.25) effectively. We can summarize the steps of mitigating (7.16) in Algorithm 1.

Algorithm 1 Distributed bandwidth optimization algorithm, DBO

Input: randomly initialize $\omega_s^{e(0)}$, $\psi_{sm}^{(0)}$, $\boldsymbol{V}_{sm}^{(0)}$, for all $i \in \mathcal{I}_s^e$, $s \in \mathcal{S}^e$ and $\boldsymbol{G}_{i,sm}^{(0)}$, for all $i \in \mathcal{I}_s^u$, $s \in \mathcal{S}^u$, $\boldsymbol{H}_{i,sm}$, for all $m \in \mathcal{M}$, let $k_{\max} = 250$.
Output: $\{\omega_s^e\}$
for $k = 1 : k_{\max}$ **do**
  for each VM $m \in \mathcal{M}$ in parallel **do**
    VM m solves the problem (7.25) to obtain $\omega_{sm}^{e(k+1)}$ and sends it to the AVM.
  end for
  After collecting all $\{\omega_{sm}^{e(k+1)}\}$, the AVM aggregates $\omega_s^{e(k+1)}$ using (7.19) and broadcasts the updated $\omega_s^{e(k+1)}$ to each VM.
  for each VM $m \in \mathcal{M}$ in parallel **do**
    VM m computes $\psi_{sm}^{(k+1)}$ using (7.20) and sends $\psi_{sm}^{(k+1)}$ to the AVM.
  end for
  if convergence or reach the maximum iteration times $k_{\max}$ **then**
    Break.
  end if
end for

7.5.5 Performance Analysis

In this subsection, we analyze the performance of DBO. We first present a lemma about the optimality of solving (7.22) and then state the computational complexity and the convergence of DBO.

If we denote $\boldsymbol{G}_{i,sm}^{\star}$ and $\boldsymbol{V}_{sm}^{\star}$ as solutions to (7.25), then the following lemma shows the tightness of exploring the SDR scheme on (7.22).

Lemma 7.5: For all $i \in \mathcal{I}_s^u$, $s \in \mathcal{S}^u$, $m \in \mathcal{M}$, the SDR for both $\boldsymbol{V}_{sm}$ and $\boldsymbol{G}_{i,sm}$ in problem (7.22) is tight, that is [30],

$$\begin{aligned} &\text{rank}(\boldsymbol{V}^{\star}_{sm}) \leq 1, \forall i \in \mathcal{I}^e_s, s \in \mathcal{S}^e, \\ &\text{rank}(\boldsymbol{G}^{\star}_{i,sm}) \leq 1, \forall i \in \mathcal{I}^u_s, s \in \mathcal{S}^u. \end{aligned} \tag{7.26}$$

Moreover, $\boldsymbol{G}^{\star}_{i,sm}$ and $\boldsymbol{V}^{\star}_{sm}$ are optimal solutions to (7.22).

The computational complexity of DBO is dominated by that of solving the SDP problem. The SDP problem has $(|\mathcal{S}^e| + I^u)$ matrices of size $JK \times JK$ and $(3|\mathcal{S}^e| + 11I^u)$ one-dimensional variables. An interior-point method can then be exploited to efficiently mitigate the SDP problem at the worst-case computational complexity of $O((|\mathcal{S}^e| + I^u)J^2K^2 + 3|\mathcal{S}^e| + 11I^u)^{3.5}$ [32]. Nevertheless, the actual complexity will usually be much smaller than the worst case.

The following lemma presents the convergence of the algorithm.

Lemma 7.6: Let $(\omega^{e\star}_{sm}, \boldsymbol{v}^{\star}_{sm}, \boldsymbol{g}^{\star}_{i,sm})$ denote the optimal solutions, under the ADMM-based distributed algorithm, $\forall k \in \mathbb{Z}^+$, $m \in \mathcal{M}$, we have that $\bar{\mathcal{L}}(\omega^{e(k)}_{sm}, \boldsymbol{v}^{(k)}_{sm}, \boldsymbol{g}^{(k)}_{i,sm})$ is bounded and [30]

$$\bar{\mathcal{L}}(\omega^{e\star}_{sm}, \boldsymbol{v}^{\star}_{sm}, \boldsymbol{g}^{\star}_{i,sm}) = \lim_{k \to \infty} \bar{\mathcal{L}}(\omega^{e(k)}_{sm}, \boldsymbol{v}^{(k)}_{sm}, \boldsymbol{g}^{(k)}_{i,sm}). \tag{7.27}$$

Besides, to show the convergence performance of the DBO algorithm intuitively, we plot its convergence curve in Figure 7.2. In this figure, we denote the loss of accuracy of the global consensus variable by $\Delta_\omega = \sum_{s \in \mathcal{S}^e} |\omega^{e(k+1)}_s - \omega^{e(k)}_s|$. According to the principle of ADMM, if the loss value Δ_ω approaches zero after a limited number of iterations, then the DBO algorithm is convergent; otherwise, the algorithm is divergent. From Figure 7.2, we can observe that the DBO algorithm can converge after several iterations.

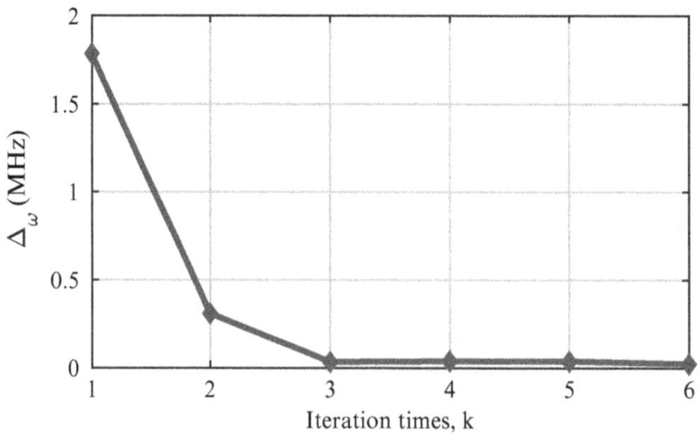

Figure 7.2 A convergence curve of the DBO algorithm. ©IEEE 2021. Reprinted with permission from [29]

7.6 Optimization of Beamforming with Imperfect Channel Gain

With the system generated channel coefficient samples, the former section obtains the approximate solution $\{\omega_s^e\}$ to (7.14). In this section, we continue to optimize minislot variables $\{\boldsymbol{V}_s(t), \boldsymbol{G}_{i,s}(t)\}$ according to sensed imperfect channel gains $\{\boldsymbol{H}_{i,s}(t)\}$, $i \in \mathcal{I}^e \cup \mathcal{I}^u$, $s \in \mathcal{S}^e \cup \mathcal{S}^u$, at the beginning of each minislot t.

Given $\{\omega_s^e\}$ and system sensed imperfect channel gains $\{\boldsymbol{H}_{i,s}(t)\}$, as the maximization of $U^e(t) + \hat{\rho}U^u(t)$ at each minislot will lead to the maximization of the time average utility over the whole time slot, the original problem (7.14) can be reduced to the following beamforming optimization problem at each minislot t

$$\underset{\{\boldsymbol{V}_s(t),\boldsymbol{G}_{i,s}(t)\}}{\text{maximize}} \quad U^e(t) + \hat{\rho}U^u(t) \tag{7.28a}$$

$$\text{s.t:} \quad \omega_s^e \log_2(1 + \frac{\text{tr}(\boldsymbol{H}_{i,s}(t)\boldsymbol{V}_s(t))}{\sigma_{i,s}^2}) \geq C_s^{th}, \forall s \in \mathcal{S}^e, i \in \mathcal{I}_s^e, \tag{7.28b}$$

$$\sum_{s\in\mathcal{S}^e} \text{tr}(\boldsymbol{Z}_j\boldsymbol{V}_s(t)) + \sum_{s\in\mathcal{S}^u}\sum_{i\in\mathcal{I}_s^u} \text{tr}(\boldsymbol{Z}_j\boldsymbol{G}_{i,s}(t)) \leq E_j, \forall j \in \mathcal{J}, \tag{7.28c}$$

$$\boldsymbol{V}_s(t) \succeq 0, \forall s \in \mathcal{S}^e, \tag{7.28d}$$

$$\boldsymbol{G}_{i,s}(t) \succeq 0, \forall i \in \mathcal{I}_s^u, s \in \mathcal{S}^u, \tag{7.28e}$$

$$\text{rank}(\boldsymbol{V}_s(t)) \leq 1, \forall s \in \mathcal{S}^e, \tag{7.28f}$$

$$\text{rank}(\boldsymbol{G}_{i,s}(t)) \leq 1, \forall i \in \mathcal{I}_s^u, s \in \mathcal{S}^u, \tag{7.28g}$$

$$\text{constraint (7.11) is satisfied.} \tag{7.28h}$$

By leveraging the presented SDR scheme and variable slack scheme in the former section, (7.28) can be equivalently transformed into a standard SDP problem that is able to be effectively mitigated by CVX or MOSEK.

Recall that the SDR for both $\boldsymbol{V}_s(t)$ and $\boldsymbol{G}_{i,s}(t)$ is tight, we therefore can perform the eigenvalue decomposition on $\boldsymbol{V}_s(t)$ and $\boldsymbol{G}_{i,s}(t)$ to obtain the optimal beamforming vectors $\boldsymbol{v}_s(t)$ and $\boldsymbol{g}_{i,s}(t)$, respectively.

Then, the bandwidth and beamforming optimization algorithm designed for the RAN slicing system can be summarized as follows.

7.7 Simulation Results

We implement the following algorithms on Python and compare them to evaluate the performance of bandwidth allocation and beamforming algorithms in the RAN slicing system: (i) the proposed B^{2}O-ADMM algorithm; (ii) the IRHS algorithm in [28], which enforces all slice requests and optimizes the same

Algorithm 2 Bandwidth and beamforming optimization algorithm based on ADMM, B^2O-ADMM

1: **Input:** $\{\boldsymbol{H}_{i,s}(t)\}$, for all $i \in \mathcal{I}^e \cup \mathcal{I}^u$, $s \in \mathcal{S}^e \cup \mathcal{S}^u$
2: **Output:** $\{\omega_s^e\}$, $\{\boldsymbol{v}_s(t)\}$, and $\{\boldsymbol{g}_{i,s}(t)\}$
3: Call Algorithm 1 to generate $\{\omega_s^e\}$, for all $s \in \mathcal{S}^e$.
4: **for** $t = 1 : T$ **do**
5: Given $\{\omega_s^e\}$, the SDRAN-C mitigates (7.28) to obtain beamformers $\{\boldsymbol{v}_s(t)\}$ for all $s \in \mathcal{S}^e$ and $\{\boldsymbol{g}_{i,s}(t)\}$ for all $i \in \mathcal{I}_s^u$, $s \in \mathcal{S}^u$.
6: **end for**

objective function as B^2O-ADMM; (iii) the proposed resource allocation algorithm without ADMM, NoADMM. Specifically, NoADMM algorithm generates the bandwidth allocated to eMBB slices based on imperfect channel gains sensed at the beginning of the first minislot.

In the simulation, we consider the CoMP-enabled RAN slicing system with three RRHs located on a circle with a radius of 0.5 km. The distance between every two RRHs is equal. eMBB and URLLC UEs are randomly and uniformly distributed in the circle. The transmit antenna gain at each RRH is set to be 5 dB, and a log-normal shadowing path loss model is utilized to simulate the path loss between a RRH and a UE. Particular, a downlink path loss is calculated by $H(\text{dB}) = 128.1 + 37.6 \log_{10} d$, where d (in km) represents the distance between a UE and a RRH. The log-normal shadowing standard deviation is set to be 10 dB. Besides, we let the maximum transmit power $E_1 = E_2 = E_3 = 1$ W, $\sigma_{i,s}^2 = -110$ dBm for all $i \in \mathcal{I}^e \cup \mathcal{I}^u$, $s \in \mathcal{S}^e \cup \mathcal{S}^u$, $L_{i,s}^u = 160$ bits, $\lambda_s = \lambda = 0.1$ packet per unit time for all $i \in \mathcal{I}_s^u$, $s \in \mathcal{S}^u$, $K = 2$, $\eta = 1000$, $\hat{\rho} = 500$, $M = 100$, $T = 60$ minislots, $\kappa = 5.12 \times 10^{-4}$ channel uses per unit time per unit bandwidth, $W = 4$ MHz, $\phi = 1.5$. Other slice configuration parameters are listed below: $\varsigma = 2 \times 10^{-5}$, $\alpha = 10^{-5}$, $\beta = 2 \times 10^{-8}$, $|\mathcal{S}^e| = 3$, $|\mathcal{S}^u| = 2$, $\{I_s^e\} = \{4, 6, 8\}$ UEs, $C_s^{th} = \{6, 4, 2\}$ Mb/s, $\{I_s^u\} = \{3, 5\}$ UEs, and $\{D_s\} = \{1, 2\}$ milliseconds [28].

The following performance indicators are adopted to evaluate the comparison algorithms: (i) system bandwidth W_u (in MHz) allocated to URLLC slices; (ii) total transmit power $E^u = \sum_{t=1}^{T} \sum_{s \in \mathcal{S}^u} \sum_{i \in \mathcal{I}_s^u} \text{tr}(\boldsymbol{G}_{i,s}(t))$ (in W) configured for URLLC slices; (iii) long-term total slice utility $\bar{U}$ that is the objective function of (7.14).

To understand the impact of packet arrival rate on the performance of comparison algorithms, we plot the relationship between the bandwidth allocated to URLLC slices and the packet arrival rate λ with $\lambda = \{0.1, 0.2, \dots, 1.0, 1.1\}$ packets per unit time in Figure 7.3. Besides, the trends of the transmit power configured for URLLC slices and the achieved total slice utility over both λ and energy efficiency coefficient $\hat{\rho}$ are illustrated in Figure 7.4.

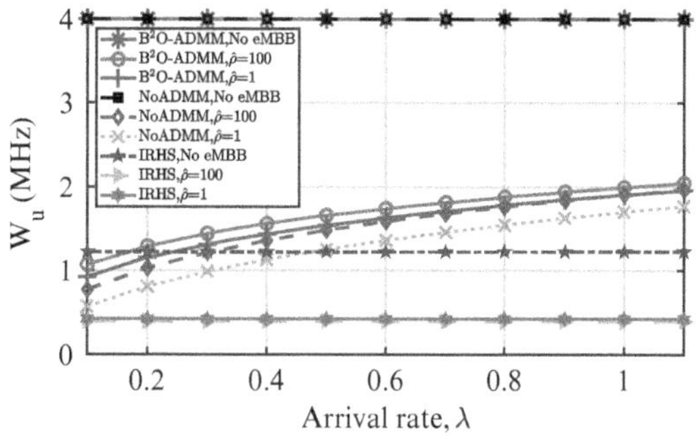

Figure 7.3 Trend of the system bandwidth allocated to URLLC slices under different arrival rates, λ. ©IEEE 2021. Reprinted with permission from [29].

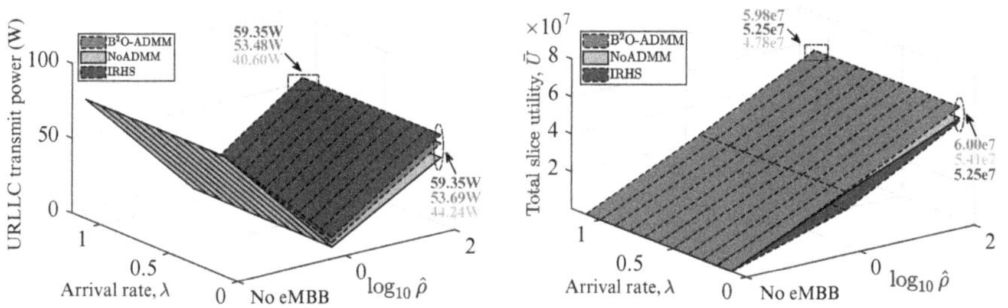

(a) Total transmit power configured for URLLC slices

(b) Achieved long-term total slice utility

Figure 7.4 Trends of transmit power configured for URLLC slices and achieved long-term total slice utility versus $\hat{\rho}$ and λ. ©IEEE 2021. Reprinted with permission from [29].

From these two figures, we have the following observations: (i) For B^2O-ADMM and NoADMM algorithms, the system bandwidth allocated to URLLC slices monotonously increases with an increasing arrival rate when the RAN slicing system provides URLLC and eMBB multiplexing services. When the system terminates eMBB services, B^2O-ADMM and NoADMM algorithms recommend to allocate the total network bandwidth to URLLC slices to guarantee more reliable URLLC transmission. (ii) For the IRHS algorithm, as it does not design some strategies to reduce the packet blocking probability of URLLC packets, it suggests keeping the amount of system bandwidth allocated to URLLC slices at a low constant value. Meanwhile, the IRHS algorithm will not allocate the total network bandwidth to URLLC slices even though in the case of end of eMBB services. (iii) Compared with B^2O-ADMM and NoADMM, although IRHS needs less network bandwidth to ensure an ultra-low URLLC decoding error probability, it desires greater transmit power E^u to satisfy QoS requirements of URLLC UEs and then consumes more energy. (iv) The B^2O-ADMM algorithm obtains the greatest long-term total slice utility. Moreover, it can be utilized to configure a slicing system, which can support URLLC transmission with higher arrival rates without significantly decreasing the achieved total slice utility $\bar{U}$.

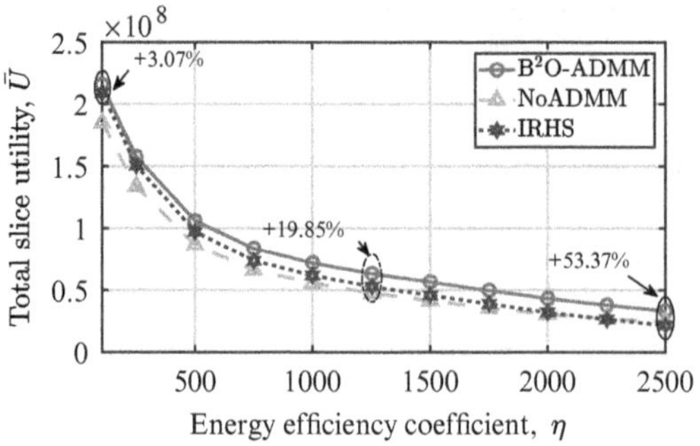

Figure 7.5 Trend of the long-term total slice utility versus η.

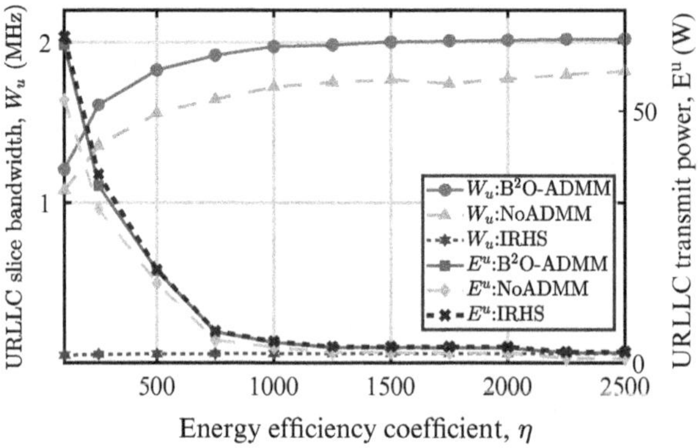

Figure 7.6 Trends of URLLC slice bandwidth W_u and transmit power E^u configured for URLLC slices versus η. ©IEEE 2021. Reprinted with permission from [29].

Besides, to understand the effect of energy efficiency coefficient η, we plot the relationship between $\bar{U}$ and η in Figure 7.5 and plot trends of W^u and E^u under different η in Figure 7.6 with $\lambda = 0.1$ and $\hat{\rho} = 500$.

We can observe the following conclusions from Figures 7.5 and 7.6: (i) The proposed B^{2}O-ADMM algorithm achieves the greatest $\bar{U}$, and the obtained total slice utilities of all comparison algorithms monotonously decrease as η increases. A great η indicates that the power consumption dominates the total slice utility; thus, all comparison algorithms reduce the power consumption and the corresponding SNR received at each end UE. (ii) For B^{2}O-ADMM and NoADMM, their obtained W_u increase with an increasing η. A large η leads to the system configuration of small transmit power; thus, the system bandwidth allocated to URLLC slices should be widened to satisfy QoS requirements of URLLC UEs. As QoS requirements of URLLC UEs are easy to be satisfied for the IRHS algorithm, a decreasing E^u does not result in a significantly increasing W_u. (iii) Compared with NoADMM, B^{2}O-ADMM suggests apportioning greater network bandwidth to URLLC slices as B^{2}O-ADMM achieves the global consensus bandwidth to ensure an ultra-low packet blocking probability over the whole time slot.

7.8 Conclusion

In this chapter, we considered a CoMP-enabled RAN slicing system simultaneously supporting URLLC and eMBB traffic transmission. In the presence of eMBB traffic, we orchestrated the shared network resources of the system to guarantee a more reliable bursty URLLC service provision from the perspectives of lowering both URLLC packet blocking probability and codeword decoding error probability. We formulated the problem of RAN slicing for bursty URLLC and eMBB service multiplexing as a resource optimization problem and developed a joint bandwidth and CoMP beamforming optimization algorithm to maximize the long-term total slice utility.

Bibliography

[1] 3GPP. Study on new radio access technology: Radio access architecture and interfaces. Technical Report 38.801, The 3rd Generation Partnership Project, Apri. 2017. https://portal.3gpp.org/desktopmodules/Specifications/SpecificationDetails. aspx?specificationId=3056.

[2] Haider D Resin Albonda and Jordi Pérez-Romero. An efficient RAN slicing strategy for a heterogeneous network with eMBB and V2X services. *IEEE Access*, 7:44771–44782, 2019.

[3] NGMN Alliance. Description of network slicing concept. *NGMN 5G P*, 1, 2016.

[4] Madyan Alsenwi, Nguyen H Tran, Mehdi Bennis, Anupam Kumar Bairagi, and Choong Seon Hong. eMBB-URLLC resource slicing: A risk-sensitive approach. *IEEE Communications Letters*, 23(4):740–743, 2019.

[5] Arjun Anand and Gustavo de Veciana. Resource allocation and HARQ optimization for URLLC traffic in 5G wireless networks. *IEEE JSAC*, 36(11): 2411–2421, 2018.

[6] MOSEK ApS. MOSEK optimization toolbox for MATLAB 8.1.0.67. https://docs.mosek.com/8.1/toolbox/index.html, 2018.

[7] Amin Azari, Mustafa Ozger, and Cicek Cavdar. Risk-aware resource allocation for URLLC: Challenges and strategies with machine learning. *IEEE Communications Magazine*, 57(3):42–48, 2019.

[8] Michela Becchi. From Poisson processes to self-similarity: a survey of network traffic models. *Washington University in St. Louis, Tech. Rep*, 2008. https://www.cse.wustl.edu/ jain/cse567-06/ftp/traffic_models1.pdf.

[9] Stephen Boyd, Neal Parikh, Eric Chu, Borja Peleato, Jonathan Eckstein, et al. Distributed optimization and statistical learning via the alternating direction method of multipliers. *Foundations and Trends® in Machine learning*, 3(1): 1–122, 2011.

[10] AB Ericsson. Common public radio interface (CPRI). Technical Report Interface Specification, V7.0, Huawei Technologies Co. Ltd, NEC Corporation, Alcatel Lucent, and Nokia Networks, 2015.

[11] Xenofon Foukas, Georgios Patounas, Ahmed Elmokashfi, and Mahesh K Marina. Network slicing in 5G: Survey and challenges. *IEEE Communications Magazine*, 55(5):94–100, 2017.

[12] Panagiotis Georgakopoulos, Tafseer Akhtar, Ilias Politis, Christos Tselios, Evangelos Markakis, and Stavros Kotsopoulos. Coordination multipoint enabled small cells for coalition-game-based radio resource management. *IEEE Network*, 33(4):63–69, 2019.

[13] M Grant and Stephen Boyd. CVX: MATLAB software for disciplined convex programming. http://cvxr.com/cvx, 2016.

[14] Mor Harchol-Balter. *Performance modeling and design of computer systems: queueing theory in action*. Cambridge University Press, 2013.

[15] Mostafa Khoshnevisan, Vinay Joseph, Piyush Gupta, Farhad Meshkati, Rajat Prakash, and Peerapol Tinnakornsrisuphap. 5G industrial networks with CoMP for URLLC and time sensitive network architecture. *IEEE JSAC*, 37(4):947–959, 2019.

[16] Sujin Kim, Raghu Pasupathy, and Shane G Henderson. A guide to sample average approximation. In *Handbook of simulation optimization*, pages 207–243. Springer, 2015.

[17] Matti Latva-aho and Kari Leppänen. White paper: Key drivers and research challenges for 6G ubiquitous wireless intelligence. (Vers. 1), 2019. http://urn.fi/urn:isbn:9789526223544.

[18] Junling Li, Weisen Shi, Peng Yang, Qiang Ye, Xuemin Sherman Shen, Xu Li, and Jaya Rao. A hierarchical soft RAN slicing framework for differentiated service provisioning. *IEEE Wireless Communications*, 2020. In press. DOI: 10.1109/MWC.001.2000010.

[19] Xin Liu, Shengqian Han, and Chenyang Yang. Energy-efficient training-assisted transmission strategies for closed-loop MISO systems. *IEEE TVT*, 64(7): 2846–2860, 2014.

[20] Wing-Kin Ken Ma. Semidefinite relaxation of quadratic optimization problems and applications. *IEEE Signal Processing Magazine*, 1053(5888/10), 2010.

[21] George R MacCartney and Theodore S Rappaport. Millimeter-wave base station diversity for 5G coordinated multipoint (CoMP) applications. *IEEE TWC*, 18(7): 3395–3410, 2019.

[22] Andrea Matera, Rahif Kassab, Osvaldo Simeone, and Umberto Spagnolini. Non-orthogonal eMBB-URLLC radio access for cloud radio access networks with analog fronthauling. *Entropy*, 20(9):661, 2018.

[23] Rui Ni, Xu Li, Jun Chen, Si Chen, Enbo Wang, Ming Zhu, Wei Zhang, and Yuhua Chen. An end-to-end demonstration for 5G network slicing. In *IEEE VTC*, pages 1–5, 2019.

[24] Yury Polyanskiy, H Vincent Poor, and Sergio Verdú. Channel coding rate in the finite blocklength regime. *IEEE TIT*, 56(5):2307, 2010.

[25] Peter Rost, Christian Mannweiler, Diomidis S Michalopoulos, Cinzia Sartori, Vincenzo Sciancalepore, Nishanth Sastry, Oliver Holland, Shreya Tayade, Bin Han, Dario Bega, et al. Network slicing to enable scalability and flexibility in 5G mobile networks. *IEEE Communications magazine*, 55(5):72–79, 2017.

[26] Sebastian Schiessl, James Gross, and Hussein Al-Zubaidy. Delay analysis for wireless fading channels with finite blocklength channel coding. In *Proceedings of the 18th ACM International Conference on Modeling, Analysis and Simulation of Wireless and Mobile Systems*, pages 13–22. ACM, 2015.

[27] M Series. IMT Vision–Framework and overall objectives of the future development of IMT for 2020 and beyond. Technical Report M.2083, Recommendation ITU, Sep. 2015. https://www.itu.int/md/R12-WP5D.AR-C-0486.

[28] Jianhua Tang, Byonghyo Shim, and Tony QS Quek. Service multiplexing and revenue maximization in sliced C-RAN incorporated with URLLC and multicast eMBB. *IEEE JSAC*, 37(4):881–895, 2019.

[29] Peng Yang, Xing Xi, Tony Q. S. Quek, Jingxuan Chen, Xianbin Cao, and Dapeng Oliver Wu. How should I orchestrate resources of my slices for bursty URLLC service provision? *IEEE TCOM*, 69(2):1134–1146, 2021.

[30] Peng Yang, Xing Xi, Yaru Fu, Tony Q. S. Quek, Xianbin Cao, and Dapeng Wu. Multicast embb and bursty URLLC service multiplexing in a comp-enabled RAN. *IEEE TWC*, 20(5):3061–3077, 2021.

[31] Qiang Ye, Weihua Zhuang, Shan Zhang, A-Long Jin, Xuemin Shen, and Xu Li. Dynamic radio resource slicing for a two-tier heterogeneous wireless network. *IEEE TVT*, 67(10):9896–9910, 2018.

[32] Yinyu Ye. *Interior point algorithms: theory and analysis*, volume 44. John Wiley & Sons, 2011.

[33] Jide Yuan, Hien Quoc Ngo, and Michail Matthaiou. Machine learning-based channel prediction in massive MIMO with channel aging. *IEEE TWC*, 19(5): 2960–2973, 2020.

8

Beamforming Design for Multi-user Downlink OFDMA-URLLC Systems

Walid R. Ghanem[1,], Vahid Jamali[2], Yan Sun[3], and Robert Schober[1]*

[1] Friedrich-Alexander-University Erlangen-Nuremberg (FAU), 91058, Erlangen, Germany
[2] Technical University Darmstadt (TU), 64283, Darmstadt, Germany
[3] Huawei Technology, Shenzhen, Guangzhou, China
* Corresponding Author

8.1 Introduction

The latest generation of wireless systems, 5G networks imposes several different system design goals including high system capacity, high spectral efficiency, massive device connectivity, and low latency. One important objective is to facilitate Ultra-Reliable Low Latency Communication (URLLC). Among the most prominent URLLC applications are tactile interaction, intelligent transportation, and factory automation, see [1]. URLLC introduces strict Quality of Service (QoS) requirements comprising a low packet error probability (e.g. 10^{-6}) [1] and an extremely low latency (e.g. 1 ms). Moreover, short data packets, e.g. around 20 bytes [2], have to be used. Unfortunately, these requirements cannot be met by current mobile communication systems. The main challenges for system design are the two conflicting requirements of ultra high reliability and low latency. Thus, new design strategies are needed to achieve URLLC.

Advanced communication systems exploit Orthogonal Frequency Division Multiple Access (OFDMA) due to its resistance to multipath fading, and its ability to utilize multi-user diversity [3]. Furthermore, multiple antenna technology facilitates diversity and multiplexing gains and provides additional degrees of freedom for the resource allocation [3]. Thus, the concepts of URLLC, OFDMA, and multiple antennas are expected to be integrated in modern communication networks.

Ultra-Reliable and Low-Latency Communications (URLLC) Theory and Practice: Advances in 5G and Beyond, First Edition. Edited by Trung Q. Duong, Saeed R. Khosravirad, Changyang She, Petar Popovski, Mehdi Bennis and Tony Q.S. Quek.
© 2023 John Wiley & Sons Ltd. Published 2023 by John Wiley & Sons Ltd.

A multi-user downlink system assisted by OFDMA transmission was investigated in [4]. In [5], resource allocation design for Multiple-Input Single-Output (MISO) systems employing OFDMA was studied, where a multiple antenna Base Station (BS) served multiple single-antenna users. However, Shannon's capacity formula for the Additive White Gaussian Noise (AWGN) channel was used for resource allocation algorithm design in [4–6]. Since URLLC systems utilize a small data packet size to minimize latency, Shannon's capacity formula cannot be used to capture the trade-off between achievable rate, transmission delay, and decoding error probability as it is based on the assumption of zero error probability and infinite block length [7]. If Shannon's capacity formula is used for URLLC resource allocation algorithm design, the QoS requirements of the URLLC users cannot be guaranteed. Therefore, current studies, such as [4–6, 8] and the related literature, are not directly applicable to resource allocation algorithm design in multi-user multiple-antenna OFDMA-URLLC systems. Hence, new resource allocation algorithms taking into account the requirements and specific properties of URLLC are needed, which is the main motivation for this chapter.

Recently, Finite Blocklength Transmission (FBT) [9] has been studied in the literature. FBT provides a relation between the achievable rate, packet length, and decoding error probability. For example, FBT for parallel Gaussian channels was considered in [9], while in [10] an asymptotic analysis for AWGN, parallel AWGN, and binary symmetric channels based on the Laplace integral was provided. The analysis in [9–11] motivated new resource allocation designs based on FBT. In particular, multi-user downlink Time Division Multiple Access (TDMA) based URLLC systems were studied in [12–14]. Energy efficiency maximization for URLLC was studied in [15]. In [16], a cross-layer framework based on the effective bandwidth was proposed for resource allocation under QoS constraints. The authors in [17] investigated the joint uplink and downlink transmission design for URLLC in MISO systems. However, the studies in [12–20] considered single-carrier transmission. Unfortunately, single-carrier transmission requires complex equalization at the receiver and suffers from poor spectrum utilization.

In this chapter, we investigate the beamforming design for multi-antenna OFDMA-URLLC systems. We formulate the beamforming algorithm design as an optimization problem to maximize the weighted sum throughput subject to QoS constraints for the URLLC users. The formulated optimization problem is highly non-convex and difficult to solve. Thus, to tackle this issue, a sub-optimal algorithm based on difference of convex programming and Sequential Convex Approximation (SCA) is developed to obtain a locally optimal solution with low-computation complexity. Numerical results reveal that the proposed algorithm enables large performance gains compared to several baseline schemes. Furthermore, our results show that equipping the BS with multiple antennas facilitates high reliability and low latency in URLLC systems.

The remainder of this chapter is organized as follows. In Section 8.2, the considered system and channel models are presented. In Section 8.3, we formulate the proposed beamforming design problem. In Section 8.4, we derive a sub-optimal algorithm to solve the formulated problem. In Section 8.5, via computer simulations, we evaluate the performance of the proposed scheme. Finally, we draw conclusions in Section 8.6.

Notation In this chapter, lower-case letters indicate scalar numbers, while bold lower and upper case letters denote vectors and matrices, respectively. $\log_2(\cdot)$ is the logarithm with base 2. $\mathrm{Tr}\,(\mathbf{A})$ and $\mathrm{Rank}\,(\mathbf{A})$ denote the trace and the rank of matrix $\mathbf{A}$, respectively. $\mathbf{A} \succeq 0$ indicates that matrix $\mathbf{A}$ is positive semi-definite. $\mathbf{A}^H$ and $\mathbf{A}^T$ denote the Hermitian transpose and the transpose of matrix $\mathbf{A}$, respectively. $\mathbb{R}_+$ denotes the set of non-negative real numbers. $\mathbb{C}$ is the set of complex numbers. $\mathbf{I}_N$ is the $N \times N$ identity matrix. $\mathbb{H}_N$ denotes the set of all $N \times N$ Hermitian matrices. $|\cdot|$ and $||\cdot||$ refer to the absolute value of a complex scalar and the Euclidean vector norm, respectively. The circularly symmetric complex Gaussian distribution with mean μ and variance σ^2 is denoted by $\mathcal{CN}(\mu, \sigma^2)$, and $\sim$ stands for "distributed as". $\mathcal{E}\{\cdot\}$ denotes statistical expectation. $\nabla_x f(\mathbf{x})$ denotes the gradient vector of function $f(\mathbf{x})$ and its elements are the partial derivatives of $f(\mathbf{x})$.

8.2 System and Channel Models

In this section, the system and channel models for multi-antenna OFDMA-URLLC are introduced.

8.2.1 System Model

A single-cell OFDMA downlink system is considered, where a BS equipped with N_T antennas serves K single-antenna URLLC users indexed by $k = \{1, \dots, K\}$, cf. Figure 8.1(b). The entire frequency band is partitioned into M sub-carriers indexed by $m \in \{1, \dots, M\}$. We assume that a resource frame has a duration of T_f seconds and consists of N_s subframes, where each subframe consists of N time slots[1] which are indexed by $n \in \{1, \dots, N\}$. Thereby, one OFDMA symbol spans one time slot, and in total $M \times N$ resource elements are available for assignment to the K users in each subframe, cf. Figure 8.1(a). All users' delay requirements are considered to be known at the BS and only users whose delay requirements

1 In existing standards such as LTE and 5G NR, the bandwidth of a typical sub-carrier is 15 kHz which corresponds to an OFDM symbol duration of $T_s = 66\ \mu$s. Therefore, to meet a URLLC delay requirement of 1 ms, N has to be smaller than 7. For larger sub-carrier spacings, such as 30, 45, 60 kHz in 5G NR, larger values of N are possible.

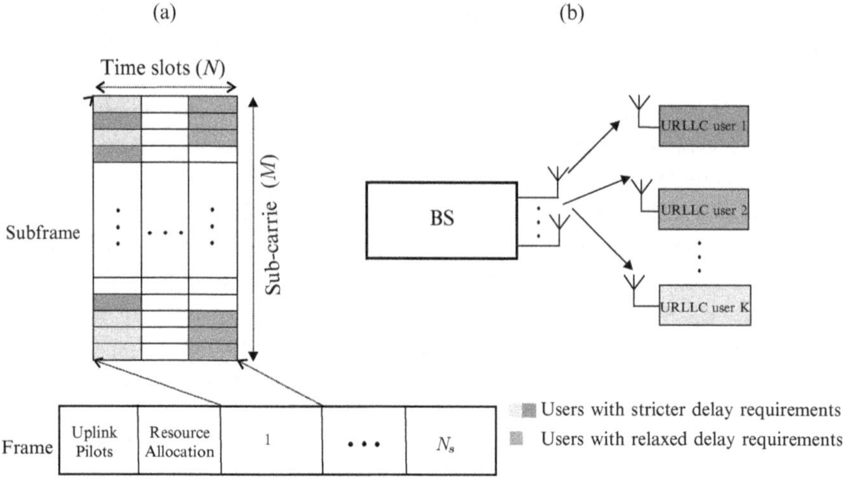

Figure 8.1 Multi-user downlink MISO OFDMA-URLLC: (a) proposed frame structure; (b) system model with N_T-antenna BS and K single-antenna users. ©IEEE 2020. Reprinted with permission from [21].

can be guaranteed in the current resource block are admitted into the system. The maximum transmit power of the BS is $P_{\max}$. Moreover, perfect channel state information (CSI) is assumed for beamforming design, which leads to a performance upper bound for multi-antenna OFDMA-URLLC systems.

8.2.2 Channel Model

The received signal at user k on sub-carrier m in time slot n is given as follows:

$$y_k[m,n] = \mathbf{h}_k^H[m]\mathbf{x}[m,n] + w_k[m,n], \tag{8.1}$$

where $\mathbf{h}_k[m] \in \mathbb{C}^{N_T \times 1}$ is the channel vector from the BS to user k on sub-carrier m, $\mathbf{x}[m,n] \in \mathbb{C}^{N_T \times 1}$ is the signal vector transmitted by the BS on sub-carrier m in time slot n, and $w_k[m,n] \sim \mathcal{CN}(0,\sigma^2)$ is the complex AWGN on sub-carrier m in time slot n at user k[2]. In this chapter, we assume linear transmit precoding at the BS, where a beamforming vector is assigned to each user. Hence, the transmit signal vector of the BS on sub-carrier m in time slot n is given by:

2 Without loss of generality, the noise variances of the K URLLC users are assumed to be identical.

$$\mathbf{x}[m,n] = \sum_{k=1}^{K} \mathbf{w}_k[m,n]u_k[m,n], \tag{8.2}$$

where $u_k[m,n] \in \mathbb{C}$ and $\mathbf{w}_k[m,n] \in \mathbb{C}^{N_T \times 1}$ are the transmit symbol and the beamforming vector of user k on sub-carrier m in time slot n, respectively. Moreover, without loss of generality, we assume that $\mathcal{E}\{|u_k[m,n]|^2\} = 1,\ \forall k \in \{1, \dots, K\}$. By substituting (8.2) into (8.1), the received signal at user k on sub-carrier m in time slot n is given by:

$$\begin{aligned} y_k[m,n] &= \mathbf{h}_k^H[m]\left(\sum_{l=1}^{K} \mathbf{w}_l[m,n]u_l[m,n]\right) + w_k[m,n] \\ &= \underbrace{\mathbf{h}_k^H[m]\mathbf{w}_k[m,n]u_k[m,n]}_{\text{desired signal}} + \underbrace{\sum_{l \neq k} \mathbf{h}_k^H[m]\mathbf{w}_l[m,n]u_l[m,n]}_{\text{multi-user interference (MUI)}} + w_k[m,n]. \end{aligned} \tag{8.3}$$

Therefore, the Signal-to-Interference-plus-Noise-Ratio (SINR) of user k on sub-carrier m in time slot n is given as follows:

$$\gamma_k[m,n] = \frac{|\mathbf{h}_k^H[m]\mathbf{w}_k[m,n]|^2}{\sum_{l \neq k} |\mathbf{h}_k^H[m]\mathbf{w}_l[m,n]|^2 + \sigma^2}, \tag{8.4}$$

i.e. the interference created by other users is treated as noise.

8.3 Resource Allocation Problem Formulation

In the following, we introduce the rate achieved via FBT, the URLLC users' QoS requirements, and the performance metric adopted for beamforming design. Furthermore, the proposed beamforming design optimization problem for multi-antenna OFDMA-URLLC systems is formulated.

8.3.1 Achievable Rate for FBT

Shannon's capacity theorem applies to the case when the probability of error approaches zero [7]. Thus, it is not suitable for the beamforming design for URLLC systems, as URLLC systems have to utilize short packets to achieve low latency, which makes decoding errors unavoidable.

The so-called normal approximation was developed for performance evaluation of FBT. For parallel complex AWGN channels, the maximum number of bits Ψ conveyed in a packet comprising L symbols can be approximated as follows [9, Equation (4.277)], [10, Figure 1]:

$$\Psi \approx \sum_{i=1}^{L} \log_2(1+\gamma[i]) - Q^{-1}(\epsilon)\sqrt{\sum_{i=1}^{L} V[i]}, \tag{8.5}$$

where ϵ is the decoding packet error probability and $Q^{-1}(\cdot)$ is the inverse of the Gaussian Q-function, defined as $Q(x) = \frac{1}{\sqrt{2\pi}} \int_x^{\infty} \exp\left(-\frac{t^2}{2}\right)\mathrm{d}t$. Moreover,

$$V[i] = a^2\left(1 - (1+\gamma[i])^{-2}\right) \tag{8.6}$$

and $\gamma[i]$ are the channel dispersion and the SINR of the ith symbol, respectively, where $a = \log_2(\mathrm{e})$ and e is the Euler's number.

In this chapter, the beamforming design for downlink multi-antenna OFDMA-URLLC systems is based on (8.5). Each resource element may carry symbols from multiple users via spatial multiplexing, and by assigning several resource elements from the available $L = M \times N$ resource elements to a given user, the number of bits received by the user with packet error probability ϵ is obtained based on (8.5).

8.3.2 QoS and System Performance Metric

The QoS requirements of URLLC user k include the minimum number of received bits, B_k, the target packet error probability, ϵ_k, and the maximum number of time slots available for transmission of the user's packet, D_k. According to (8.5), the total number of bits transmitted over the resources allocated to user k can be written as:

$$\Psi_k(\mathbf{w}_k) = F_k(\mathbf{w}_k) - V_k(\mathbf{w}_k), \tag{8.7}$$

where

$$F_k(\mathbf{w}_k) = \sum_{m=1}^{M}\sum_{n=1}^{N} \log_2(1+\gamma_k[m,n]), \tag{8.8}$$

$$V_k(\mathbf{w}_k) = Q^{-1}(\epsilon_k)\sqrt{\sum_{m=1}^{M}\sum_{n=1}^{N} V_k[m,n]}. \tag{8.9}$$

Here, the channel dispersion $V_k[m,n]$ is given by:

$$V_k[m,n] = a^2\left(1 - (1+\gamma_k[m,n])^{-2}\right). \tag{8.10}$$

Furthermore, $\mathbf{w}_k$ is the collection of all beamforming vectors $\mathbf{w}_k[m,n]$, $\forall m, n$, of user k.

The maximum delay of user k can be guaranteed by allocating all symbols of user k to the first D_k time slots. Therefore, users demanding very low latency are allocated resource elements at the start of a subframe, cf. Figure 8.1(a). Moreover, the decoding of a user's message can start after all OFDMA symbols that contain its data have been received, i.e. after D_k time slots.

The fairness between the URLLC users can be controlled by adopting the weighted sum throughput as performance metric. The weighted sum throughput is defined as follows:

$$U(\mathbf{w}) = \sum_{k=1}^{K} \mu_k \Psi_k(\mathbf{w}_k) = F(\mathbf{w}) - V(\mathbf{w}), \tag{8.11}$$

where

$$F(\mathbf{w}) = \sum_{k=1}^{K} \mu_k F_k(\mathbf{w}_k), \quad V(\mathbf{w}) = \sum_{k=1}^{K} \mu_k V_k(\mathbf{w}_k), \tag{8.12}$$

and μ_k is the weight specified for user k. A user with higher priority is assigned a larger value of μ_k, thereby, it will achieve a higher throughput (i.e. more bits are transmitted to the user) compared to the other users. The values of the μ_k may be selected in the Medium Access Control (MAC) layer and are assumed to be given in the following. Moreover, $\mathbf{w}$ is the collection of the beamforming vectors $\mathbf{w}_k$ of all users.

8.3.3 Optimization Problem Formulation

In this subsection, the resource allocation optimization problem is formulated with the goal of maximizing the weighted sum throughput of the system subject to the QoS requirements of all users in terms of the received number of bits, reliability, and latency. In particular, the proposed resource allocation policies are calculated by solving the following optimization problem:

$$\begin{aligned} &\underset{\mathbf{w}}{\text{maximize}}\ F(\mathbf{w}) - V(\mathbf{w}) \\ &\text{s.t. C1: } F_k(\mathbf{w}_k) - V_k(\mathbf{w}_k) \geq B_k,\ \forall k, \\ &\quad \text{C2: } \sum_{k=1}^{K} \sum_{m=1}^{M} \sum_{n=1}^{N} \|\mathbf{w}_k[m,n]\|^2 \leq P_{\max}, \\ &\quad \text{C3: } \mathbf{w}_k[m,n] = 0, \quad \forall n > D_k, \forall k. \end{aligned} \tag{8.13}$$

In (8.13), constraint C1 ensures the transmission of a minimum number of B_k bits to user k. C2 constrains the BS power budget. Finally, constraint C3 guarantees that user k is served within the first D_k time slots to meet its delay requirements. The problem in (8.13) is a non-convex optimization problem. The non-convexity is caused by the form of the SINR in (8.4) and the non-convex normal approximation in (8.7), which appear in the cost function and constraint C1.

There is no systematic approach for solving general non-convex optimization problems in polynomial time. Hence, in the next section, we focus on developing a sub-optimal solution based on the SCA method. This ensures computational efficiency and real-time applicability.

8.4 Sub-optimal Resource Allocation Algorithm

In this section, a sub-optimal resource allocation algorithm based on SCA is proposed to obtain a locally optimal solution of the formulated optimization problem. The problem in (8.13) is first transformed into a more tractable equivalent form. In particular, we employ Semi-Definite Programming (SDP) and some transformations.

8.4.1 Problem Transformation

To facilitate the application of SDP, we define new variables $\mathbf{W}_k[m,n] = \mathbf{w}_k[m,n]\mathbf{w}_k^H[m,n]$ and $\mathbf{H}_k[m] = \mathbf{h}_k[m]\mathbf{h}_k^H[m]$, $\forall k, m, n$, and rewrite (8.13) in equivalent form as follows:

$$
\begin{aligned}
&\underset{\mathbf{W}}{\text{maximize}}\; F(\mathbf{W}) - V(\mathbf{W}) \\
&\text{s.t.}\; \overline{\text{C1}} : \; F_k(\mathbf{W}_k) - V_k(\mathbf{W}_k) \geq B_k,\; \forall k, \\
&\quad \overline{\text{C2}} : \; \sum_{k=1}^{K}\sum_{m=1}^{M}\sum_{n=1}^{N} \text{Tr}(\mathbf{W}_k[m,n]) \leq P_{\text{max}}, \\
&\quad \overline{\text{C3}} : \; \text{Tr}(\mathbf{W}_k[m,n]) = 0, \forall n > D_k, \forall k, \\
&\quad \text{C4:} \quad \mathbf{W}_k[m,n] \succeq 0, \quad \forall k, m, n, \\
&\quad \text{C5:} \; \text{Rank}(\mathbf{W}_k[m,n]) \leq 1, \quad \forall k, m, n,
\end{aligned}
\tag{8.14}
$$

where

$$
F(\mathbf{W}) = \sum_{k=1}^{K} \mu_k \sum_{m=1}^{M}\sum_{n=1}^{N} \log_2(1 + \gamma_k[m,n]), \tag{8.15}
$$

$$
V(\mathbf{W}) = \sum_{k=1}^{K} \mu_k a Q^{-1}(\epsilon_k) \sqrt{\sum_{m=1}^{M}\sum_{n=1}^{N} (1 - (1 + \gamma_k[m,n])^2)}\,, \tag{8.16}
$$

and

$$\gamma_k[m,n] = \frac{\mathrm{Tr}(\mathbf{H}_k[m]\mathbf{W}_k[m,n])}{\sum_{l\neq k}\mathrm{Tr}(\mathbf{H}_k[m]\mathbf{W}_l[m,n]) + \sigma^2}. \tag{8.17}$$

We note that $\mathbf{W}_k[m,n] \succeq 0$ and $\mathrm{Rank}(\mathbf{W}_k[m,n]) \leq 1, \forall k, m, n$, in constraints C4 and C5 are imposed to ensure that $\mathbf{W}_k[m,n] = \mathbf{w}_k[m,n]\mathbf{w}_k^H[m,n]$ holds after optimization. Moreover, for simplicity of notation, we define $\mathbf{W}_k$ as the collection of all Hermitian matrices $\mathbf{W}_k[m,n] \in \mathbb{H}_{N_T}, \forall m, n$, and $\mathbf{W}$ as the collection of all $\mathbf{W}_k, \forall k$.

The objective function and constraint $\overline{\text{C1}}$ in (8.14) have a complicated structure. In fact, the objective function in (8.14) and constraint $\overline{\text{C1}}$ are differences of two functions that are monotonic in the SINR, i.e. $\gamma_k[m,n]$, which in turn is a function of optimization variable $\mathbf{W}$, cf. (8.17). To handle the related complexity and to facilitate the application of SCA, a set of auxiliary variables $z_k[m,n], \forall k, m, n$ are introduced to bound the SINR from below, i.e.

$$0 \leq z_k[m,n] \leq \gamma_k[m,n] = \frac{f_k[m,n](\mathbf{W})}{g_k[m,n](\mathbf{W})}, \forall k, m, n, \tag{8.18}$$

where $f_k[m,n](\mathbf{W})$ and $g_k[m,n](\mathbf{W})$ are the numerator and denominator of the SINR in (8.17) and are given respectively by

$$f_k[m,n](\mathbf{W}) = \mathrm{Tr}(\mathbf{H}_k[m]\mathbf{W}_k[m,n]), \forall k, m, n, \tag{8.19}$$

$$g_k[m,n](\mathbf{W}) = \sum_{l\neq k}\mathrm{Tr}(\mathbf{H}_k[m]\mathbf{W}_l[m,n]) + \sigma^2, \forall k, m, n. \tag{8.20}$$

Let us replace $\gamma_k[m,n]$ by $z_k[m,n]$ in $F(\mathbf{W})$, $V(\mathbf{W})$, $F_k(\mathbf{W}_k)$, and $V_k(\mathbf{W}_k)$ and denote the resulting functions by $F(\mathbf{z})$, $V(\mathbf{z})$, $F_k(\mathbf{z}_k)$, and $V_k(\mathbf{z}_k)$, respectively, i.e.

$$F(\mathbf{z}) = \sum_{k=1}^{K} \mu_k F_k(\mathbf{z}_k), \qquad V(\mathbf{z}) = \sum_{k=1}^{K} \mu_k V(\mathbf{z}_k), \tag{8.21}$$

$$F_k(\mathbf{z}_k) = \sum_{m=1}^{M}\sum_{n=1}^{N} \log_2(1 + z_k[m,n]), \forall k, \tag{8.22}$$

$$V_k(\mathbf{z}_k) = aQ^{-1}(\epsilon_k)\sqrt{\sum_{m=1}^{M}\sum_{n=1}^{N}(1 - (1 + z_k[m,n])^{-2})}, \tag{8.23}$$

where $\mathbf{z}_k$ denotes the collection of optimization variables $z_k[m,n]$, $\forall m, n$, and $\mathbf{z}$ denotes the collection of optimization variables $\mathbf{z}_k, \forall k$. Using these notations, and after dropping rank constraint C5 in (8.14), we formulate a new optimization problem as follows:

$$\begin{aligned}
&\underset{\mathbf{W},\mathbf{z}}{\text{maximize}}\ F(\mathbf{z}) - V(\mathbf{z}) \\
&\text{s.t.}\ \widetilde{\text{C1}}:\ F_k(\mathbf{z}_k) - V_k(\mathbf{z}_k) \geq B_k, \forall k, \\
&\quad \overline{\text{C2}}, \overline{\text{C3}}, \text{C4}, \\
&\quad \text{C6: } z_k[m,n] \leq \frac{f_k[m,n](\mathbf{W})}{g_k[m,n](\mathbf{W})}, \forall k, m, n, \\
&\quad \text{C7: } z_k[m,n] \geq 0.
\end{aligned} \tag{8.24}$$

Lemma 8.1: Optimization problems (8.24) and (8.13) are equivalent and share the same solution **W**.

Proof. Optimization problem (8.24) can be reformulated as a monotonic optimization problem. As a result, constraint C6 in (8.24) has to hold with equality. See [21] for a similar proof. Moreover, one can show that the solution to (8.24) yields a beamforming matrix that has a rank equal to or smaller than one, i.e. $\text{Rank}(\mathbf{W}_k[m,n]) \leq 1, \forall k, m, n$. The corresponding proof is similar to the one presented in [21, Appendix B].

8.4.2 Difference of Convex Programming Reformulation

In the following, we solve optimization problem (8.24), as (8.24) is equivalent to (8.13). We solve optimization problem (8.24) in two steps. First, we transform the problem into the canonical form required for application of difference of convex programming. Second, we employ a Taylor series expansion to obtain a convex approximation of the non-convex terms. Thereby, we obtain a convex optimization problem that can be efficiently solved using standard convex optimization software. In the following, these two steps are explained in detail.

Step 1 We rewrite non-convex constraint C6 in (8.24) as follows:

$$\begin{aligned}
&\widetilde{\text{C6}}\text{: } z_k[m,n] g_k[m,n](\mathbf{W}) \\
&= z_k[m,n](I_k[m,n](\mathbf{W}) + \sigma^2) \leq f_k[m,n](\mathbf{W}),\ \forall k, m, n,
\end{aligned} \tag{8.25}$$

where $g_k[m,n] = I_k[m,n](\mathbf{W}) + \sigma^2$. We note that $z_k[m,n]I_k[m,n](\mathbf{W})$ in (8.25) is a bilinear term which is non-convex. In fact, the Hessian matrix of a bilinear function is neither positive nor negative semi-definite. Thus, bilinear functions are neither convex nor concave in general, which is a barrier for designing computationally efficient resource allocation algorithms. The product of two convex function $f_1(x)$ and $f_2(x)$ can be written as a difference of two convex functions as follows [22]:

$$f_1(x)f_2(x) = 0.5(f_1(x) + f_2(x))^2 - 0.5f_1(x)^2 - 0.5f_2(x)^2. \tag{8.26}$$

Exploiting (8.26), where we substitute $z_k[m,n]$ and $I_k[m,n](\mathbf{W})$ as f_1 and f_2, respectively, we can express the product term $z_k[m,n]I_k[m,n](\mathbf{W})$ in (8.25) as follows:

$$z_k[m,n]I_k[m,n](\mathbf{W}) = Q(z_k[m,n],\mathbf{W}) - T(z_k[m,n],\mathbf{W}), \tag{8.27}$$

where

$$Q(z_k[m,n],\mathbf{W}) = \frac{1}{2}(z_k[m,n] + I_k[m,n](\mathbf{W}))^2, \forall k,m,n, \tag{8.28}$$

$$T(z_k[m,n],\mathbf{W}) = \frac{1}{2}(z_k[m,n])^2 + \frac{1}{2}(I_k[m,n](\mathbf{W}))^2, \forall k,m,n. \tag{8.29}$$

Furthermore, substituting (8.27) into (8.25), we obtain an equivalent representation of constraint $\widetilde{\text{C6}}$ in (8.25) as follows:

$$\begin{aligned}\widetilde{\widetilde{\text{C6}}}\text{: } & Q(z_k[m,n],\mathbf{W}) - T(z_k[m,n],\mathbf{W}) \\ & \leq f_k[m,n](\mathbf{W}) - \sigma^2 z_k[m,n], \forall k,m,n,\end{aligned} \tag{8.30}$$

where the left-hand-side is a difference of two convex functions. Hence, optimization problem (8.24) can now be rewritten as follows:

$$\begin{aligned}\underset{\mathbf{W},\mathbf{z}}{\text{minimize}} \quad & -[F(\mathbf{z}) - V(\mathbf{z})] \\ \text{s.t. } & \widetilde{\text{C1}}, \overline{\text{C2}}, \overline{\text{C3}}, \text{C4}, \text{C7}, \\ & \widetilde{\widetilde{\text{C6}}}\text{: } Q(z_k[m,n],\mathbf{W}) - T(z_k[m,n],\mathbf{W}) \\ & \leq f_k[m,n](\mathbf{W}) - \sigma^2 z_k[m,n], \forall k,m,n.\end{aligned} \tag{8.31}$$

The optimization problem in (8.31) belongs to the class of difference of convex programming problems, since its objective function can be written as a difference of two convex functions and constraints $\widetilde{\text{C1}}$ and $\widetilde{\widetilde{\text{C6}}}$ can also be expressed as the differences of two convex functions. In particular, functions $-F(\mathbf{z})$, $-V(\mathbf{z})$, $Q(z_k[m,n],\mathbf{W})$, and $T(z_k[m,n],\mathbf{W})$ are convex functions.

Step 2 To obtain a convex optimization problem that can be efficiently solved, we have to address the non-convex objective function and non-convex constraints $\widetilde{\text{C1}}$ and $\widetilde{\widetilde{\text{C6}}}$. To this end, we determine the first order approximations of functions

$V_k(\mathbf{z}_k)$ and $T(z_k[m,n],\mathbf{W})$ using Taylor series as follows:

$$V_k(\mathbf{z}_k) \leq \tilde{V}_k(\mathbf{z}_k) = V_k(\mathbf{z}_k^{(j)}) + \nabla_{\mathbf{z}_k} V_k(\mathbf{z}_k^{(j)})^T(\mathbf{z}_k - \mathbf{z}_k^{(j)}), \tag{8.32}$$

and

$$\begin{aligned} T(z_k[m,n],\mathbf{W}) \geq \tilde{T}(z_k[m,n],\mathbf{W}) &= T(z_k^{(j)}[m,n],\mathbf{W}^{(j)}) \\ &+ \nabla_{z_k[m,n]} T(z_k^{(j)}[m,n],\mathbf{W}^{(j)})(z_k[m,n] - z_k^{(j)}[m,n]) \\ &+ \mathrm{Tr}(\nabla_{\mathbf{W}} T(z_k^{(j)}[m,n],\mathbf{W}^{(j)})^T)(\mathbf{W} - \mathbf{W}^{(j)}), \forall k,m,n, \end{aligned} \tag{8.33}$$

where $\mathbf{W}^{(j)}$, $\mathbf{z}_k^{(j)}$, and $z_k^{(j)}[m,n]$ are initial feasible points, and

$$\nabla_{\mathbf{z}_k} \mathbf{V}_k(\mathbf{z}_k) = \frac{a^2 Q^{-1}(\epsilon_k)}{\sqrt{\sum_{m=1}^{M}\sum_{n=1}^{N} V_k[m,n]}} \begin{pmatrix} \frac{1}{(1+z_k[1,1])^3} \\ \frac{1}{(1+z_k[2,1])^3} \\ \vdots \\ \frac{1}{(1+z_k[M,N])^3} \end{pmatrix}, \tag{8.34}$$

$$\nabla_{z_k[m,n]} T(z_k[m,n],\mathbf{W}) = z_k[m,n], \tag{8.35}$$

and

$$\nabla_{\mathbf{W}} T(z_k[m,n],\mathbf{W}) = I_k[m,n](\mathbf{W})\mathbf{H}_k[m]. \tag{8.36}$$

The right-hand-sides of (8.32) and (8.33) are affine functions, and by substituting them in (8.31), we obtain the following convex optimization problem:

$$\begin{aligned} &\underset{\mathbf{W},\mathbf{z}}{\text{minimize}} \; -[F(\mathbf{z}) - \tilde{V}(\mathbf{z})] \\ &\text{s.t. } \widetilde{\widetilde{\text{C1}}} : \; F_k(\mathbf{z}_k) - \tilde{V}_k(\mathbf{z}_k) \geq B_k, \; \forall k, \\ &\quad \overline{\text{C2}}, \overline{\text{C3}}, \text{C4}, \text{C7}, \\ &\quad \widetilde{\widetilde{\widetilde{\text{C6}}}} : \; Q(z_k[m,n],\mathbf{W}) - \tilde{T}(z_k[m,n],\mathbf{W}) \\ &\qquad\qquad \leq f_k[m,n](\mathbf{W}) - \sigma^2 z_k[m,n], \forall k,m,n. \end{aligned} \tag{8.37}$$

We can efficiently solve optimization problem (8.37) by standard convex solvers such as CVX [23]. Problem (8.37) can be solved iteratively where the solution of (8.37) in iteration j is used as the initial point for the next iteration $j+1$. The algorithm generates a sequence of improved feasible solutions until convergence to a local optimum point of problem (8.37) or equivalently problem (8.13) in polynomial time [24, 25].

Algorithm 1 Sequential Convex Approximation (SCA)

1: Initialize: Maximum number of iterations $J_{\max}$, iteration index $j = 1$, initial points $\mathbf{W}^{(1)}$, $\mathbf{z}^{(1)}$.
2: **Repeat**
3: Solve convex problem (8.37) for given $\mathbf{W}^{(j)}$ and $\mathbf{z}^{(j)}$ and store the intermediate resource allocation policy $\{\mathbf{W}, \mathbf{z}\}$
4: Set $j = j + 1$ and update $\mathbf{W}^{(j)} = \mathbf{W}$, $\mathbf{z}^{(j)} = \mathbf{z}$.
5: **Until** convergence or $j = J_{\max}$
6: $\mathbf{W}^* = \mathbf{W}^{(j)}$,

8.5 Performance Evaluation

In this section, the performance of the developed resource allocation scheme is evaluated via simulations. We summarize the relevant simulation parameters in Table 8.1. In our simulations, we consider a single cell with inner and outer radii of $r_1 = 50$ m and $r_2 = 250$ m, respectively. We position the BS at the center of the cell. We calculate the path loss by using $35.3 + 37.6 \log_{10}(d_k)$ [16], where d_k(m) is the distance from the BS to user k. We model the small scale fading gains between the BS and the users as independent and identically Rayleigh distributed. For simplicity, all user weights are set to $\mu_k = 1, \forall k$. We average the simulation results over different channel realizations of the multipath fading and path loss.

8.5.1 Performance Metric

To evaluate the system performance, the sum throughput of the system is defined for a given channel realization as follows:

Table 8.1 System parameters [21].

Parameter	Value
Number of sub-carriers	64
Bandwidth of each sub-carrier	15 kHz
Noise power density	-174 dBm/Hz
Number of bits per packet	160 bits
Maximum BS transmit power $P_{\max}$	45 dBm
Packet error probability for user k	$\epsilon_k = 10^{-6}$

$$\bar{R} = \begin{cases} \frac{1}{MN} \sum_{k=1}^{K} \Psi_k(\mathbf{W}_k), & \text{if } \mathbf{W} \text{ is feasible} \\ 0 & \text{otherwise.} \end{cases} \tag{8.38}$$

If the optimization problem is infeasible for a given channel realization, we set the corresponding sum throughput to zero. The Average System Sum Throughput (ASST) is obtained by averaging $\bar{R}$ over all considered channel realizations.

8.5.2 Performance Bound and Benchmark Schemes

We compare the performance of the proposed resource allocation algorithm design with the following upper bound and baseline schemes:

- **Upper bound** To get a performance upper bound, Shannon's capacity formula is used in problem (8.13), i.e. $V(\mathbf{w})$ and $V_k(\mathbf{w}_k)$ are set to zero in the objective function and in constraint C1, respectively, and all other constraints are retained. The resulting optimization problem is solved using a modified version of **Algorithm** 1. The corresponding sum throughput is obtained from (8.38), where $\Psi_k(\mathbf{w}_k) = F_k(\mathbf{w}_k)$ is used.
- **Baseline scheme 1** For this scheme, as for the performance upper bound, resource allocation is performed based on Shannon's capacity formula. However, now $\Psi_k(\mathbf{w}_k) = F_k(\mathbf{w}_k) - V_k(\mathbf{w}_k)$ is used in (8.38), and $\Psi_k(\mathbf{w}_k) = F_k(\mathbf{w}_k) - V_k(\mathbf{w}_k) \geq B_k$ is required for feasibility of the solution.
- **Baseline scheme 2** For this scheme, we use Maximum Ratio Transmission (MRT) beamforming, where $\mathbf{w}_k[m,n] = \sqrt{p_k[m,n]}\frac{\mathbf{h}_k[m]}{\|\mathbf{h}_k[m]\|}$, and optimize the powers $p_k[m,n]$. The resulting optimization problem is solved using sequential convex approximation based on a similar approach as for deriving **Algorithm** 1.

8.5.3 Simulation Results

In Figure 8.2, we show the ASST versus the maximum BS transmit power. It is apparent that by increasing the maximum BS transmit power $P_{\max}$, the ASST can be improved. The reason for this behaviour is that the users' SINRs can be boosted by assigning more power. In Figure 8.2, we also compare the proposed algorithm with the two baseline schemes. For the baseline schemes, we observe lower throughputs compared to the proposed scheme. The non-optimality of using a fixed beamformer in baseline scheme 2 causes a performance loss which results in performance saturation for transmit powers above 25 dBm. For baseline scheme 1, the resource allocation policies obtained for Shannon's capacity formula may violate constraint C1 in (8.13), especially for small $P_{\max}$, which

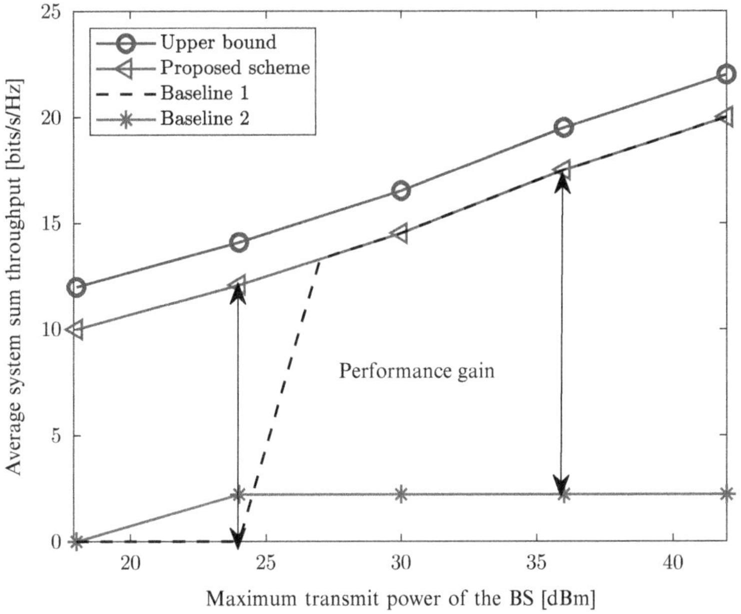

Figure 8.2 ASST [bits/s/Hz] versus BS transmit power, P_{max}, for different resource allocation schemes. $M = 16$, $N_T = 2$, $N = 2$, $K = 2$, $d_k = 50$ m, $\forall k \in \{1,2\}$, and $D_1 = 1, D_2 = 2$. ©IEEE 2020. Reprinted with permission from [21].

results in an infeasible solution. Therefore, we cannot use Shannon's capacity formula for URLLC resource allocation design, since the QoS requirements of the users cannot always be met. For high P_{max}, for the proposed scheme, all non-zero $\|\mathbf{w}_k[m,n]\|^2$ assume large values. Hence, the corresponding $\gamma_k[m,n]$ in (8.7) are large and $V_k(\mathbf{w}_k)$ becomes small compared to $F_k(\mathbf{w}_k)$. Therefore, in this case, baseline scheme 1, which assumes $V_k(\mathbf{w}_k)$ is zero, yields a similar performance as the proposed scheme.

In Figure 8.3, we depict the ASST versus the number of antennas equipped at the BS, N_T, for two delay scenarios S_0 and S_1, and different numbers of URLLC users. For delay scenario S_0 all URLLC users have no delay restrictions, i.e. $D_k = N = 4, \forall k$. For delay scenario S_1, two users have stringent constraints on the delay, while we do not impose constraints on the delays of the other URLLC users, i.e. $D_1 = D_2 = 2, D_k = N = 4, \forall k \neq \{1,2\}$. Figure 8.3 reveals that the ASST can be improved by increasing the number of BS antennas. This results from the fact that additional antennas provide additional degrees of freedom for resource allocation which gives rise to higher SINRs at the URLLC users' receivers. Moreover, the proposed scheme approaches the performance upper bound as the number of BS antennas grows as the value of $V_k(\mathbf{w}_k)$ in (8.7) becomes small compared to that

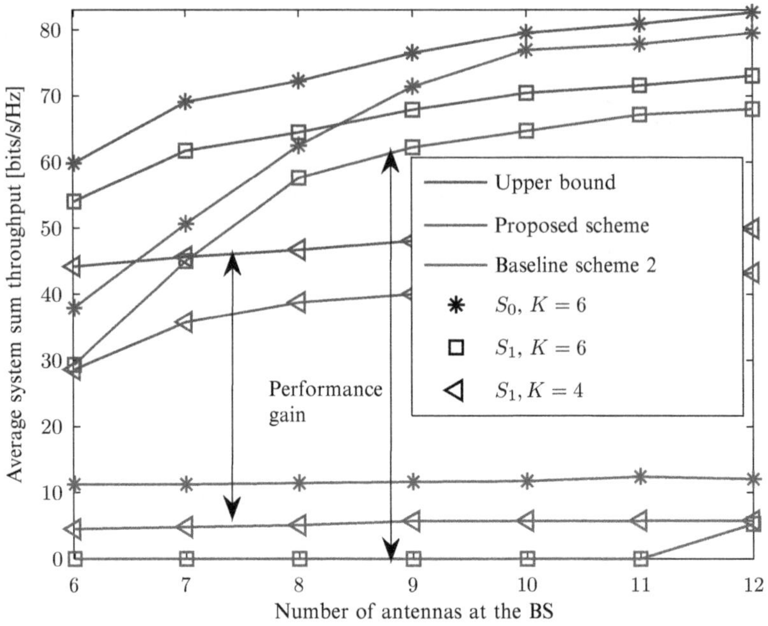

Figure 8.3 ASST [bits/s/Hz] versus the number of antennas at the BS. $P_{max} = 45$ dBm, $M = 64$, and $N = 4$. The users are randomly distributed within the inner and the outer radius. ©IEEE 2020. Reprinted with permission from [21].

of $F_k(\mathbf{w}_k)$. Therefore, the impact of FBT coding on the ASST can be mitigated by utilizing a BS equipped with a large number of antennas. Furthermore, altering the delay requirements from S_0 to S_1 reduces the throughput for all considered schemes. To overcome this effect, the BS can deploy additional antennas in order to be able to serve the users with stringent delay requirements in a more efficient manner. Figure 8.3 also shows the impact of the numbers of users on the ASST. Since the proposed scheme is able to exploit multi-user diversity, enlarging the number of users from $K = 4$ to $K = 6$ boosts the throughput. In contrast, baseline scheme 2 cannot support $K = 6$ users for delay scenario S_1 because this scheme cannot exploit all available degrees of freedom for resource allocation, and hence, the two users with stringent delay requirements cause infeasible solutions, which negatively affects the ASST.

8.6 Conclusions

In this chapter, the beamforming design for multiple-antenna OFDMA-URLLC systems was studied. We formulated the design as a non-convex optimization problem to maximize the weighted system sum throughput of the URLLC users. Moreover, the problem formulation took into account the QoS requirements of the URLLC users. Due to the non-convexity of the problem, solving the problem optimally is challenging. Thus, we proposed a sub-optimal algorithm based on SCA. Numerical results showed the potential gains in weighted system sum throughput enabled by the proposed algorithm.

Bibliography

[1] G. Durisi, T. Koch, and P. Popovski, "Toward massive, ultra-reliable, and low-latency wireless communication with short packets," *Proc. IEEE*, vol. 104, pp. 1711–1726, Sept 2016.

[2] P. Popovski, "Ultra-reliable communication in 5G wireless systems," in *Proc. IEEE Int. Conf. 5G Ubiq. Connect*, pp. 146–151, Nov 2014.

[3] D. W. K. Ng, E. S. Lo, and R. Schober, "Energy-efficient resource allocation in OFDMA systems with large numbers of base station antennas," *IEEE Trans. Wireless. Commun*, vol. 11, no. 9, pp. 3292–3304, Sept. 2012.

[4] K. Seong, M. Mohseni, and J. M. Cioffi, "Optimal resource allocation for OFDMA downlink systems," in *Proc. IEEE Intern. Sympos. on Inf. Theory*, pp. 1394–1398, July 2006.

[5] V. D. Papoutsis, I. G. Fraimis, and S. A. Kotsopoulos, "User selection and resource allocation algorithm with fairness in MISO-OFDMA," *IEEE Commun. Lett.*, vol. 14, pp. 411–413, May 2010.

[6] W. Dang, M. Tao, H. Mu, and J. Huang, "Subcarrier-pair based resource allocation for cooperative multi-relay OFDM systems," *IEEE Trans. Wireless Commun*, vol. 9, pp. 1640–1649, May 2010.

[7] C. E. Shannon, "A mathematical theory of communication," *Bell Syst. Tech. J*, vol. 56, pp. 2307–2359, May 2010.

[8] D. W. K. Ng, E. S. Lo, and R. Schober, "Energy-efficient resource allocation in multi-cell OFDMA systems with limited backhaul capacity," *IEEE Trans. Wireless. Commun*, vol. 11, no. 10, pp. 3618–3631, Oct. 2012.

[9] Y. Polyanskiy, *Channel coding: Non-asymptotic fundamental limits.* PhD thesis, Princeton University.

[10] T. Erseghe, "Coding in the finite-blocklength regime: Bounds based on Laplace integrals and their asymptotic approximations," *IEEE Trans. Inf. Theory*, vol. 62, pp. 6854–6883, Dec 2016.

[11] Y. Polyanskiy, H. V. Poor, and S. Verdu, "Channel coding rate in the finite blocklength regime," *IEEE Trans. Inf. Theory*, vol. 56, pp. 2307–2359, May 2010.

[12] Y. Hu, M. Ozmen, M. C. Gursoy, and A. Schmeink, "Optimal power allocation for QoS-constrained downlink multi-user networks in the finite blocklength regime," *IEEE Trans. Wireless Commun*, vol. 17, pp. 5827–5840, Sept 2018.

[13] J. Chen, L. Zhang, Y. Liang, X. Kang, and R. Zhang, "Resource allocation for wireless-powered IoT networks with short packet communication," *IEEE Trans. Wireless Commun*, vol. 18, pp. 1447–1461, Feb 2019.

[14] S. Xu, T. H. Chang, S. C. Lin, C. Shen, and G. Zhu, "Energy-efficient packet scheduling with finite blocklength codes: convexity analysis and efficient algorithms," *IEEE Trans. Wireless Commun*, vol. 15, pp. 5527–5540, Aug 2016.

[15] C. Sun, C. She, C. Yang, T. Q. S. Quek, Y. Li, and B. Vucetic, "Optimizing resource allocation in the short blocklength regime for ultra-reliable and low-latency communications," *IEEE Trans. Wireless Commun*, vol. 18, pp. 402–415, Jan 2019.

[16] C. She, C. Yang, and T. Q. S. Quek, "Cross-layer optimization for ultra-reliable and low-latency radio access networks," *IEEE Trans. Commun*, vol. 17, pp. 127–141, Jan 2018.

[17] C. Shen, T. Chang, H. Xu, and Y. Zhao, "Joint uplink and downlink transmission design for URLLC using finite blocklength codes," in *Proc. ISWCS 2018, Lisbon, Portugal, August 28-31, 2018*, pp. 1–5, 2018.

[18] A. Avranas, M. Kountouris, and P. Ciblat, "Energy-latency tradeoff in ultra-reliable low-latency communication with retransmissions," *IEEE J. Sel. Areas Commun*, vol. 36, pp. 2475–2485, Nov 2018.

[19] D. Qiao, M. C. Gursoy, and S. Velipasalar, "Throughput-delay tradeoffs with finite blocklength coding over multiple coherence blocks," *IEEE Trans. Commun*, pp. 5892–5904, Aug. 2019.

[20] M. Haghifam, M. Robat Mili, B. Makki, M. Nasiri-Kenari, and T. Svensson, "Joint sum rate and error probability optimization: Finite blocklength analysis," *IEEE Wireless Commun. Lett*, vol. 6, pp. 726–729, Dec 2017.

[21] W. Ghanem, V. Jamali, Y. Sun, and R. Schober, "Resource Allocation for Multi-User Downlink MISO OFDMA-URLLC Systems," *IEEE Trans. Commun*, vol. 68, pp. 7184–7200, August 2020.

[22] H. Tuy, *Convex Analysis and Global Optimization.* Springer Publishing Company, Incorporated, 2nd ed., 2016.

[23] M. Grant and S. Boyd, "CVX: Matlab software for disciplined convex programming, version 2.1." http://cvxr.com/cvx, Mar. 2014.

[24] Y. Sun, D. W. K. Ng, Z. Ding, and R. Schober, "Optimal joint power and subcarrier allocation for full-duplex multicarrier non-orthogonal multiple access systems," *IEEE Trans. Commun*, vol. 65, pp. 1077–1091, March 2017.

[25] E. Che, H. D. Tuan, and H. H. Nguyen, "Joint optimization of cooperative beamforming and relay assignment in multi-user wireless relay networks," *IEEE Trans. Wirel. Commun*, vol. 13, pp. 5481–5495, Oct 2014.

9

A Full-Duplex Relay System for URLLC with Adaptive Self-Interference Processing

Hanjun Duan, Yufei Jiang, Xu Zhu, and Fu-Chun Zheng*

The School of Electronic and Information Engineering, Harbin Institute of Technology, Shenzhen, 518055, Guangdong, Shenzhen, Nanshan District, China
** Corresponding Author*

9.1 Introduction

Low latency and high reliability have become two major challenges in future wireless communication design. However, more and more emerging devices and applications such as industrial automation, Tactile Internet, virtual reality, Internet of vehicles, telemedicine and so on [2, 25] (as shown in Figure 9.1) require strict constraints of high reliability and low latency. In the 5th Generation (5G) cellular networks, Ultra-Reliable and Low-Latency Communication (URLLC) is introduced to achieve reliability of 99.999 % and latency of 1 ms. The future networks are expected to require even more stringent reliability and latency (e.g. below 1 ms).

It is very challenging to achieve two such ambitious targets for cell-edge user terminals that encounter wireless deep fading, dramatically reducing the link reliability. The application of a relay [8], as an efficient way to mitigate wireless fading, can be utilized to guarantee high reliability of a communication link highly separated by a cell-edge user and base station. However, the latency is very high for a traditional Half-Duplex (HD) relay system, as two time slots are required to complete a single data transmission. The source transmits data to an HD relay in the first time slot, and the HD relay forwards them to their destination in the second time slot. Meanwhile, the source is not allowed to transmit data in the

Ultra-Reliable and Low-Latency Communications (URLLC) Theory and Practice: Advances in 5G and Beyond, First Edition. Edited by Trung Q. Duong, Saeed R. Khosravirad, Changyang She, Petar Popovski, Mehdi Bennis and Tony Q.S. Quek.
© 2023 John Wiley & Sons Ltd. Published 2023 by John Wiley & Sons Ltd.

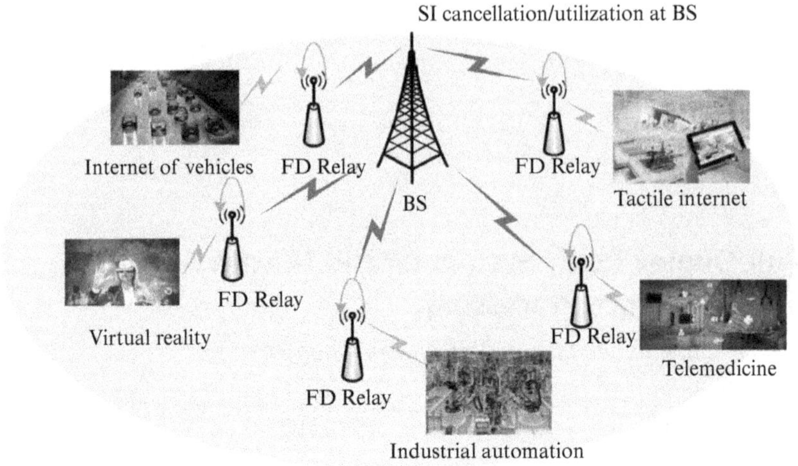

Figure 9.1 Emerging devices and applications that require ultra-reliable and low-latency demands. User terminals far from the base station are assisted by a full-duplex relay [1] to guarantee high reliability and low latency, allowing a full-duplex relay to work in the AF mode and pushing digital SI cancellation/utilization at base station to reduce relaying latency (BS: base station). ©IEEE 2021. Reprinted with permission from [2].

second time slot. Thus, the latency of the HD relay system is twice, as far as possible, that of direct transmissions with no relay and no retransmission during data transmission completion. In order to reduce the latency, a Full-Duplex (FD) relay, equipped with transmit antennas and receive antennas, has been widely studied in the literature, to enable simultaneous transmission and reception in the same frequency band, with theoretically doubled throughput [22]. The source and relay can successively transmit data in consecutive time slots. Thus, the FD relay system is a promising solution to provide latency lower than the HD relay system, and close to direct transmissions with no relay and no retransmission. In the literature, there is only one overview of an HD relay-enabled URLLC system in the finite blocklength regime [8]. So far, there has been no overview of an FD relay system from the URLLC perspective.

In an FD relay system, reliability is limited by Self-Interference (SI) due to signal leakage from the transmit antennas to the receive antennas at relay, seriously affecting Bit Error Rate (BER) performance [21]. Recent breakthroughs have revealed up to 120 dB SI cancellation capability [26], and have facilitated the real application of FD communications at relay rather than at base station. The SI suppression is as low as approximately −97 dBm at relay with transmission power of 23 dBm, and −74 dBm at base station with transmission power of 46 dBm. The significant SI suppression at FD relay can be as low as the noise power level of −90 dBm. To guarantee high reliability, traditional works [21, 22, 24, 26] require the FD relay to work in the Decode-and-Forward (DF) mode with a number of SI cancellation processes, such as Radio Frequency (RF) cancellation, analog cancellation and digital cancellation. However, the SI cancellation increases processing latency, referred to as relaying latency, which should be maintained at a low level as much as possible, without compromising reliability in terms of SI cancellation capability, to achieve URLLC.

We present an insightful investigation of reliability and latency together for an FD relay assisted URLLC system. We provide an overview of the end-to-end latency, where the relaying latency plays a significant part. This has not been presented in the existing literature [8, 13, 26]. We discuss possible relaying latency reduction solutions. An efficient solution is to allow the FD relay to work in the Amplify-and-Forward (AF) mode with low-complexity RF cancellation and analog cancellation. This is different to existing work [8, 26], where the DF mode is required with all three SI cancellation processes, yielding high relaying latency. We also investigate the residual SI cancellation and utilization conducted at base station in the digital domain, not necessarily being canceled as much as possible at the FD relay as in [24, 26]. The residual SI can be utilized to improve reliability and enhance the degrees of freedom in signal processing, which has not been introduced in the recent literature [8, 26].

We evaluate the end-to-end latency and reliability in terms of BER. The FD relay system with the AF mode provides latency significantly lower than that with the DF mode and the HD relay system, respectively, and close to direct transmissions with no relay and no retransmission. Also, the AF FD relay system with SI utilization at base station provides better BER performance, compared to the DF FD relay system with all SI cancellations at FD relay.

9.2 FD Relay in URLLC

For FD relay systems, as shown in Figure 9.2, the overall latency is addressed on the physical layer, and can generally be divided into transmission latency, propagation latency, pre-processing latency, processing latency, and relaying latency.

Transmission Latency. This corresponds to the time duration between the beginning of a packet transmission and the end of the same packet transmission at the transmitter. The packet design is a key issue to minimize the transmission latency in URLLC. In 5G New Radio systems, a non-square packet stretched in frequency is used to reduce transmission latency [13]. As shown in Figure 9.2, there are two parts of transmission latency between source to relay and between relay to destination for relay-based systems. The transmission latency can be minimized at an FD relay with simultaneous reception and transmission. Thus, the FD relay system provides transmission latency close to the direct transmission system with no relay and no retransmission.

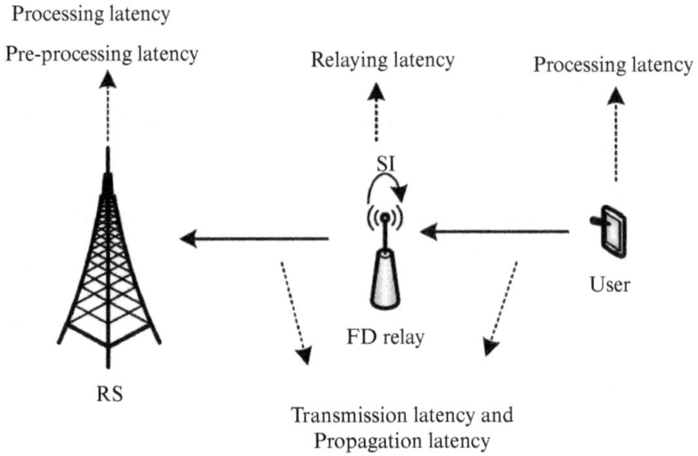

Figure 9.2 Latency diagram for FD relay systems in the physical layer with the main components: pre-processing latency, relaying latency, processing latency, transmission latency, and propagation latency (BS: base station). ©IEEE 2021. Reprinted with permission from [2].

Propagation Latency. We consider an urban scenario, where the transmission distance is about 500 meters between base station and cell-edge users [21]. There are two courses of signal propagation: between source and relay, between relay and destination. Given a user terminal far from the serving base station, as shown in Figure 9.3, it is important to select a proper relay to secure two signal propagations in the line-of-sight scenario with strong channel gains and short propagation distances, achieving simultaneous high reliability and low latency.

Pre-processing Latency. This corresponds to signaling feedback and exchange at base station, including a number of feedback information such as Channel State Information (CSI), quality of service requirement, bandwidth and capacity requirements. Also, the control signaling is incorporated, e.g. Hybrid Automatic Repeat Request (HARQ), connection request message, scheduling grant message, queuing latency and so on.

Processing Latency. This includes optimization latency and signal processing latency. In URLLC, short-frame transmissions are allowed, with a number of resources available, such as frequency, bandwidth, channel, time and power resources, the number of relays, the number of antennas at relay and at base station, the number of symbols in a data frame, queue state information, and packet loss probability. With the given resources, a low-complexity cross-layer optimization scheme is preferable to achieve low-latency and high-reliability requirements, formulated with tractable solutions in a closed form. It has been shown in [25]

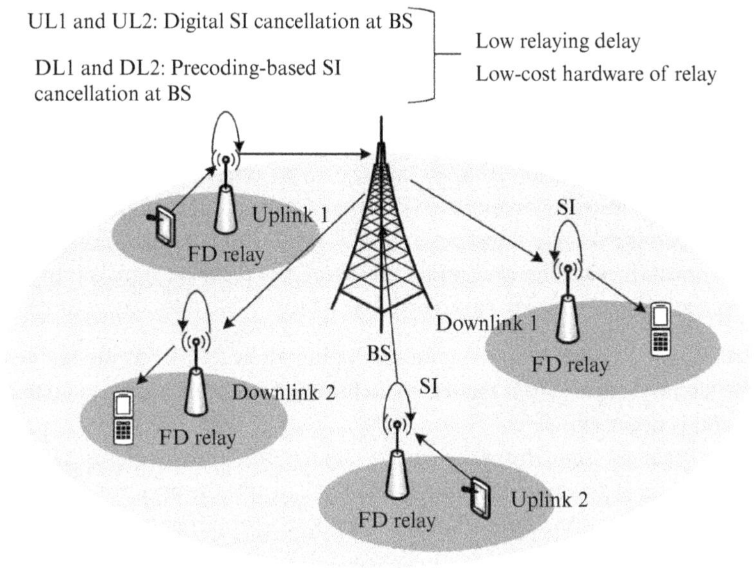

Figure 9.3 Diagram of residual SI cancellation and utilization in digital domain at base station with uplink, downlink, and relay selection. ©IEEE 2021. Reprinted with permission from [2].

that a cross-layer optimization problem is formulated to minimize transmission power in a closed form under the required quality of service, subject to a number of constraints on physical and media access control layers, such as packet dropping and joint power, bandwidth and subcarrier allocations. In signal processing, we are not able to employ traditional long-frame-based channel coding, such as turbo codes with long redundancy check, which reduces data rates and increases latency in finite block length. It is possible to use control-signaling-based channel coding in URLLC, for example, polar codes in control channels in 5G [11]. It is worth investigating advanced channel coding for FD relay in URLLC to be robust against the SI. The other signal processing parts are related to channel estimation, Carrier Frequency Offset (CFO) estimation, I/Q imbalance compensation, phase noise estimation, requiring a number of pilots. In the literature, many estimation approaches have been proposed which have pros and cons. However, the number of pilots is limited in short-frame transmissions. Therefore, the same pilots can be utilized efficiently as much as possible to accomplish multi-tasks. It has been shown in [20] that a single pilot can be used to jointly estimate SI, CFO, and channels for FD systems, providing high reliability and low latency.

Relaying Latency. Compared with direct transmissions, there is an extra latency, referred to as relaying latency, including RF SI cancellation, analog SI cancellation, and digital SI cancellation, which should be maintained at a low level for FD relay systems. *RF SI cancellation* is performed via antenna shielding and isolation plus directional or dual-polarized antennas. The SI can be suppressed by around 45 dB, providing a negligible latency in the propagation domain due to very short signal propagation between receive and transmit antennas at relay. If there is a strong line-of-sight component of SI, *analog SI cancellation* must be operated to cancel the significant component before the received signals transfer into a power amplifier. There are two reasons for this:

- First, the received power from SI is much higher than that from the source. Thus, the signal received from the source is hardly decoded correctly, if analog cancellation is not used.
- Second, the received signal from the source and SI could provide power higher than the limit of a power amplifier. The signal distortion occurs when the received signal reaches the saturation area of a power amplifier.

The residual SI is the non-line-of-sight component that can be reduced in the digital domain. Traditionally, *digital SI cancellation* allows the relay to work in the DF mode, requiring a complex signal processing based on an equivalent baseband model in a chip with a complex design, which increases processing latency at the relay, unable to meet the low-latency requirement. An efficient solution is to allow digital SI cancellation at the base station rather than at the relay, reducing relaying latency. The base station is equipped with high-quality electronic components and chips, and provides high computation capability in digital SI cancellation with low signal processing latency.

9.3 Self-Interference Cancellation with Low Latency and Low-Cost Hardware at Relay

Assume that Orthogonal Frequency Division Multiplexing (OFDM) modulation is employed. There is an integer OFDM symbol processing latency at the FD relay. The latency is at least one OFDM symbol to ensure that the received symbol is not correlated with the transmitted symbol within one OFDM symbol at the FD relay [24]. Traditionally, as shown in Figure 9.4, digital SI cancellation is conducted at the FD relay to mitigate the residual SI signals from the output of analog cancellation. The CSI of SI is estimated from transmit antennas to receive antennas at the FD relay by a number of pilots. The SI signal is regenerated using the estimated channel and the decoded signal in the previous OFDM symbol. The regenerated

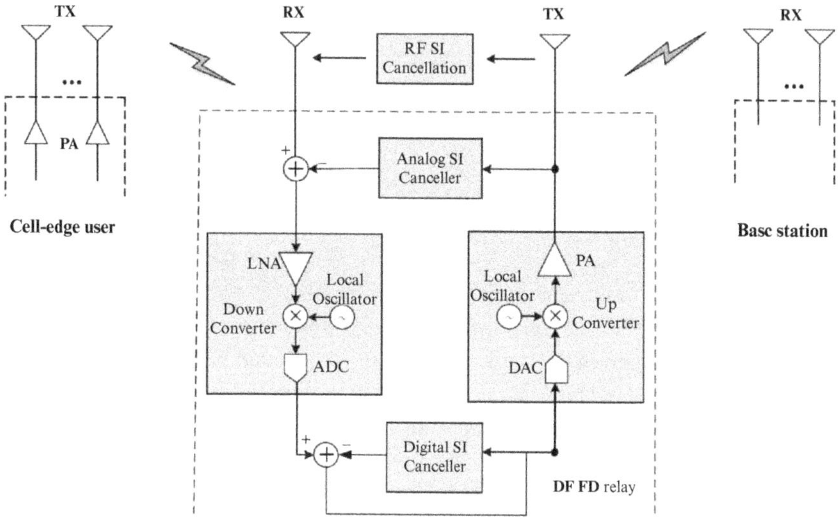

Figure 9.4 For traditional DF FD relay systems, the SI is canceled at the relay, using RF, analog, and digital cancellations, which leads to an increase in relaying latency. (LNA: low-noise amplifier, PA: power amplifier, ADC: analog-to-digital converter, DAC: digital-to-analog converter, RX: receiver, TX: transmitter). ©IEEE 2021. Reprinted with permission from [2].

SI signal is subtracted from the received signal in the current OFDM symbol to obtain a clear signal with no SI before transmission from the transmit antenna at the FD relay.

In order to achieve the goal of URLLC, digital SI cancellation can be processed at a base station rather than at a relay [10-12]. The main reasons can be listed as follows:

- **Low relaying latency**: Compared to direct transmissions, relaying latency plays a significant part in the end-to-end latency for the FD relay system, and should be minimized, by allowing the FD relay to work in the AF mode with no complex decoding and encoding process. Complex digital SI cancellation can be pushed as much as possible to the base station with high calculation capability to achieve low latency.
- **Low-cost hardware of relay**: Generally, a base station is surrounded by a number of relays to extend the coverage range and improve the reliability of cell-edge users. Digital SI cancellation that works at a number of DF FD relays requires multiple sets of complex chips and hardware, while just one set is probably

required at the base station to cancel residual SI digitally with the AF FD relay used. Therefore, the AF FD relay can reduce the cost of hardware, and a large number of low-cost AF FD relays can be deployed.

However, there is a lack of research investigating digital SI cancellation at the base station.

9.3.1 System Model

We consider an AF FD relay assisted OFDM system in the uplink, where the source is equipped with a single antenna, the AF relay has N_t transmit antennas and N_m receive antennas, and the destination is equipped with N_d antennas, as shown in Figure 9.5. In the FD mode, the relay transmits and receives signals at the same time and the same frequency, and hence SI is introduced. We apply an adaptive SI processing mode selection to maintain the SI energy at a reasonable level, which is determined by a threshold of SIR at the relay. In order to measure the residual SI after SI cancellation, we define β as the ratio of the SI power before and after suppression/cancellation. α_{SR} and α_{RD} denote the path loss from source to relay and the path loss from relay to destination, respectively. All the channels are modeled as Rayleigh frequency-selective fading channels, where the channel of L paths, remains constant for a frame duration of N_s OFDM blocks each with

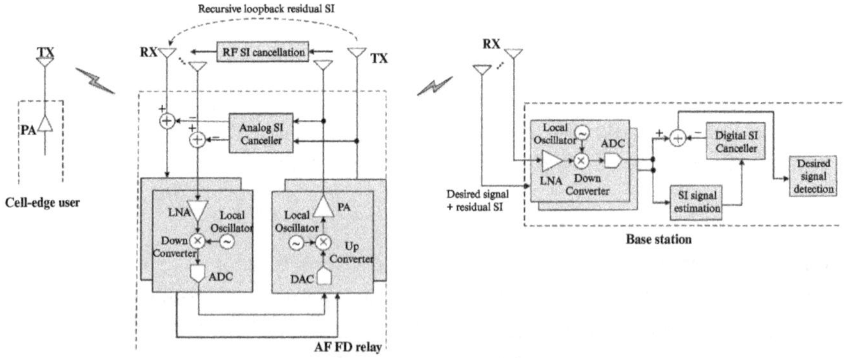

Figure 9.5 An AF FD relay system model is presented, with a single transmit antenna and a single receive antenna at the relay as well as a single antenna at the base station. In order to meet the low latency requirement, analog and RF cancellations are necessary to perform at the relay. The residual SI is transmitted together with the desired signal to the base station for further processing.(LNA: low-noise amplifier, PA: power amplifier, ADC: analog-to-digital converter, DAC: digital-to-analog converter, RX: receiver, TX: transmitter). ©IEEE 2021. Reprinted with permission from [2].

N subcarriers. Each OFDM block is prepended with a Cyclic Prefix (CP) of length L_{cp} ($L_{\text{cp}} \geq L-1$) before transmission, which is removed at the destination to avoid inter-block interference.

Let $s_{\text{U}}(n,i)$ denote the transmitted Quadrature Phase Shift Keying (QPSK) symbol on the nth ($n=0,1,\dots,N-1$) subcarrier in the ith ($i=0,1,\dots,N_{\text{s}}-1$) OFDM block. A non-redundant precoding mechanism [7] is expressed as $s_{\text{U}}(n,i)=\frac{1}{\sqrt{1+a^2}}[d(n,i)+ad_{\text{ref}}(n,i)]$, where $d_{\text{ref}}(n,i)$ is the reference symbol, which is used at the destination to eliminate the ambiguity caused by ICA, $d(n,i)$ is the source symbol, and a ($0 \leqslant a \leqslant 1$) is the precoding constant which makes a trade-off in power allocation between the source symbol and the reference symbol.

Let $H_m^{[\text{SR}]}(n)$ and $H_{m,t}^{[\text{RR}]}(n)$ denote the channel frequency response matrices on the nth subcarrier, between the user and the mth ($m=0,1,\dots,N_{\text{m}}-1$) receive antenna at the relay, and between the mth receive antenna and the tth ($t=0,1,\dots,N_{\text{t}}-1$) transmit antenna at the relay, respectively. The received signal in the frequency domain on the nth subcarrier and the mth receive antenna at the relay is given by

$$r_m(n,i) = \underbrace{\sqrt{P_{\text{s}}\alpha_{\text{SR}}}H_m^{[\text{SR}]}(n)s_{\text{U}}(n,i)}_{\text{Desired Signal}} + \underbrace{\sqrt{\frac{1}{\beta}}\sum_{t=0}^{N_{\text{t}}-1} H_{m,t}^{[\text{RR}]}(n)r_t(n,i-1)+z_m(n,i)}_{\text{Residual SI}}, \tag{9.1}$$

where P_{s} is the transmitted power at the source, and $z_m(n,i)$ is the Additive White Gaussian Noise (AWGN) with zero mean and variance of N_0. $r_t(n,i) = \sqrt{\beta_{\text{PA}}}r_m(n,i)$ is the transmitted signal on the nth subcarrier at the tth transmit antenna of the relay, where β_{PA} is the amplification power at the relay. The received signal $y_d(n,i)$ on the nth subcarrier at the dth ($d=0,1,\dots,N_{\text{d}}-1$) received antenna of the destination is written as

$$y_d(n,i) = \sum_{t=0}^{N_{\text{t}}-1} \sqrt{\beta_{\text{PA}}\alpha_{\text{RD}}}H_{d,t}^{[\text{RD}]}(n)r_m(n,i) + z_d(n,i), \tag{9.2}$$

where $H_{d,t}^{[\text{RD}]}(n)$ is the channel frequency response from the tth transmit antenna at the relay to the dth receive antenna at the destination, and $z_d(n,i)$ is the noise. Substituting (9.1) into (9.2) yields

$$\tilde{y}_d(n,i) = \underbrace{H_{\text{U},d}(n)s_{\text{U}}(n,i)}_{\text{Desired Signal}} + \underbrace{H_{\text{I},d}(n)s_{\text{I}}(n,i)}_{\text{Residual SI}} + \underbrace{\tilde{z}_d(n,i)}_{\text{Equivalent Noise}}, \tag{9.3}$$

where $s_{\mathrm{I}}(n,i) = s_{\mathrm{U}}(n,i-1)$ is the SI, $H_{\mathrm{U},d}(n) = \sqrt{\beta_{\mathrm{PA}} P_{\mathrm{s}} \alpha_{\mathrm{SR}} \alpha_{\mathrm{RD}}} \sum_{t_1=0}^{N_{\mathrm{t}}-1} H_{d,t_1}^{[\mathrm{RD}]}(n)$ $H_{t_1}^{[\mathrm{SR}]}(n)$ is the equivalent channel frequency response of the desired signal on the dth antenna of the destination. We assume that the number of transmit antennas is equal to the number of receiver antennas at the relay, $N_{\mathrm{t}} = N_{\mathrm{m}}$. $H_{\mathrm{I},d}(n) = \sqrt{\frac{\beta_{\mathrm{PA}}^2 P_{\mathrm{s}} \alpha_{\mathrm{SR}} \alpha_{\mathrm{RD}}}{\beta}} \sum_{t_1=0}^{N_{\mathrm{t}}-1} \sum_{t=0}^{N_{\mathrm{t}}-1} H_{d,t_1}^{[\mathrm{RD}]}(n) H_{t_1,t}^{[\mathrm{RR}]}(n) H_{t}^{[\mathrm{SR}]}(n)$ is the equivalent channel frequency response of SI, and $\tilde{z}_d(n,i)$ is the equivalent noise, as expressed in (9.4).

$$
\begin{aligned}
\tilde{z}_d(n,i) =& \sqrt{\frac{\beta_{\mathrm{PA}}^2 \alpha_{\mathrm{RD}}}{\beta}} \sum_{t_1=0}^{N_{\mathrm{t}}-1} \sum_{t=0}^{N_{\mathrm{t}}-1} \sum_{t_2=0}^{N_{\mathrm{t}}-1} H_{d,t_1}^{[\mathrm{RD}]}(n) H_{t_1,t}^{[\mathrm{RR}]}(n) H_{t,t_2}^{[\mathrm{RR}]}(n) r_{t_2}(n,i-2) \\
&+ \sqrt{\frac{\beta_{\mathrm{PA}}^2 \alpha_{\mathrm{RD}}}{\beta}} \sum_{t_1=0}^{N_{\mathrm{t}}-1} \sum_{t=0}^{N_{\mathrm{t}}-1} H_{d,t_1}^{[\mathrm{RD}]}(n) H_{t_1,t}^{[\mathrm{RR}]}(n) z_t(n,i-1) \\
&+ \sqrt{\beta_{\mathrm{PA}} \alpha_{\mathrm{RD}}} \sum_{t_1=0}^{N_{\mathrm{t}}-1} H_{d,t_1}^{[\mathrm{RD}]}(n) z_{t_1}(n,i) + z_d(n,i),
\end{aligned}
\tag{9.4}
$$

9.3.2 SI Power Attenuation

The previous OFDM symbol from the transmit antenna at relay is taken as an SI interfering with the current OFDM symbol from cell-edge users, which forms an SI loop with cumulative effect over all previous OFDM symbols. In other words, the SI consists of summing up all previous OFDM symbols, if the residual SI is not canceled at the relay. However, the power of the residual SI consecutively fades exponentially with the number of transmitted OFDM symbols in the past, as the residual SI channel gain is relatively low [5, 15, 19], with the line-of-sight component being mitigated by analog cancellation. Therefore, the current OFDM symbol only interferes with a small number of previously successive OFDM symbols [5, 15, 19] that play a significant part in the SI power. The residual SI power of the other left OFDM symbols is too low, and can be treated as noise [5, 15, 19].

9.3.3 Digital Cancellation Procedure

Because of the high complexity of pseudo-inverse and convolution operations in the time domain, digital SI cancellation should be implemented in the frequency domain by utilizing the element-wise calculations. As shown in Figure 9.5, the digital cancellation procedure at the base station is described as follows:

- The SI includes equivalent channels from the transmit antenna to the receive antenna at the relay, and further to the base station. There is no need to estimate channels separately, as the equivalent channels can be estimated by pilots transmitted together with source data. The received signal includes equivalent channels from cell-edge users to the relay, and further to the base station. The equivalent channels can also be estimated by the pilots.
- The SI signal is reconstructed via a number of previously decoded OFDM symbols and SI channels.
- The SI is subtracted digitally from the current received OFDM symbol at the base station.

To some extent, the channel estimation error destroys the reconstruction of the SI signal, and gives rise to additional interference and system degradation. Also, the number of OFDM symbols considered as SI depends on the Signal-to-Residual-Self-Interference Ratio (SRSIR). It has been shown in [19] that two previous OFDM symbols dominate 99.9 percent power of the residual SI when the ratio of the SI to relay-to-destination channel variances is as low as −40 dB.

9.3.4 Precoding-based SI Cancellation

For a downlink, digital SI cancellation is not preferable at a user terminal, since most mobile phones and devices are composed of cheap electronic components and are incapable of dealing with complex signal processing. Hence, it is preferable to perform precoding-based SI cancellation at a base station with high computation capability, allowing low relaying latency. The reconstructed SI signal with a known CSI is subtracted from the desired signal before transmission, which does not dramatically reduce the signal power, because the desired signal provides a channel gain higher than the SI signal. The propagation of the desired signal between the base station and the relay is line-of-sight, while the SI channels are non-line-of-sight components after analog cancellation at the relay. The SI can be canceled automatically at the receive antennas of the relay, when the received signal is superimposed with the SI from the transmit antennas of the relay.

9.4 An Adaptive Self-Interference Processing with High Reliability and Low Latency

As SI originates from previous OFDM symbols, the residual SI can be utilized to enhance reliability and the degree of freedom in signal processing rather than being canceled as much as possible in the digital domain, when being transmitted together with the desired signal to the base station.

9.4.1 System Model

We consider an FD AF relay assisted OFDM system in the uplink depicted in Figure 9.6. The source is equipped with a single antenna. The AF relay has N_t transmit antennas and N_m receive antennas, $N_t = N_m$. The destination is equipped with N_d receive antennas. In the FD mode, the relay transmits and receives signals at the same time and the same frequency, which causes SI. We apply an adaptive SI processing mode selection to maintain the SI energy at a reasonable level. The details of adaptive SI processing mode selection strategy are presented in Section 3. The Information of Mode Selection (IMS) is transmitted with the transmitted signal from relay to destination, to assist ICA based adaptive signal separation or antenna regrouping. In order to measure the residual SI after SI cancellation, we define β as the ratio of the SI power before and after suppression/cancellation. α_{SR} and α_{RD} denote the path loss from source to relay and the path loss from relay to destination, respectively. All the channels are modeled as Rayleigh frequency-selective fading channels [16][6], where the channel of L paths remains constant for a frame duration of N_s OFDM blocks with N subcarriers. Each OFDM block is prepended with a Cyclic Prefix (CP) of length L_{cp}

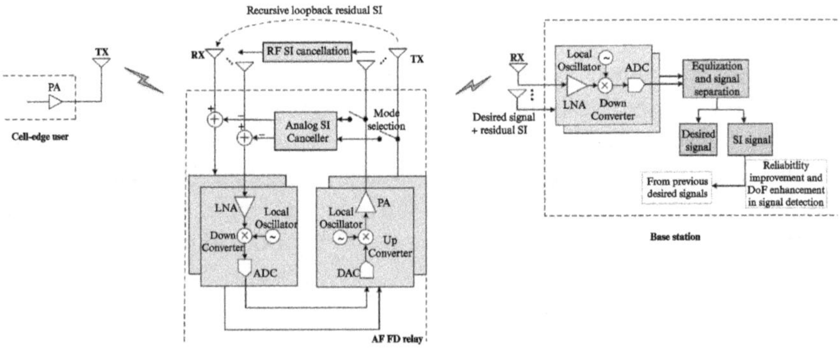

Figure 9.6 An AF FD relay system model with multiple transmit and multiple receive antennas at the relay as well as a multi-antenna base station. The residual SI after analog and RF cancellations is transmitted together with the desired signal to base station in the digital domain, and is utilized to improve reliability and enhance the degree of freedom in signal processing. (LNA: low-noise amplifier, PA: power amplifier, ADC: analog-to-digital converter, DAC: digital-to-analog converter, RX: receiver, TX: transmitter, DoF: degree of freedom). ©IEEE 2021. Reprinted with permission from [2].

($L_{\text{cp}} \geq L - 1$) before transmission, which is removed at the destination to avoid inter-block interference.

9.4.2 Adaptive SI Processing Mode Selection

In this section, we apply an adaptive SI processing mode selection to maintain the SI energy at a reasonable level. There are four SI processing modes described as follows: (a) PSAC-S: SI is canceled by Passive Suppression (PS) and Analog Cancellation (AC) at the relay and the residual SI is treated as the source at the destination; (b) PSAC-N: SI is canceled by PS and AC at the relay and the residual SI is treated as noise at the destination; (c) PS-S: SI is canceled by PS only at the relay and the residual SI is treated as the source at the destination; (d) PS-N: SI is canceled by PS at the relay and the residual SI is treated as noise at the destination. Two thresholds of SRSIR are applied at the relay to help determine the SI processing modes.

9.4.2.1 SRSIR at Relay

Let $q_m(n) = \frac{\beta_{\text{PA}}}{\beta} \sum_{t=0}^{N_t-1} |H_{m,t}^{[\text{RR}]}(n)|^2$ denote the power coefficient of the loop channel of the SI on the nth subcarrier. The SRSIR on the nth subcarrier at the mth antenna of the relay can be derived as

$$\gamma_{\text{R},m}(n) = 10\,\text{lg}\left(\frac{P_{\text{s}}\alpha_{\text{SR}}|H_m^{[\text{SR}]}(n)|^2}{\sum_{j=1}^{N_s-1} (q_m(n))^j\, P_{\text{s}}\alpha_{\text{SR}}|H_t^{[\text{SR}]}(n)|^2}\right). \tag{9.5}$$

Equation (9.5) can be summarized as

$$\gamma_{\text{R},m}(n) = 10\,\text{lg}\left(\frac{1 - q_m(n)}{q_m(n) - (q_m(n))^{N_s-1}}\right), \tag{9.6}$$

where $(q_m(n))^i$ can be estimated as

$$(\hat{q}_m(n))^i = \frac{P_m(n,i) - P_m(n,i-1)}{P_m(n,1)}, \tag{9.7}$$

where $P_m(n,i)$ is the power of the ith ($i = 0, 1, \ldots, N_s - 1$) OFDM block on the nth subcarrier at the mth receive antenna of the relay. Therefore, the system can realize switching of SI cancellation mode of PS and PSAC to ensure BER performance and reduce the power consumption of SI cancellation. The SRSIR on the nth subcarrier at the mth antenna of the relay can be estimated as

$$\hat{\gamma}_{\text{R},m}(n) = 10\,\text{lg}\left(\frac{1}{\sum_{j=1}^{N_s-1} (\hat{q}_m(n))^j}\right). \tag{9.8}$$

9.4.2.2 SRSIR Thresholds at the Relay for SI Processing Mode Selection

The SI processing mode selection depends on the power ratio of the desired signal and the residual SI signal after PS. Let $g_m(n) = P_s\alpha_{SR}|H_m^{[SR]}(n)|^2$ denote the power coefficient of the desired signal on the nth subcarrier, $P_{R,U}(n,i) = g_m(n)|s_U(n,i)|^2$ denote the power of the desired signal of current block on the nth subcarrier, and $P_{R,I}(n,i) = q_m(n)g_m(n)|s_U(n,i-1)|^2$ denote the power of the current SI signal block on the nth subcarrier, $P_{R,I}(n,i)$ and $P_{R,U}(n,i)$ are estimated as

$$\hat{P}_{R,I}(n,i) = (|P_m(n,i) - P_m(n,i-1)|), \tag{9.9}$$

$$\hat{P}_{R,U}(n,i) = \left(|P_m(n,i) - \sum_{k=1}^{i} \hat{P}_m(n,k)| \right). \tag{9.10}$$

When the power of the desired signal is equal to or less than that of the current SI signal block $P_{R,U}(n,i) \le P_{R,I}(n,i)$, the SI is canceled by PSAC. When $P_{R,U}(n,i) > P_{R,I}(n,i)$ the SI is suppressed only by PS. Note that the power of the desired signal is normalized $|s_U(n,0)|^2 = |s_U(n,1)|^2 = \dots = |s_U(n,N_s-1)|^2 = 1$. Therefore, a threshold of the SI processing mode selection is applied to determine whether the SI is canceled by PS or PSAC at the relay, which can be given as

$$\gamma_{Th1}(n,i) = \gamma_R(n) + 10\lg\left(\frac{P_{R,I}(n,i)}{P_{R,U}(n,i)}\right). \tag{9.11}$$

At the destination, the ICA assisted adaptive signal processing algorithm depends on the power of the SI at the relay. When the power of the SI signal is relatively high, the residual SI is regarded as a useful signal. When the power of the SI signal is relatively low, the residual SI is regarded as noise. When $P_{R,U}(n,i) \le \sum_{j=1}^{\infty}(q_m(n))^{j-1}P_{R,I}(n,i-j)$, the residual SI is regarded as a useful signal. When $P_{R,U}(n,i) > \sum_{j=1}^{\infty}(q_m(n))^{j-1}P_{R,I}(n,i-j)$, the residual SI is regarded as noise. Therefore, another higher threshold of the SI processing mode selection is applied to determine whether the residual SI is regarded as a useful signal or noise at the destination, which can be given as

$$\gamma_{Th2}(n,i) = \gamma_R(n) + 10\lg\left(\frac{\sum_{j=1}^{\infty}(q_m(n))^{j-1}P_{R,I}(n,i-j)}{P_{R,U}(n,i)}\right). \tag{9.12}$$

The conditions for SI processing mode selection are shown in Table 9.1. Where $\gamma_R^{[PS]}$ denotes the SRSIR at relay after PS, $\gamma_R^{[PSAC]}$ denotes the SRSIR at relay after PSAC. $\gamma_{Th1}(n,i)$ and $\gamma_{Th2}(n,i)$ are abbreviated as γ_{Th1} and γ_{Th2}. When SRSIR is less than the lower threshold, the SI is canceled by PSAC, which is suitable for PSAC-S and PSAC-N modes. When SRSIR is larger than the lower threshold, the SI is suppressed only by PS, which is suitable for PS-S and PS-N modes. When SRSIR is between the lower and higher threshold, the residual SI is treated as a

Table 9.1 Conditions of Adaptive Mode Selection for SI Processing.

SI processing Mode	Condition
PSAC-S	$\gamma_R^{[PS]} < \gamma_{Th1}, \gamma_{Th1} \leq \gamma_R^{[PSAC]} < \gamma_{Th2}$
PSAC-N	$\gamma_R^{[PS]} < \gamma_{Th1}, \gamma_R^{[PSAC]} \geq \gamma_{Th2}$
PS-S	$\gamma_{Th1} \leq \gamma_R^{[PS]} < \gamma_{Th2}$
PS-N	$\gamma_R^{[PS]} \geq \gamma_{Th2}$

useful signal to enhance the degree of freedom in signal detection, which is suitable for PSAC-S and PS-S modes. When SRSIR is larger than the higher threshold, the residual SI is treated as a green source for Energy Harvesting (EH), which is suitable for PSAC-N and PS-N modes.

9.4.3 ICA Assisted Adaptive Signal Separation, Detection, and Energy Recycling

We propose an adaptive signal detection and energy recycling structure for different SI processing modes. According to the SI processing mode selection, we propose two approaches. In the modes of PSAC-S and PS-S, ICA is employed on at least two receive antennas at the destination to allow separation of the desired signal and SI, and then the desired signal is decoded via ambiguity elimination. In the modes of PSAC-N and PS-N, at least one receive antenna with higher SRSIR is selected for signal detection, while the other antennas are used for energy recycling. This is essentially different from the conventional FD transmission methods [27] [17], where SI is canceled as much as possible. Moreover, the proposed structure provides higher resource utilization over the Time-Splitting (TSP) [9] and Power-Splitting (PSP) [28] based self-energy recycling approaches, since the TSP and PSP based approaches fail to achieve consecutive FD transmission all the time due to the use of partial resources (time or power) for EH.

9.4.3.1 ICA Based Signal Separation and Detection for Low SRSIR Case

In the SI processing mode of PSAC-S or PS-S, the SRSIR of the relay is less than the higher threshold $\gamma_R < \gamma_{Th2}$, ICA is employed on at least two antennas to separate the desired signal and SI. The cross correlation between ICA separated signals and reference signals is explored to detect the desired signals.

ICA is an efficient blind source separation technique by maximizing the non-Gaussianity of received signals. Since ICA requires no training for channel estimation, it is more spectrum efficient than the conventional channel estimation methods [10]. Among different ICA-based methods, the Joint Approximate Diagonalization of Eignmatrices (JADE) [4], a well established batch algorithm based on joint diagonalization of the cumulant matrices of the received components, requires shorter data sequences than other ICA methods. Thus, JADE is employed in this chapter to perform semi-blind joint signal separation and detection.

Let $\mathbf{s}(n,i) = [s_{\mathrm{U}}(n,i), s_{\mathrm{I}}(n,i)]^{\mathrm{T}}$ denote the transmitted signal vector. Let $\mathbf{y}(n,i) = [y_0(n,i), y_1(n,i), \dots, y_{N_{\mathrm{d}}-1}(n,i)]^{\mathrm{T}}$ denote all received signals from N_{d} receive antennas of the destination on the nth subcarrier in the ith OFDM block, which is calculated as

$$\mathbf{y}(n,i) = \mathbf{H}(n)\mathbf{s}(n,i) + \tilde{\mathbf{z}}(n,i), \tag{9.13}$$

where $\mathbf{H}(n) = [\mathbf{h}_{\mathrm{U}}(n), \mathbf{h}_{\mathrm{I}}(n)]$ with $\mathbf{h}_{\mathrm{U}}(n) = [H_{\mathrm{U},0}(n), H_{\mathrm{U},1}(n), \dots, H_{\mathrm{U},N_{\mathrm{d}}-1}(n)]^{\mathrm{T}}$ and $\mathbf{h}_{\mathrm{I}}(n) = [H_{\mathrm{I},0}(n), H_{\mathrm{I},1}(n), \dots, H_{\mathrm{I},N_{\mathrm{d}}-1}(n)]^{\mathrm{T}}$, and $\tilde{\mathbf{z}}(n,i) = [\tilde{z}_0(n,i), \tilde{z}_1(n,i), \dots, \tilde{z}_{N_{\mathrm{d}}-1}(n,i)]^{\mathrm{T}}$. As the received signals $\mathbf{y}(n,i)$ in (9.13) are a linear mixture of the desired signal $s_{\mathrm{U}}(n,i)$ and the SI $s_{\mathrm{I}}(n,i)$ on each subcarrier. Thus, JADE is employed on $\mathbf{y}(n,i)$ in (9.13) to perform separation of the desired signal and SI. For the ICA approach, we can obtain the equalized signals as $\tilde{\mathbf{s}}_{\mathrm{LSIR}}(n,i) = [\tilde{s}_{\mathrm{LSIR,U}}(n,i), \tilde{s}_{\mathrm{LSIR,I}}(n,i)]^{\mathrm{T}}$ [4, 10, 14], de-rotated by the phase of each substream as follows [14]

$$\check{\mathbf{s}}_{\mathrm{LSIR}}(n,i) = \mathbf{G}(n)\tilde{\mathbf{s}}_{\mathrm{LSIR}}(n,i), \tag{9.14}$$

where $\check{\mathbf{s}}_{\mathrm{LSIR}}(n,i) = [\check{s}_{\mathrm{LSIR,U}}(n,i), \check{s}_{\mathrm{LSIR,I}}(n,i)]^{\mathrm{T}}$, $\mathbf{G}(n) = \mathrm{diag}\{[g_{\mathrm{U}}(n), g_{\mathrm{I}}(n)]^{\mathrm{T}}\}$, with $g_{\mathrm{U}}(n) = \alpha_{\mathrm{U}}(n)/|\alpha_{\mathrm{U}}(n)|$, $\alpha_{\mathrm{U}}(n) = \{(1/N_{\mathrm{s}})\sum_{i=0}^{N_{\mathrm{s}}-1}[\tilde{s}_{\mathrm{LSIR,U}}(n,i)]^4\}^{-\frac{1}{4}}e^{j\frac{\pi}{4}}$ and $g_{\mathrm{I}}(n) = \alpha_{\mathrm{I}}(n)/|\alpha_{\mathrm{I}}(n)|$, $\alpha_{\mathrm{I}}(n) = \{(1/N_{\mathrm{s}})\sum_{i=0}^{N_{\mathrm{s}}-1}[\tilde{s}_{\mathrm{LSIR,I}}(n,i)]^4\}^{-\frac{1}{4}}e^{j\frac{\pi}{4}}$. $\alpha_{\mathrm{U}}(n)$ and $\alpha_{\mathrm{I}}(n)$ denote the factors obtained from $\tilde{s}_{\mathrm{LSIR,U}}(n,i)$ and $\tilde{s}_{\mathrm{LSIR,I}}(n,i)$ for QPSK modulation, respectively [14].

In the next step, we need to find the desired signal and the SI. Define $\rho_{\mathrm{U}}(n)$ and $\rho_{\mathrm{I}}(n)$ as the cross-correlations between two equalized signals and the reference signal, respectively, which are given by

$$\rho_{\mathrm{U}}(n) = \frac{1}{N_{\mathrm{s}}}\sum_{i=0}^{N_{\mathrm{s}}-1}\{\check{s}_{\mathrm{LSIR,U}}(n,i)d_{\mathrm{ref}}^{*}(n,i)\}. \tag{9.15}$$

$$\rho_{\mathrm{I}}(n) = \frac{1}{N_{\mathrm{s}}}\sum_{i=0}^{N_{\mathrm{s}}-1}\{\check{s}_{\mathrm{LSIR,I}}(n,i)d_{\mathrm{ref}}^{*}(n,i)\}. \tag{9.16}$$

By applying permutation ambiguity elimination, the order of the desired signals can be identified by

$$\hat{\mathrm{U}} = \max\{|\rho_{\mathrm{U}}(n)|, |\rho_{\mathrm{I}}(n)|\}. \tag{9.17}$$

By applying quadrant ambiguity elimination [7], the desired signal is given by

$$\hat{s}_{\mathrm{LSIR},\hat{\mathrm{U}}}(n,i) = \left[e^{-j\frac{\pi}{4}} \mathrm{sign}\left(\frac{\rho_{\hat{\mathrm{U}}}(n)}{|\rho_{\hat{\mathrm{U}}}(n)|} e^{j\frac{\pi}{4}}\right)\right]^{-1} \check{s}_{\mathrm{LSIR},\hat{\mathrm{U}}}(n,i). \tag{9.18}$$

The order of SI can be identified by

$$\hat{\mathrm{I}} = \min\{|\rho_{\mathrm{U}}(n)|, |\rho_{\mathrm{I}}(n)|\}. \tag{9.19}$$

With the identified SI order of $\hat{\mathrm{I}}$, the SI $\check{s}_{\mathrm{LSIR},\hat{\mathrm{I}}}(n,i)$ are extracted from the received signals. The quadrant ambiguity elimination is not required for the SI.

9.4.3.2 Space-splitting Based Signal Separation and ICA Based Signal Detection for the High SRSIR Case

In the SI processing mode of PSAC-N or PS-N, when the SRSIR of the relay is equal to or larger than the higher threshold $\gamma_{\mathrm{R}} \geq \gamma_{\mathrm{Th2}}$, the SI power is low. There is no need to separate the SI from the received signals. The SI power can be incorporated into the noise. At least one of the antennas with the highest SIRs can be used for signal detection, while the other antennas are utilized for space-splitting based energy recycling.

From (9.3), the SRSIR on the nth subcarrier of the destination can be derived as

$$\gamma_d = 10\lg\left(\frac{P_{\mathrm{U},d}}{P_{\mathrm{I},d}}\right), \tag{9.20}$$

where $P_{\mathrm{U},d} = \sum_{n=0}^{N-1}\sum_{i=0}^{N_s-1} P_{\mathrm{U},d}(n,i)$ with $P_{\mathrm{U},d}(n,i) = |H_{\mathrm{U},d}(n)s_{\mathrm{U}}(n,i)|^2$ denotes the power of the desired signal on the dth receive antenna at the destination, and $P_{\mathrm{I},d} = \sum_{n=0}^{N-1}\sum_{i=0}^{N_s-1} P_{\mathrm{I},d}(n,i)$ with $P_{\mathrm{I},d}(n,i) = |H_{\mathrm{I},d}(n)s_{\mathrm{I}}(n,i)|^2$ denotes the power of SI on the dth receive antenna at the destination.

The transmitted signals are assumed to have constant power. The power difference between a number of consecutive OFDM blocks can be used to estimate the SRSIR, as the SI is from the previous blocks. Thus, QPSK modulation is used in this chapter. The power of the received signal in the ith block at the dth antenna of the destination is given by

$$P_d(n,i) = \begin{cases} P_{\mathrm{U},d}(n,i) + P_{\mathrm{Z},d}(n,i), & \text{if } \ i = 0 \\ P_{\mathrm{U},d}(n,i) + P_{\mathrm{I},d}(n,i) & \\ +P_{\mathrm{Z},d}(n,i), & \text{if } \ i = 1, \dots, N_s - 1 \end{cases} \tag{9.21}$$

where $P_{\mathrm{Z},d}(n,i) = |\tilde{z}_d(n,i)|^2$ is the equivalent noise power. Due to the constant power of the transmitted signals, we have $P_{\mathrm{U},d}(n,0) = P_{\mathrm{U},d}(n,1)\cdots = P_{\mathrm{U},d}(n, N_s - 1)$. Since the SI is from the previous block, the SI power can be estimated from the power difference between received signals over two consecutive OFDM blocks as

$$\hat{P}_{\mathrm{I},d}(n,i) = (|P_d(n,i) - P_d(n,i-1)|). \tag{9.22}$$

The power of the desired signal in the ith block is estimated as

$$\hat{P}_{\mathrm{U},d}(n,i) = \left(|P_d(n,i) - \sum_{k=1}^{i} \hat{P}_{\mathrm{I},d}(n,k)| \right). \tag{9.23}$$

The estimation of SRSIR is given by

$$\hat{\gamma}_d = 10\lg\left(\frac{\hat{P}_{\mathrm{U},d}}{\hat{P}_{\mathrm{I},d}}\right). \tag{9.24}$$

where $\hat{P}_{\mathrm{U},d} = \sum_{n=0}^{N-1}\sum_{i=0}^{N_s-1} \hat{P}_{\mathrm{U},d}(n,i)$ is the estimated power of the desired signal and $\hat{P}_{\mathrm{I},d} = \sum_{n=0}^{N-1}\sum_{i=0}^{N_s-1} \hat{P}_{\mathrm{I},d}(n,i)$ is the estimated power of the SI at the dth antenna.

We select a number of antennas with estimated N_q highest SIRs for signal detection, i.e. ($\hat{\gamma}_0 > \hat{\gamma}_1, \dots > \hat{\gamma}_{N_\mathrm{q}-1}$), while the rest of N_b antennas are used for space-splitting based energy recycling. (9.3) can be re-expressed as

$$y_d(n,i) = \underbrace{H_{\mathrm{U},d}(n)s_\mathrm{U}(n,i)}_{\text{Desired Signal}} + \underbrace{\check{z}_d(n,i)}_{\text{Equivalent Noise}}, \tag{9.25}$$

where $\check{z}_d(n,i) = \tilde{z}_d(n,i) + H_{\mathrm{I},d}(n)s_\mathrm{I}(n,i)$ is the equivalent noise including SI. Assume that N_q antennas at the destination are selected for signal detection by the proposed space-splitting approach. Let $\mathbf{y}_{\mathrm{HSIR}}(n,i) = [y_0(n,i), y_1(n,i), \dots, y_{N_\mathrm{q}-1}(n,i)]^\mathrm{T}$ denote the received signals of N_q receive antennas from a total number of N_d antennas at the destination on the nth subcarrier in the ith OFDM block, expressed as

$$\mathbf{y}_{\mathrm{HSIR}}(n,i) = \mathbf{h}_\mathrm{U}(n)s_\mathrm{U}(n,i) + \check{\mathbf{z}}(n,i), \tag{9.26}$$

where $\mathbf{h}_\mathrm{U}(n) = [H_{\mathrm{U},0}(n), H_{\mathrm{U},1}(n), \dots, H_{\mathrm{U},N_\mathrm{d}-1}(n)]^\mathrm{T}$, and $\check{\mathbf{z}}(n,i) = [\check{z}_0(n,i), \check{z}_1(n,i), \dots, \check{z}_{N_\mathrm{q}-1}(n,i)]^\mathrm{T}$. JADE [4] is employed on $\mathbf{y}_{\mathrm{HSIR}}(n,i)$ in (9.26) for signal detection to obtain the equalized signal $\tilde{s}_{\mathrm{HSIR}}(n,i)$, de-rotated by the phase of each substream as follows

$$\check{s}_{\mathrm{HSIR}}(n,i) = \frac{\alpha(n)}{|\alpha(n)|}\tilde{s}_{\mathrm{HSIR}}(n,i), \tag{9.27}$$

where $\alpha(n) = \{(1/N_\mathrm{s})\sum_{i=0}^{N_s-1}[\tilde{s}_{\mathrm{HSIR}}(n,i)]^4\}^{-\frac{1}{4}}e^{j\frac{\pi}{4}}$ [14][7] is obtained from $\tilde{s}_{\mathrm{HSIR}}(n,i)$. The cross-correlation $\rho_{\mathrm{HSIR}}(n)$ between equalized signal $\check{s}_{\mathrm{HSIR}}(n,i)$ and reference signal $d_{\text{ref}}(n,i)$ on the nth subcarrier is defined as

$$\rho_{\text{HSIR}}(n) = \frac{1}{N_{\text{s}}} \sum_{i=0}^{N_{\text{s}}-1} \{\check{s}_{\text{HSIR}}(n,i) d_{\text{ref}}^{*}(n,i)\}. \tag{9.28}$$

The remaining quadrant ambiguity is solved by [7]

$$\hat{s}_{\text{HSIR}}(n,i) = \left[e^{-j\frac{\pi}{4}} \text{sign} \left(\frac{\rho_{\text{HSIR}}(n)}{|\rho_{\text{HSIR}}(n)|} e^{j\frac{\pi}{4}} \right) \right]^{-1} \check{s}_{\text{HSIR}}(n,i). \tag{9.29}$$

When a number of N_{q} antennas are selected for signal detection, the rest of the N_{b} antennas are used for space-splitting based energy recycling. The recycled power collected from N_{b} antennas can be expressed as

$$P_{\text{HSIR,SS}} = \sum_{d=0}^{N_{\text{b}}-1} \sum_{i=0}^{N_{\text{s}}-1} \sum_{n=0}^{N-1} P_d(n,i). \tag{9.30}$$

A logistic function based model is adopted [1] [3], which well describes the saturation effect at high input power level as well as the breakdown effect low input power level. The logistic function based efficiency can be calculated as

$$P_{\text{EH}} = \frac{\Psi_{\text{EH}} - P_{\text{sat}}\Omega}{1-\Omega}, \tag{9.31}$$

where $\Psi_{\text{EH}} = \frac{P_{\text{sat}}}{1+\exp(-b_1(P_{\text{in}}-b_2))}$ is the conventional logistic function with respect to the input power of the harvester P_{in}. P_{sat} denotes the maximum harvested power when the harvesting circuit is saturated. Parameter Ω is calculated as $\Omega = \frac{1}{1+\exp(b_1 b_2)}$, where b_1 and b_2 are parameters accounting for physical hardware phenomena, such as the turn-on voltage of the diode and the maximum output power of the rectifier. The values of P_{sat}, b_1 and b_2 are related to the specific circuit design and can be determined by the curve fitting method. Using the EH model in (9.31), the overall harvested energy based on space-splitting is given by

$$P_{\text{HSIR}} = TP_{\text{EH}}(P_{\text{HSIR,SS}}). \tag{9.32}$$

9.4.4 Virtual MISO Systems

The residual SI is a delayed version of the desired signal, and can be modelled as a virtual Multiple-Input Single-Output (MISO) system [19] together with the desired signal, if there is a single receive antenna at the FD relay and at the base station, respectively. Thus, the residual SI can be treated as a self-coding to build a structure of space time code at the received signals at the base station to enhance the degree of freedom in signal processing. It has been shown in [15] that a spatial diversity is built at the base station with two virtual antennas, where one received signal is from the relay and the other one is from the cell-edge user via direct transmissions. The desired signal is mixed with the SI signal, to improve reliability in terms of BER, based on bit-interleaved coded modulation [15].

9.4.5 Virtual MIMO Systems

The residual SI signal and the desired signal can be viewed as multiple input signals to build a virtual Multiple-Input Multiple-Uutput (MIMO) system, if multiple antennas are equipped at transmit and receive sides of the FD relay as well as at the base station. As shown in Figure 9.6, a receive and a transmit antenna are grouped into a transmission link at the relay. The received signal at AF FD relay is first operated by analog SI cancellation, and then re-transmitted to a multi-antenna base station. By building an equivalent MIMO channel model, the desired and SI signals are equalized on each subcarrier in the frequency domain at the base station. The equalized SI is a replica of previous OFDM symbols, and is utilized to combine with the equalized signal in previous OFDM symbols to maximize the desired signal power for enhancement of the degree of freedom in signal processing as well as reliability improvement. It has been shown that up to 10 dB signal power enhancement can be yielded [5]. The procedure of SI utilization is described as follows:

- Build an equivalent virtual MIMO system model with respect to the SI and desired signals.
- Estimate SI and the desired signal channels.
- Employ zero-forcing or minimum mean squared error criterion to equalize the desired and SI signals.
- Apply the equalized SI signals to enhance the power of the desired signal in previous OFDM symbols.

In order to successfully equalize the desired and SI signals, the number of antennas required is the same as or larger than the number of OFDM symbols taken as SI plus one desired signal. For example, if the SI power is dominated by the previous one OFDM symbol, the SI signal generated from the previous OFDM symbol and the desired signal are used to form two virtual inputs, which requires at least two receive and two transmit antennas at the FD relay along with at least two receive antennas at the base station. The degree of freedom in signal processing is enhanced at the expense of a number of additional antennas used. This is not a big problem, as current and future devices are equipped with multiple antennas.

9.4.6 Adaptive Blind Source Separation

Since the desired and SI signals can be formulated as a linear MIMO model, a number of blind source separation approaches [5, 20] can be employed to separate the received signals blindly, such as subspace [20] and independent component analysis [5]. Blind source separation provides high spectral efficiency to recover the desired and SI signals with no knowledge of CSI and therefore no pilot is required, which can reduce the impact of channel estimation errors resulting from a limited number of pilots used in short-frame transmission in URLLC. It has been shown in [5] that the previous one OFDM symbol is considered as a significant component in the SI and can be extracted by blind source separation, with the residual OFDM symbols in the SI being treated as noise, when SIR is at least around 20 dB. However, the separated signals present some ambiguity drawbacks in terms of phase and permutation. Phase ambiguity is referred to as a number of phase shifts in the equalized signals, while permutation ambiguity is to sort out the disorder problem in the separated signals. In other words, the blindly separated signals should recognize which is the desired signal and which is the SI signal. The ambiguities can be eliminated by precoding or short pilots. Nevertheless, blind source separation is based on data statistics, and requires a data frame that is not too short.

9.4.7 Millimeter-wave Transmissions

FD relay is applied for millimeter-wave transmissions to extend network coverage, as the propagation attenuation is significantly stronger in small wavelength. Due to the fact that millimeter-wave transmissions are easily blocked by obstacles, SI power can be significantly reduced by employing antenna shielding between the receive and transmit sides of relay and highly directional transmit antennas with narrow beamwidth directly to the destination [27]. The SI channels between receive and transmit antennas of the relay can be modelled as Rayleigh fading collected from reflected waves, with strong line-of-sight path being blocked. Thus, analog cancellation is not required at the relay, and the recursive loopback residual SI can be canceled in the digital domain at the base station if the relay works in the AF mode. Also, the residual SI can be utilized to enhance the degree of freedom in signal processing at the base station with multiple antennas.

9.5 Reliability and Latency Evaluation

System parameters are set as follows: the source and destination are equipped with a single transmit antenna and $N_\mathrm{d} = 2$ receive antennas respectively; the relay is equipped with $N_\mathrm{t} = 2$ transmit and $N_\mathrm{m} = 2$ receiver antennas; QPSK modulation scheme is utilized; the channel follows an exponential delay profile with a normalized Root Mean Square (RMS) delay spread of 1.4; a slot level is considered with $N_s = 7$ OFDM symbols [13]; The subcarrier spacing is 60 kHz; the bandwidth is set as 20 MHz. RF and analog cancellations are up to 30 dB and 40 dB, respectively;

transmission power ranges from 10 dBm to 30 dBm; noise floor is −95 dBm; the precoding constant is set as $a = 0.26$; the PA gain is $\beta_{\mathrm{PA}} = 36$ dB; the noise figure of each LNA is $F = 1.5$ dB; the TSP and PSP coefficients ω_{T} and ω_{P} of the existing methods [9] [28] are both set as 0.5; regarding the non-linear EH model parameters, we assume $P_{\mathrm{sat}} = 300$ mW, $b_1 = 10$, and $b_2 = 0.14$ [1]. All the simulation results are averaged over Monte-Carlo runs with independent source data, noise, and channel realizations.

9.5.1 Reliability Evaluation

Block error rate is used as a performance metric to describe the performance of the encoding and decoding processes using a cyclic redundancy check in the literature [12], which can correct some error bits. However, a cyclic redundancy check is not used in this chapter. Error bits cannot be detected and corrected. Thus, it is better to use BER as a performance metric to reflect the performance of the algorithm design at the receiver. Therefore, BER is used as a performance metric.

Figure 9.7 shows BER performance of the AF FD relay system with adaptive digital SI cancellation or utilization at the base station. The AF FD relay system with adaptive SI processing provides BER lower than 10^{-5}, from Signal to Noise Ratio (SNR) being 22 dB to 30 dB, which meets the 99.999 % reliability requirement of URLLC [13], equivalent to 10^{-5} in terms of BER [5]. Also, its reliability improves by about 4 dB, compared to the DF FD relay system with all SI cancellations at the relay. The AF FD relay with digital SI cancellation at the base station is shown to provide BER performance close to the DF FD relay system.

Figure 9.8 shows the BER performance of the AF FD relay system with adaptive SI processing in comparison with the existing methods [9] [28] at SNR = 20 dB. The AF FD relay system with adaptive SI processing achieves a much better BER performance than the PSP assisted self-energy recycling [28] and the TSP assisted self-energy recycling [9], especially in the low SRSIR range of $-30 \sim -15$ dB. The unsmooth transition in the curves is caused by the mode change between PSAC and PS, dependent on SRSIR and the threshold. There is an additional process of analog cancellation in PSAC, compared to PS. When SRSIR after PS is lower than the threshold, PSAC is used. When SRSIR after PS is higher than the threshold, PS is used with no additional AC process. When $\gamma_{\mathrm{R}} \leq -2$ dB, the ICA based signal detection method can obtain better BER performance than the space-splitting method. However, when $\gamma_{\mathrm{R}} \geq 0$ dB, the BER performance of the ICA based signal separation method is worse than that of the space-splitting method. This indicates the ICA based signal separation method is more suitable for low SRSIR and the space-splitting based method is more suitable for high SRSIR. Thus, the AF FD relay system with adaptive SI processing is proposed to adapt to both low SRSIR and high SRSIR cases. Since the proposed scheme achieves the

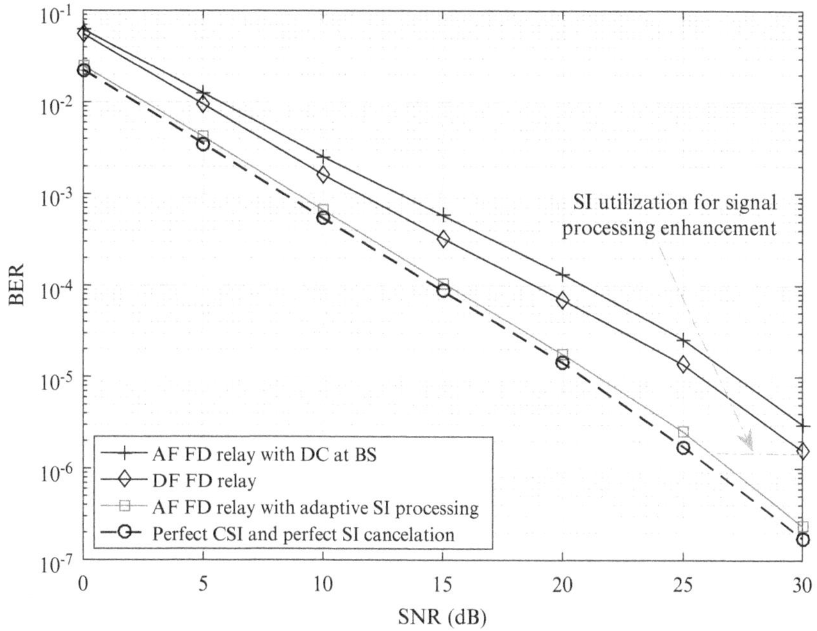

Figure 9.7 BER performance of the AF FD relay system with adaptive digital SI cancellation or utilization at base station, in comparison with the traditional DF FD relay system with all SI cancellations at relay. SRSIR is 20 dB (DC: digital cancellation). ©IEEE 2021. Reprinted with permission from [2].

adaptive SI processing mode selection to balance utilizing SI and canceling SI, it is shown to be more robust against SRSIR than the existing methods [9] [28]. It is noteworthy that the proposed scheme has a small fluctuation from SRSIR = −16 dB to SRSIR = −14 dB owing to the adaptive mode switching between PSAC and PS.

Figure 9.9 demonstrates the BER performance of the AF FD relay system with adaptive SI processing with different SNR. They have an inflection point at −15 dB, which is due to the switch between PSAC and PS mode. In addition, as the increase of SNR, the BER performance of the proposed method is improved accordingly. BER of the AF FD relay system with adaptive SI processing scheme is close to 10^{-5} at $\gamma_{\mathrm{R}} = 20$ dB.

Figure 9.10 shows the probability of SI processing mode selection for SNR = 20 dB. The probability of using PSAC mode is much higher than that of PS mode for $\gamma_{\mathrm{R}}^{[\mathrm{PS}]} < -15$ dB. Conversely, the probability of using PS mode is much higher than that of PSAC mode for $\gamma_{\mathrm{R}}^{[\mathrm{PS}]} > 15$ dB. This is determined by the first threshold of the SRSIR γ_{Th1} at the relay. The probability of using PSAC-S mode is the largest

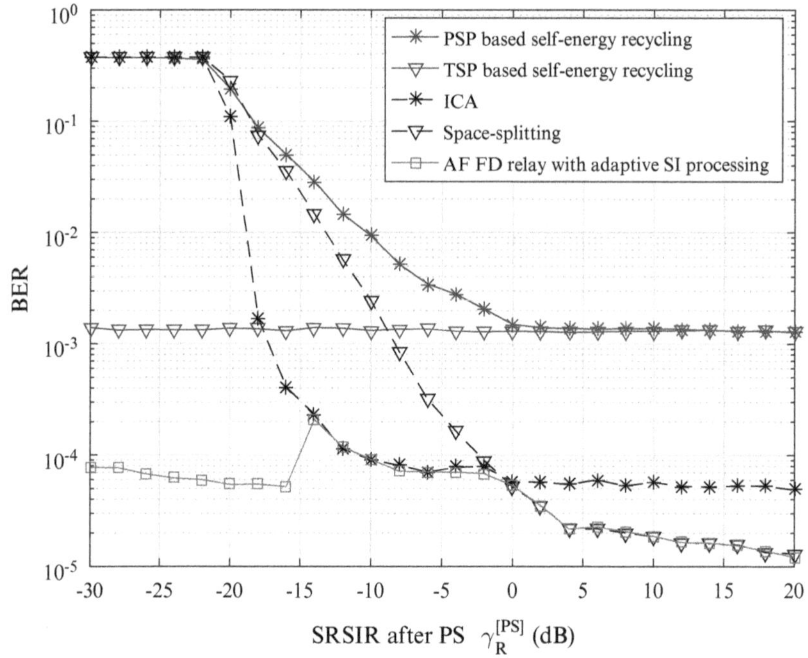

Figure 9.8 BER performance of the AF FD relay system with adaptive digital SI cancellation or utilization at base station, in comparison to the PSP assisted self-energy recycling and the TSP assisted self-energy recycling. SNR is 20 dB. ©IEEE 2020. Reprinted with permission from [5].

for $\gamma_R^{[PS]} < -23$ dB and the probability of using PS-S mode is the largest for -15 dB $< \gamma_R^{[PS]} < -3$ dB. The probability of using PSAC-N mode is the largest for $\gamma_R^{[PS]} > -3$ dB. This is determined by the higher threshold of the SRSIR γ_{Th2} at the relay.

9.5.2 Latency Evaluation

In Table 9.2, the latency evaluation for the AF FD relay is shown and is compared to the DF HD relay, the DF FD relay, and direct transmissions with no relay and no retransmission. There are three significant latency components: transmission latency, relaying latency and processing latency. The transmission latency is the time-to-transmit, corresponding to an OFDM symbol of 17.86 μs in URLLC [13]. Due to simultaneous transmission and reception at FD relay, the FD relay system provides transmission latency close to direct transmissions, and lower than the HD relay system that requires one time slot for data transmission and the other

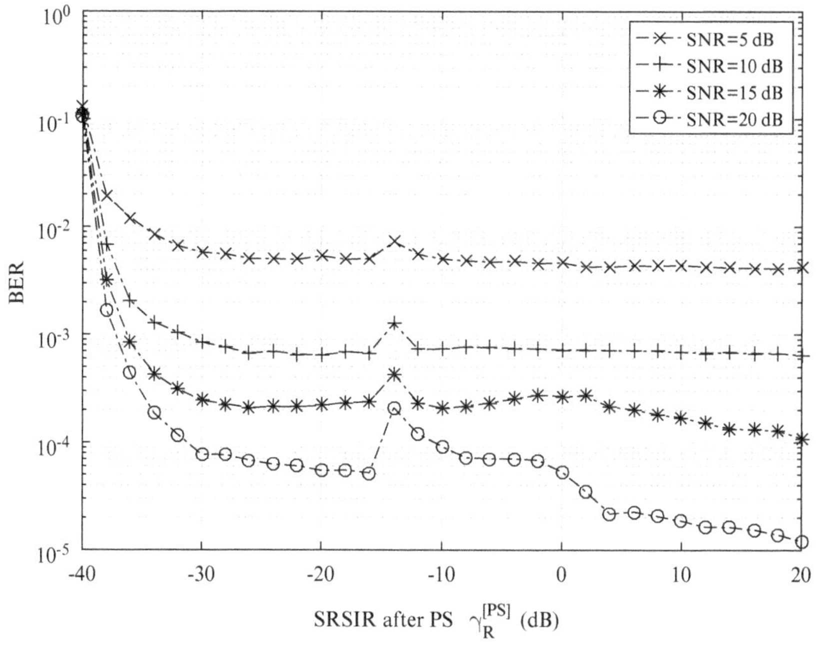

Figure 9.9 BER performance of the AF FD relay system with adaptive digital SI cancellation or utilization at base station, with different SNR. ©IEEE 2020. Reprinted with permission from [5].

time slot for data reception. Thus, the transmission latency of the FD relay system is approximately as much as that of direct transmissions, and a half of that of the HD relay system, as shown in Table 9.2. The relaying latency and the processing latency are evaluated by simulation, and are determined by the computational complexity of the signal processing algorithms, such as the channel estimation algorithm, signal detection algorithm, and SI cancellation method, requiring a number of matrix multiplications and matrix inversions. The computational complexity is represented as the latency evaluation, characterized by the time running on CPU. In simulation, the associated calculations are simulated to run over 10 000 OFDM symbols to obtain the average time consumption on each OFDM symbol for each algorithm [23]. Also, the relaying latency and the processing latency are affected by the baseband chip. The baseband chip at base station generally provides a computation capability higher than that at the relay. In URLLC, the retransmission is allowed in the HARQ protocol when block error rate is lower than 10^{-3}, and the maximum number of retransmissions is two [13]. Hence, when cell-edge users are far from the base station, the AF FD relay system with SI cancellation or utilization at the base station can guarantee higher reliability with

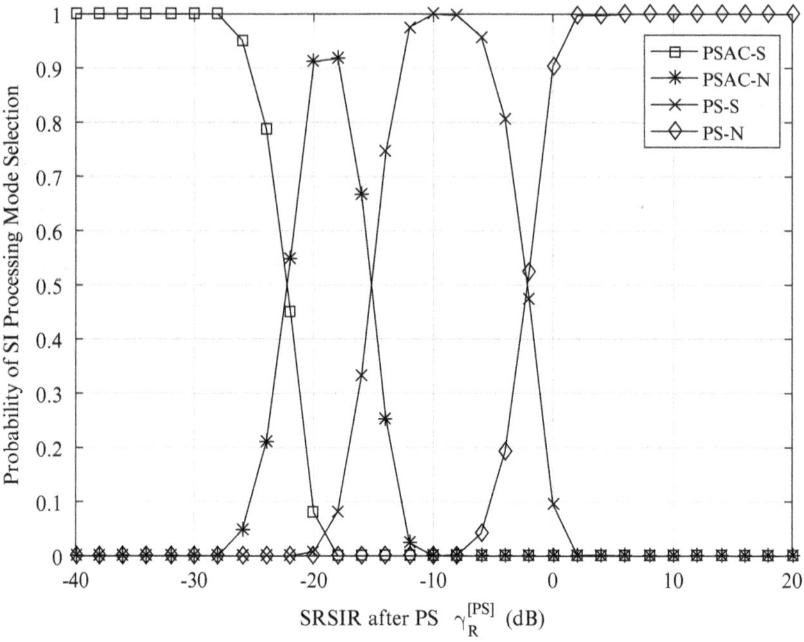

Figure 9.10 The probability of SI processing mode selection, with SNR = 20 dB. ©IEEE 2020. Reprinted with permission from [5].

lower block error rate, requiring less retransmissions, compared to direct transmissions in deep fading. The queuing latency in URLLC can be modelled as a statistical queuing requirement, characterized by the maximum queuing latency of 0.8 ms and a small latency violation probability [25]. It is demonstrated in Table 9.2 that the AF FD relay system with digital SI cancellation or utilization at the base station provides the overall latency of an OFDM symbol close to the direct transmission with no relay and no retransmission, and significantly lower than the DF HD relay system and the DF FD relay system, respectively.

9.6 Conclusion

In this chapter an overview of the reliability and latency of an FD relay has been provided for URLLC-enabled systems. The relaying latency is a significant part in the end-to-end latency, and can be minimized, while guaranteeing high reliability. To reduce the relaying latency, the FD relay to work in the AF mode with low-complexity RF SI cancellation and analog SI cancellation has been discussed. The residual SI can be processed together with the desired signals at a base station in

Table 9.2 Evaluation of latency. System level parameters are set as follows: the subcarrier spacing is 60 kHz; the bandwidth is set as 20 MHz; RF and analog cancellations are up to 30 dB and 40 dB, respectively; transmission power ranges from 10 dBm to 30 dBm, and the noise floor is −95 dBm (DC: Digital cancellation, BS: Base station). ©IEEE 2021. Reprinted with permission from [2].

Transmission category	Accomplished steps		Average time (μs)	Latency of each OFDM symbol (μs)
DF HD relay [18]	Transmission latency [13]		35.72	99.609
	Relaying latency	Channel estimation	25.669	
		Signal detection	36.565	
	Processing latency at BS	Channel estimation Signal detection	1.655	
DF FD relay [22]	Transmission latency [13]		17.86	46.028
	Relaying latency	Analog SI cancellation	0.122	
		SI channel estimation	26.359	
		SI signal reconstruction and mitigation	0.032	
	Processing latency at BS	Channel estimation Signal detection	1.655	

(Continued)

Table 9.2 (Continued)

Transmission category	Accomplished steps		Average time (μs)	Latency of each OFDM symbol (μs)
AF FD relay with DC at BS	Transmission latency [13]		17.86	20.551
	Relaying latency	Analog SI cancellation	0.122	
	Processing latency at BS	Channel estimation SI signal reconstruction and mitigation Signal detection	2.569	
AF FD relay with SI utilization at BS [5]	Transmission latency [13]		17.86	19.747
	Relaying latency	Analog SI cancellation	0.122	
	Processing latency at BS	Equalization and signal separation SI utilization	1.765	
Direct transmissions with no relay and no retransmission	Transmission latency [13]		17.86	19.515
	Processing latency at BS	Channel estimation Signal detection	1.655	

the digital domain. Also, the residual SI can be utilized to improve the reliability and enhance the degree of freedom in signal processing. The FD relay system with the AF mode provides latency significantly lower than that with the DF mode and the HD relay system, respectively, and close to direct transmissions with no relay and no retransmission. Also, the AF FD relay system with SI utilization at the base station provides better BER performance, compared to the DF FD relay system with all SI cancellations at FD relay.

Bibliography

[1] E. Boshkovska, D. W. K. Ng, N. Zlatanov, and R. Schober. Practical non-linear energy harvesting model and resource allocation for SWIPT systems. *IEEE Communications Letters*, 19(12):2082–2085, Dec. 2015.

[2] Y. Jiang, H. Duan, X. Zhu, Z. Wei, T. Wang, F. Zheng, and S. Sun. Toward URLLC: A full duplex relay system with self-interference utilization or cancellation. *IEEE Wireless Communications*, 28(1):74–81, Feb. 2021.

[3] E. Boshkovska, D. W. K. Ng, N. Zlatanov, A. Koelpin, and R. Schober. Robust resource allocation for MIMO wireless powered communication networks based on a non-linear EH model. *IEEE Transactions on Communications*, 65(5): 1984–1999, May 2017.

[4] J. F. Cardoso. High-order contrasts for independent component analysis. *Neural Computation*, 11(1):157–192, Jan. 1999.

[5] H. Duan, X. Zhu, Y. Jiang, Z. Wei, and S. Sun. An adaptive self-interference cancelation/utilization and ica-assisted semi-blind full-duplex relay system for llhr iot. *IEEE Internet of Things Journal*, 7(3):2263–2276, Mar. 2020.

[6] E. Everett, A. Sahai, and A. Sabharwal. Passive self-interference suppression for full-duplex infrastructure nodes. *IEEE Transactions on Communications*, 13(2): 680–694, Feb. 2014.

[7] J. Gao, X. Zhu, and A. K. Nandi. Non-redundant precoding and PAPR reduction in MIMO OFDM systems with ICA based blind equalization. *IEEE Transactions on Wireless Communications*, 8(6):3038–3049, Jun. 2009.

[8] Y. Hu, M. C. Gursoy, and A. Schmeink. Relaying-enabled ultra-reliable low-latency communications in 5g. *IEEE Network*, 32(2):62–68, Apr. 2018.

[9] D. Hwang, K. C. Hwang, D. I. Kim, and T. J. Lee. Self-energy recycling for RF powered multi-antenna relay channels. *IEEE Transactions on Wireless Communications*, 16(2):812–824, Feb. 2017.

[10] A. Hyvarinen, J. Karhunen, and E. Oja. *Independent Component Analysis*. John Wiley & Sons, New York, USA, May 2002.

[11] A. Jalali and Z. Ding. Joint detection and decoding of polar coded 5g control channels. *IEEE Transactions on Wireless Communications*, 19(3):2066–2078, Jan. 2020.

[12] W. Kim, S. K. Bandari, and B. Shim. Enhanced sparse vector coding for ultra-reliable and low latency communications. *IEEE Transactions on Vehicular Technology*, 69(5):5698–5702, May 2020.

[13] H. Ji, S. Park, J. Yeo, Y. Kim, J. Lee, and B. Shim. Ultra-reliable and low-latency communications in 5g downlink: Physical layer aspects. *IEEE Wireless Communications*, 25(3):124–130, Jun. 2018.

[14] Y. Jiang, X. Zhu, E. Lim, Y. Huang, and H. Lin. Low-complexity semiblind multi-CFO estimation and ICA-based equalization for CoMP OFDM systems. *IEEE Transactions on Vehicular Technology*, 63(4):1928–1934, May 2014.

[15] Y. Jin, X. Xia, Y. Chen, and R. Li. Full-duplex delay diversity relay transmission using bit-interleaved coded ofdm. *IEEE Transactions on Communications*, 65 (8):3250–3258, Aug. 2017.

[16] J. Lee and O. S. Shin. Full-duplex relay based on distributed beamforming in multiuser MIMO systems. *IEEE Transactions on Vehicular Technology*, 62(4): 855–1860, May 2013.

[17] R. Li, Y. Chen, G. Y. Li, and G. Liu. Full-duplex cellular networks. *IEEE Communications Magazine*, 55(4):184–191, Apr. 2017.

[18] W. Liu, C. Li, and J. Li. Achieving maximum degrees of freedom of two-hop mimo alternate half-duplex relaying system for linear transceivers: A unified transmission framework for df and af protocols. *IEEE Transactions on Vehicular Technology*, 64(5):2144–2148, Jul. 2015.

[19] Y. Liu, X. Xia, and H. Zhang. Distributed linear convolutional space-time coding for two-relay full-duplex asynchronous cooperative networks. *IEEE Transactions on Wireless Communications*, 12(12):6406–6417, Dec. 2013.

[20] Y. Liu, X. Zhu, E. G. Lim, Y. Jiang, and Y. Huang. Fast iterative semi-blind receiver for urllc in short-frame full-duplex systems with cfo. *IEEE Journal on Selected Areas in Communications*, 37(4):839–853, Apr. 2019.

[21] B. Ma, H. Shah-Mansouri, and V. W. S. Wong. Full-duplex relaying for d2d communication in millimeter wave-based 5g networks. *IEEE Transactions on Wireless Communications*, 17(7):4417–4431, Jul. 2018.

[22] M. Mohammadkhani Razlighi and N. Zlatanov. Buffer-aided relaying for the two-hop full-duplex relay channel with self-interference. *IEEE Transactions on Wireless Communications*, 17(1):477–491, Jan. 2018.

[23] X. Quan, Y. Liu, D. Chen, S. Shao, Y. Tang, and K. Kang. Blind nonlinear self-interference cancellation for wireless full-duplex transceivers. *IEEE Access*, 6:37725–37737, Jul. 2018.

[24] T. Riihonen, S. Werner, and R. Wichman. Hybrid full-duplex/half-duplex relaying with transmit power adaptation. *IEEE Transactions on Wireless Communications*, 10(9):3074–3085, Sep. 2011.

[25] C. She, C. Yang, and T. Q. S. Quek. Cross-layer optimization for ultra-reliable and low-latency radio access networks. *IEEE Transactions on Wireless Communications*, 17(1):127–141, Jan. 2018.

[26] M. S. Sim, M. Chung, D. Kim, J. Chung, D. K. Kim, and C. Chae. Nonlinear self-interference cancellation for full-duplex radios: From link-level and system-level performance perspectives. *IEEE Communications Magazine*, 55(9): 158–167, Sep. 2017.

[27] Z. Wei, X. Zhu, S. Sun, Y. Huang, L. Dong, and Y. Jiang. Full-duplex versus half-duplex amplify-and-forward relaying: Which is more energy efficient in 60-GHz dual-hop indoor wireless systems? *IEEE Journal on Selected Areas in Communications*, 33(12):2936–2947, Dec. 2015.

[28] W. Wu, B. Wang, Y. Zeng, H. Zhang, Z. Yang, and Z. Deng. Robust secure beamforming for wireless powered full-duplex systems with self-energy recycling. *IEEE Transactions on Vehicular Technology*, 66(11):10055–10069, Nov. 2017.

10

Mobility Prediction for Reducing End-to-End Delay in URLLC

Zhanwei Hou, Changyang She, Yonghui Li, and Branka Vucetic*

School of Electrical and Information Engineering, The University of Sydney, 2006, NSW, Sydney, City road, Australia
** Corresponding Author*

10.1 Introduction

Ultra-reliable and low-latency communications (URLLC) is one of the new application scenarios in 5th Generation (5G) communications [2]. By achieving ultra-high reliability (e.g. 10^{-5} to 10^{-8} packet loss probability) and ultra-low end-to-end (E2E) delay (e.g. 1 ms), URLLC lays the foundation for several mission-critical applications, such as industrial automation, Tactile Internet, remote driving, virtual reality (VR), and tele-surgery [8, 15, 41]. How these two conflicting requirements on delay and reliability can be achieved remains an open problem.

To improve reliability, several technologies have been proposed in the existing literature and specifications, such as K-repetition [1], frequency hopping [30], large-scale antenna systems [39], and multi-connectivity [24]. With these technologies, different kinds of diversities are exploited to improve reliability at the cost of more radio resources. On the other hand, to reduce latency in the air interface, the short frame structure was proposed in 5G New Radio (NR)[20], and fast uplink grant schemes were proposed to reduce access delay [14, 29]. However, there are some other delay components in the networks, such as buffers of devices, computing systems, backhauls, and core networks. As a result, the user experienced delay can hardly meet the requirements of URLLC. Novel concepts and technologies that can reduce the user experienced delay and improve overall

Ultra-Reliable and Low-Latency Communications (URLLC) Theory and Practice: Advances in 5G and Beyond, First Edition. Edited by Trung Q. Duong, Saeed R. Khosravirad, Changyang She, Petar Popovski, Mehdi Bennis and Tony Q.S. Quek.
© 2023 John Wiley & Sons Ltd. Published 2023 by John Wiley & Sons Ltd.

reliability (i.e. total packet losses and errors in different parts of the system) are in urgent need.

To tackle these challenges, we aim to meet the requirements of URLLC by jointly optimizing prediction and communication. The basic idea is to predict the future system states in the transmitter, such as locations and force feedback, and then send them to the receiver in advance. In this way, the user experienced delay can be reduced significantly. For example, consider a communication and prediction system, where the E2E delay is 10 time slots and the prediction horizon is 9 slots. In the current slot, the transmitter predicts its future state in the 9th slot and sends it to the receiver. After 10 slots of communication delay, the state in the 9th slot is received by the receiver in the 10 th slot. As a result, the user experienced delay is 1 slot. If the prediction horizon is equal to the E2E delay, the user experienced delay will be zero[1]. However, predictions are not error-free, and a long prediction horizon will lead to a large prediction error probability. Intuitively, there is a trade-off between the user experienced delay and the overall reliability. To satisfy the two conflicting requirements of URLLC, we need to jointly optimize the prediction and communication systems. Specifically, in this chapter, we will address the following questions: *(1) how can we characterize the trade-off between user-experienced delay and overall reliability with prediction and communication co-design? (2) Is it possible to satisfy the requirements of URLLC by prediction and communication co-design? (3) If yes, how can we maximize the number of URLLC services that can be supported by the system?*

These questions are challenging to answer since multiple components of delays and errors are involved in prediction and communication systems. As such, we need a prediction and communication co-design framework which takes different delay components and errors into account. Moreover, the complicated constraints on the user experienced delay and the overall reliability are non-convex in general, and hence it is very difficult to find the optimal solution.

10.1.1 Our Contributions

The main contributions of this chapter are summarized as follows:

- We establish a framework for prediction and communication co-design, where the time and frequency resource allocation in the communication system and the prediction horizon in the prediction system are jointly optimized to maximize the number of devices that can be supported in the system.
- We derive the closed-form expressions of the decoding error probability, the queueing delay violation probability, prediction error probability, and analyzed

1 It should be noted that the user experienced delay is different from the E2E delay. The E2E delay of a communication system can never be zero because of the limitation of physical laws, e.g. the propagation delay due to the electromagnetic wave propagation in the air or fiber cannot be zero.

their properties. From these results, the trade-off between user experienced delay and overall reliability can be obtained.

- We propose an algorithm to find a near optimal solution of the optimization problem. The performance loss of the near optimal solution is studied and further validated via numerical results. In addition, we analyze the complexity of the algorithm, which linearly increase with the number of devices.

Furthermore, to evaluate the performance of the proposed method, we compare it with a benchmark solution without prediction. Simulation results show that the trade-off can be improved remarkably with prediction and communication co-design. In addition, an experiment is carried out to validate the accuracy of mobility prediction in practical remote-control scenarios.

10.2 Related Work

10.2.1 Communications in URLLC

There are some existing solutions to reduce latency in communication systems for URLLC [14, 20, 26, 29, 36]. With the 5G NR [20], the notion of the "mini-slot" is introduced to support transmissions with the delay as low as the duration of a few symbols. The queueing delay is analyzed and optimized in [36], where the trade-off among throughput, delay and reliability was studied. To reduce the access delay in uplink transmissions, a Semi-Persistent Scheduling (SPS) scheme was developed in [29]. A grant-free protocol was proposed in [14] to further avoid the delay caused by scheduling requests and transmission grants. With the preemptive scheduling scheme in [26], the short packets with high priority can preempt an ongoing long packet transmission without waiting for the next scheduling period. With this scheme, the scheduling delay of short packets is reduced.

To improve the reliability for the low latency communications, different kinds of diversities were introduced [1, 24, 30, 39]. In [1], K-repetition was proposed to avoid retransmission feedback. The basic idea is to send multiple copies of each packet without waiting for the acknowledgment feedback. Considering that the required delay is shorter than channel coherence time, frequency hopping was adopted in [30] to improve reliability. In [39], a Lyapunov optimization problem was formulated to improve the reliability with guaranteed latency, where spatial diversity was used to improve reliability. In [24], interface diversity was proposed to achieve URLLC without modifications in the baseband designs by providing multiple communication interfaces. However, by introducing diversities, the reliability is improved at the cost of low resource utilization efficiency.

This trade-off between delay and reliability has been exhaustively studied in communication systems [4, 5, 32, 40]. To reduce the transmission delay, the blocklength of channel codes is short, and the decoding error probability is non-zero for arbitrary Signal-to-Noise Ratio (SNR). The fundamental trade-off between transmission delay and decoding error probability in the short blocklength regime was derived in [40]. The trade-off between the queueing delay and the delay bound violation probability was studied in [4]. To achieve a lower delay bound, the violation probability increases. Moreover, grant-free schemes can help reduce latency, but introduce extra packet losses due to transmission collisions. How to achieve ultra-high reliability with grant-free schemes was studied in [5]and it was shown that the proposed stop-and-wait protocol can achieve 10^{-5} outage probability.

10.2.2 Predictions in URLLC

To achieve satisfactory delay and reliability in URLLC, different kinds of prediction have been studied in the existing literature [11, 19, 21, 35, 37, 38].

In [38], the predicted control commands were sent to the receiver and were stored in the buffer. When a control command is lost in communication, predicted commands in the receiver's buffer will be executed. The length of predictive control commands was optimized to minimize the resource consumption. The idea of a model-mediated tele-operation approach was mentioned in [35]. By predicting the movement or the force feedback, the user experienced delay can be reduced. In both [38] and [35], prediction errors were not considered, and whether we can achieve ultra-high reliability in the systems remains unclear.

Different from command or mobility predictions in control systems, predicting some other features of traffic or performance of communications is also helpful. In [11], based on the predicted traffic state, a bandwidth reservation scheme was proposed to improve the spectral efficiency of URLLC. By exploiting the correlation among different nodes, the behavior of different users can be predicted [19]. Then, by reserving resources according to the predicted behavior, the access delay can be reduced. A fast Hybrid Automatic Repeat Request (HARQ) protocol was proposed in [21], prediction is used to omit some HARQ feedback signals and successive message decodings, so that the expected delay can be improved by 27% to 60% compared with standard HARQ. In [37], the outcome of the decoding was predicted before the end of the transmission. With the predicted result, there is no need to wait for the acknowledgment feedback, and thus the E2E delay can be reduced.

10.3 System Model

As shown in Figure 10.1, we consider a joint prediction and communication system, where N mobile devices send packets to a receiver, which could be a data

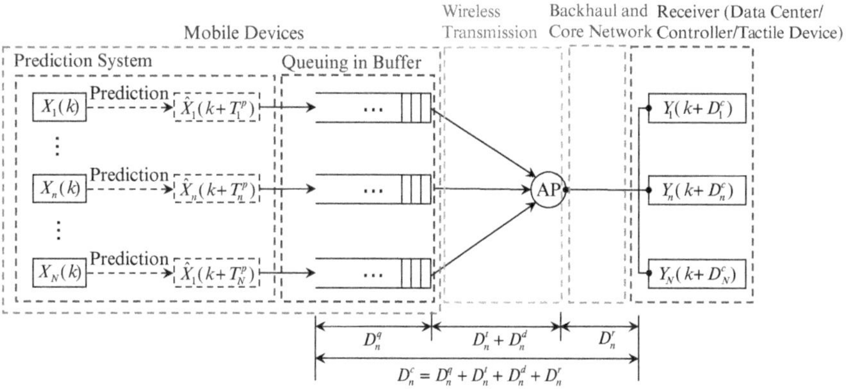

Figure 10.1 Illustration of network structure. ©IEEE 2020. Reprinted with permission from [12].

center, controller, or tactile device. The function of the receiver depends on specific applications. In remote driving [15], a human driver can remotely control a vehicle based on the feedback from various sensors installed on the vehicle. In factory automation [41], sensors update information to the controller to perform better closed-loop control, or to a data center for monitoring or fault detection. In Tactile Internet [8], force and torques are sent to a tactile device to render the sense of touch, and thus can enable haptic communications. The packets generated by each device may include different features, such as the location, velocity, and acceleration of a device in remote driving or industrial automation, or the force and torques in Tactile Internet.

The receiver can be deployed at a Mobile Edge Computing (MEC) server or a cloud center. In our framework, we consider a general wireless communication system, where mobile devices send packets to a cloud center via wireless links, backhauls, and core networks. The framework is also suitable for an MEC system, where the delays and packet losses in backhauls and core networks are set to be zero [33].

10.3.1 User Experienced Delay

Time is discretized into slots. The duration of each slot is denoted as T_S. Let $X_n(k) = [x_n^1(k), x_n^2(k), ..., x_n^F(k)]^T$ be the state of the nth device in the kth slot, where F is the number of features. The state of the nth device that is received by the receiver in the kth slot is denoted as $Y_n(k)$. In traditional communication systems, each device sends its current state $X_n(k)$ to the data center. Let D_n^{c} (slots) be the nth device's end-to-end (E2E) delay in the communication system. If the packet that conveys $X_n(k)$ is decoded successfully in the $(k + D_n^c)$th slot, then $Y_n(k + D_n^c) = X_n(k)$, and the user experienced delay is D_n^c.

As shown in Figure 10.2, to improve the user experienced delay, each device predicts its future state. T_n^p is denoted as the prediction horizon. In the kth slot, the device generates a packet based on the predicted state $\hat{X}_n(k + T_n^p)$. After D_n^c slots, the packet is received by the data center. Then, we have $Y_n(k + D_n^c) = \hat{X}_n(k + T_n^p)$, which is equivalent to $Y_n(k) = \hat{X}_n\left[k - (D_n^c - T_n^p)\right], \forall k$. Therefore, the delay experienced by the user is $D_n^e = D_n^c - T_n^p$.[2]

Remark 1. *It is worth noting that the states of adjacent slots could be correlated. Thus, source coding schemes that compress the information in multiple slots can achieve higher compression ratio. On the other hand, channel coding schemes that encode the packets to be transmitted in multiple slots into one block, can achieve higher reliability. However, both of them will lead to a longer decoding delay. To achieve ultra-low latency, in this chapter we assume that the source coding and channel coding in the kth slots only depend on $\hat{X}_n(k + T_n^p)$ and the data to be transmitted in this slot.*

10.3.2 Delay and Reliability Requirements

The delay and reliability requirements are characterized by a maximum delay bound and a maximum tolerable error probability, $D_{\max}$ and $\varepsilon_{\max}$. This means that $X_n(k)$ should be received by the data center before the $(k + D_{\max})$th slot with probability $1 - \varepsilon_{\max}$.

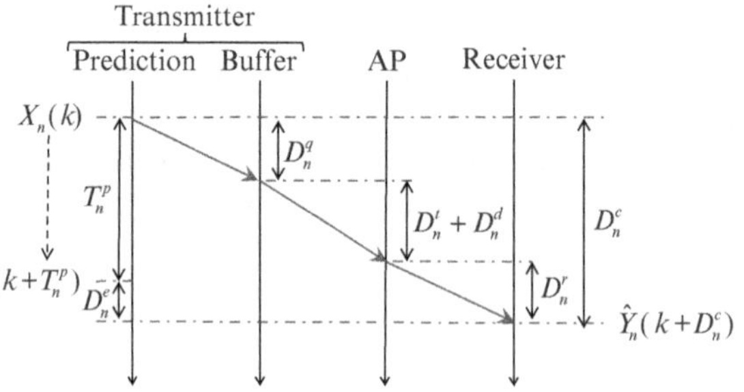

Figure 10.2 Illustration of prediction structure. ©IEEE 2020. Reprinted with permission from [12].

2 If D_n^c is smaller than T_n^p, D_n^e is negative. This means that the receiver can predict the states of devices. In this chapter, we only consider the scenario that $D_n^e \geq 0$.

To satisfy the delay requirement, the user experienced delay should not exceed a maximal delay bound, i.e.

$$D_n^{\mathrm{e}} = D_n^{\mathrm{c}} - T_n^p \leq D_{max}. \tag{10.1}$$

In the considered communication system, the E2E communication delay D_n^{c} includes queueing delay D_n^{q}, transmission delay D_n^{t}, decoding delay D_n^{d}, and delay in backhauls and core networks D_n^{r} [13, 42].

Thus, the constraint in (10.1) can be re-expressed as follows,

$$D_n^{\mathrm{q}} + D_n^{\mathrm{t}} + D_n^{\mathrm{d}} + D_n^{\mathrm{r}} - T_n^p \leq D_{max}, \tag{10.2}$$

where $D_n^{\mathrm{d}} = \kappa D_n^{\mathrm{t}}, \kappa > 0$.

The overall reliability depends on prediction errors and packet losses in communications. In the control system, if the difference between the actual state of the device and the received state does not exceed a required threshold, the user cannot notice the difference. For example, in Tactile Internet, the minimum Difference of the force stimulus intensity that our hands can perceive is referred to as Just Noticeable difference (JND) [9]. We define the difference between $\hat{X}_n(k)$ and $X_n(k)$ as $E_n(k) = [e_n^1(k), e_n^2(k), \ldots, e_n^F(k)]^T$, where $e_n^j(k) = \hat{x}_n^j(k + D_n^{\mathrm{e}}) - x_n^j(k)$. The JND of this system is denoted as $\Delta = [\delta_1, \delta_2, \ldots, \delta_N]^T$. Then, the prediction error probability is given by

$$\varepsilon_n^{\mathrm{p}} = 1 - \prod_{j=1}^{N} \Pr\{|e_n^j(k)| \leq \delta_j\}, \tag{10.3}$$

Even if $\hat{X}_n(k)$ is accurate enough, it will be useless if it is not received by the data center before the $(k + D_{\mathrm{max}})$th slot. Denote the queueing delay bound violation probability and the packet loss probability of the nth device as $\varepsilon_n^{\mathrm{q}}$ and $\varepsilon_n^{\mathrm{t}}$, respectively. Then, the overall reliability of the device can be expressed as follows,

$$\varepsilon_n^{\mathrm{o}} = 1 - (1 - \varepsilon_n^{\mathrm{q}})(1 - \varepsilon_n^{\mathrm{t}})(1 - \varepsilon_n^{\mathrm{p}}). \tag{10.4}$$

To achieve ultra-high reliability, all of $\varepsilon_n^{\mathrm{q}}$, $\varepsilon_n^{\mathrm{t}}$ and $\varepsilon_n^{\mathrm{p}}$ should be small (i.e. less than 10^{-5}). Thus, (10.4) can be accurately approximated by $\varepsilon_n^{\mathrm{o}} \approx \varepsilon_n^{\mathrm{q}} + \varepsilon_n^{\mathrm{t}} + \varepsilon_n^{\mathrm{p}}$, and the reliability requirement can be satisfied if

$$\varepsilon_n^{\mathrm{q}} + \varepsilon_n^{\mathrm{t}} + \varepsilon_n^{\mathrm{p}} \leq \varepsilon_{max}. \tag{10.5}$$

10.4 Trade-off Analyses for Predictions and Communications

In this section, we first consider a general linear prediction framework, and derive the relation between the prediction error probability and the prediction horizon in a closed form. Then, we characterize the trade-off between communication reliability and E2E delay for short packet transmissions in a closed form. Based on the analysis, we further study how to maximize the number of URLLC services that can be supported by the system.

10.4.1 State Transition Function

We assume that the state of the nth device, $X_n(k)$, changes according to the following state transition function [16]

$$X_n(k+1) = \Phi_n X_n(k) + W_n(k), \tag{10.6}$$

where $\Phi_n = [\phi_n^{i,j}]_{F\times F}$, $i, j = 1, 2, \cdots, F$, is the state transition matrix and $W_n(k) = [w_n^i(k)]_{F\times 1}$, $i = 1, 2, \cdots, F$, is the transition noise. We assume that Φ_n is constant, and thus it can be obtained from measurements or physical laws. The elements of $W_n(k)$ are independent random variables that follow Gaussian distributions with zero mean and variances $\sigma_1^2, \sigma_2^2, \cdots, \sigma_F^2$, respectively.

Remark 2. *This model is widely adopted in kinematics systems or control systems [16, 17]. Here we consider a general prediction method for a linear system. This is because for non-linear system, the relation between the prediction horizon and the prediction error probability can hardly be derived in a closed-form expression. To implement our framework in non-linear systems, data-driven prediction methods such as neural networks should be applied. These methods do not rely on system models, and will be considered in future work.*

According to (10.6), the state in the $(k + T_n^p)$th slot is given by

$$X_n(k + T_n^p) = (\Phi_n)^{T_n^p} X_n(k) + \sum_{i=1}^{T_n^p} (\Phi_n)^{T_n^p - i} W_n(k + i - 1). \tag{10.7}$$

10.4.2 Prediction Horizon and Prediction Error Probability

Inspired by the Kalman filter, we consider a general linear prediction method [16]. Based on the system state in the kth slot, we can predict the state in the $(k + 1)$th slot according to the following expression,

$$\hat{X}_n(k+1) = \Phi_n X_n(k). \tag{10.8}$$

From (10.8), we can further predict the state in the $(k + T_n^p)$th slot,

$$\hat{X}_n(k+T_n^p) = (\Phi_n)^{T_n^p} X_n(k). \tag{10.9}$$

After T_n^p steps of prediction, the difference between $X_n(k+T_n^p)$ and $\hat{X}_n(k+T_n^p)$ can be derived as follows,

$$\begin{aligned} E_n(k+T_n^p) &\triangleq X_n(k+T_n^p) - \hat{X}_n(k+T_n^p) \\ &= W_n(k+T_n^p-1) \\ &+ \sum_{i=1}^{T_n^p-1} (\Phi_n)^{T_n^p-i} W_n(k+i-1). \end{aligned} \tag{10.10}$$

The jth element of $E_n(k+T_n^p)$ is given by

$$\begin{aligned} e_n^j(k+T_n^p) &= w_n^j(k+T_n^p-1) \\ &+ \sum_{i=1}^{T_n^p-1} \sum_{m=1}^{F} \phi_{n,j,m,T_n^p-i} w_n^m(k+i-1), \end{aligned} \tag{10.11}$$

where ϕ_{n,j,m,T_n^p-i} is the element of $(\Phi_n)^{T_n^p-i}$ at the jth row and the mth column.

Since the state transition noises follow independent Gaussian distributions, and $e_n^j(k+T_n^p)$ is a linear combination of them, $e_n^j(k+T_n^p)$ follows a Gaussian distribution with zero mean. The variance of $e_n^j(k+T_n^p)$ is denoted as $\rho_{n,j}^2(T_n^p)$, which is given by

$$\rho_{n,j}^2(T_n^p) = \sigma_j^2 + \sum_{i=1}^{T_n^p-1} \sum_{m=1}^{F} (\phi_{n,j,m,T_n^p-i})^2 \sigma_m^2. \tag{10.12}$$

Therefore, $\Pr\{|e_n^j(k+T_n^p)| \le \delta_j\}$ can be derived as follows,

$$\begin{aligned} \Pr\{|e_n^j(k+T_n^p)| \le \delta_j\} =& 1 - \Pr\{|e_n^j(k+T_n^p)| > \delta_j\} \\ =& 1 - \psi_{T_n^p,j}(-\delta_j) \\ =& 1 - \psi\left(\frac{-\delta_j}{\rho_{n,j}(T_n^p)}\right), \end{aligned} \tag{10.13}$$

where $\psi_{T_n^p,j}(\cdot)$ is the Cumulative Distribution Function (CDF) of $e_n^j(k+T_n^p)$, and $\psi(\cdot)$ is the CDF of the standard Gaussian distribution with zero mean and unit variance.

By substituting (10.13) into (10.3), $\varepsilon_n^{\mathrm{p}}$ can be expressed as follows,

$$\varepsilon_n^{\mathrm{p}} = 1 - \prod_{j=1}^{F}\left[1 - \psi\left(\frac{-\delta_j}{\rho_{n,j}(T_n^p)}\right)\right]. \tag{10.14}$$

From the expression in (10.14), we can obtain the following property of $\varepsilon_n^{\mathrm{p}}$.

Lemma 10.1: $\varepsilon_n^{\mathrm{p}}$ strictly increases with the prediction horizon T_n^p.

Proof. To prove this lemma, we need to prove that for any $T_n^{p,1} < T_n^{p,2}$, $\varepsilon_n^{\mathrm{p}}(T_n^{p,1}) < \varepsilon_n^{\mathrm{p}}(T_n^{p,2})$ holds. From (10.12), we have

$$\sigma_j^2(T_n^p + 1) - \sigma_j^2(T_n^p) = \sum_{m=1}^{F} \phi_{n,j,m,n}\sigma_m^2 > 0.$$

As such, we can conclude that $\sigma_j(T_n^p), j = 1, 2, \cdots, N$, increases with T_n^p. Moreover, from (10.14), we can see that $\varepsilon_n^{\mathrm{p}}$ increases with $\sigma_j(T_n^p), j = 1, 2, \cdots, N$. Therefore, $\varepsilon_n^{\mathrm{p}}$ increases with the prediction horizon T_n^p. This completes the proof.

Lemma 10.1 indicates that a longer prediction horizon leads to a larger prediction error probability. This is in accordance with intuition. For example, predicting the mobility of a device in the next 100 ms will be much harder than predicting the mobility in the next 10 ms.

10.4.3 Queueing Delay Bound Violation Probability

To derive the queueing delay bound violation probability, $\varepsilon_n^{\mathrm{q}}$, we can use the concept of effective bandwidth [32]. Effective bandwidth is defined as the minimal constant service rate of the queueing system that is required to ensure the maximum queueing delay bound and the delay bound violation probability [6].[3]

The number of packets generated in each slot depends on the mobility of the device and the random events detected by the device. According to the observation in [7], packet arrival processes in the Tactile Internet are very bursty. To capture the burstiness of the packet arrival process, a Switched Poisson Process (SPP) can be applied [11][4]. A SPP includes two traffic states. In each state, the packet arrival process follows a Poisson process. The average packet arrival rates are different in the two states, and the SPP switches between the two states according to a Markov

3 To analyze the upper bound of the delay bound violation probability, a widely used tool is network calculus [3]. However, with network calculus, one can hardly obtain a closed-form expression of the delay bound violation probability. Since we are interested in the asymptotic scenarios that $\varepsilon_n^{\mathrm{q}}$ is very small, effective bandwidth can be used [6].
4 In standardizations of 3GPP, queueing models are not specified since they depend on specific applications.

chain. With the traffic state classification methods in [11], the AP knows the average packet arrival rate in the current state, λ_n (packets/slot). According to [32], the effective bandwidth of the Poisson process is given by

$$E_n^{\mathrm{B}} = \frac{\ln(1/\varepsilon_n^{\mathrm{q}})}{D_n^{\mathrm{q}} \ln\left[\frac{\ln(1/\varepsilon_n^{\mathrm{q}})}{\lambda_n D_n^{\mathrm{q}}} + 1\right]} \text{ packets/slot,} \tag{10.15}$$

which is the minimal constant service rate required to ensure D_n^{q} and $\varepsilon_n^{\mathrm{q}}$. Since the transmission delay of each packet is fixed as D_n^{t}, to guarantee the queueing delay violation probability, the following constraint should be satisfied,

$$\frac{1}{D_n^{\mathrm{t}}} = E_n^{\mathrm{B}}. \tag{10.16}$$

Then, the queueing delay violation probability can be derived as

$$\varepsilon_n^{\mathrm{q}} = e^{D_n^{\mathrm{q}} \phi(\lambda_n, E_n^{\mathrm{B}})}, \tag{10.17}$$

where

$$\phi(\lambda_n, E_n^{\mathrm{B}}) = E_n^{\mathrm{B}} \mathbb{W}_{-1}\left(-\frac{\lambda_n}{E_n^{\mathrm{B}}} e^{-\frac{\lambda_n}{E_n^{\mathrm{B}}}}\right) + \lambda_n, \tag{10.18}$$

where $\mathbb{W}_{-1}(\cdot)$ is the "−1" branch of the Lambert W-function, which is defined as the inverse function of $f(x) = xe^x$.

The derivations of (10.17) and (10.18) are given as follows. The equation (10.15) can be re-expressed as

$$\frac{\ln(1/\varepsilon_n^{\mathrm{q}}) + \lambda_n D_n^{\mathrm{q}}}{\lambda_n D_n^{\mathrm{q}}} = \exp\left[\frac{\ln(1/\varepsilon_n^{\mathrm{q}}) + \lambda_n D_n^{\mathrm{q}}}{D_n^{\mathrm{q}} E_n^{\mathrm{B}}} - \frac{\lambda_n}{E_n^{\mathrm{B}}}\right], \tag{10.19}$$

and

$$\begin{aligned} -\frac{\lambda_n}{E_n^{\mathrm{B}}} \exp\left(-\frac{\lambda_n}{E_n^{\mathrm{B}}}\right) = & \frac{\ln(1/\varepsilon_n^{\mathrm{q}}) + \lambda_n D_n^{\mathrm{q}}}{-D_n^{\mathrm{q}} E_n^{\mathrm{B}}} \\ & \cdot \exp\left[\frac{\ln(1/\varepsilon_n^{\mathrm{q}}) + \lambda_n D_n^{\mathrm{q}}}{-D_n^{\mathrm{q}} E_n^{\mathrm{B}}}\right]. \end{aligned} \tag{10.20}$$

According to the definition of the Lambert function, (10.20) can be written as

$$\frac{\ln(1/\varepsilon_n^{\mathrm{q}}) + \lambda_n D_n^{\mathrm{q}}}{-D_n^{\mathrm{q}} E_n^{\mathrm{B}}} = \mathbb{W}\left[-\frac{\lambda_n}{E_n^{\mathrm{B}}} \exp\left(-\frac{\lambda_n}{E_n^{\mathrm{B}}}\right)\right]. \tag{10.21}$$

It should be noted that when $-\frac{\lambda_n}{E_n^{\mathrm{B}}}\exp\left(-\frac{\lambda_n}{E_n^{\mathrm{B}}}\right) < 0$, the Lambert function has two branches according to the range of $\frac{\ln(1/\varepsilon_n^{\mathrm{q}})+\lambda_n D_n^{\mathrm{q}}}{D_n^{\mathrm{q}} E_n^{\mathrm{B}}}$. Specifically, we have

$$\frac{\ln(1/\varepsilon_n^{\mathrm{q}})+\lambda_n D_n^{\mathrm{q}}}{-D_n^{\mathrm{q}} E_n^{\mathrm{B}}} = \begin{cases} \mathbb{W}_0\left[-\frac{\lambda_n}{E_n^{\mathrm{B}}}\exp\left(-\frac{\lambda_n}{E_n^{\mathrm{B}}}\right)\right], \\ \qquad -1 < \frac{\ln(1/\varepsilon_n^{\mathrm{q}})+\lambda_n D_n^{\mathrm{q}}}{-D_n^{\mathrm{q}} E_n^{\mathrm{B}}} < 0 \\ \mathbb{W}_{-1}\left[-\frac{\lambda_n}{E_n^{\mathrm{B}}}\exp\left(-\frac{\lambda_n}{E_n^{\mathrm{B}}}\right)\right], \\ \qquad \frac{\ln(1/\varepsilon_n^{\mathrm{q}})+\lambda_n D_n^{\mathrm{q}}}{-D_n^{\mathrm{q}} E_n^{\mathrm{B}}} \geq 0. \end{cases} \tag{10.22}$$

In the first case in (10.22), $\mathbb{W}_0\left[-\frac{\lambda_n}{E_n^{\mathrm{B}}}\exp\left(-\frac{\lambda_n}{E_n^{\mathrm{B}}}\right)\right] = -\frac{\lambda_n}{E_n^{\mathrm{B}}} = \frac{\ln(1/\varepsilon_n^{\mathrm{q}})+\lambda_n D_n^{\mathrm{q}}}{-D_n^{\mathrm{q}} E_n^{\mathrm{B}}}$. We can obtain that $\varepsilon_n^{\mathrm{q}} = 1$, which does not satisfy the reliability requirement. Thus, the first case in (10.22) can be removed. As such, we have

$$\frac{\ln(1/\varepsilon_n^{\mathrm{q}})+\lambda_n D_n^{\mathrm{q}}}{-D_n^{\mathrm{q}} E_n^{\mathrm{B}}} = \mathbb{W}_{-1}\left[-\frac{\lambda_n}{E_n^{\mathrm{B}}}\exp\left(-\frac{\lambda_n}{E_n^{\mathrm{B}}}\right)\right], \tag{10.23}$$

and

$$\varepsilon_n^{\mathrm{q}} = \exp\left\{D_n^{\mathrm{q}} E_n^{\mathrm{B}} \mathbb{W}_{-1}\left[-\frac{\lambda_n}{E_n^{\mathrm{B}}}\exp\left(-\frac{\lambda_n}{E_n^{\mathrm{B}}}\right)\right] + D_n^{\mathrm{q}}\lambda_n\right\}. \tag{10.24}$$

With the expressions in (10.17) and (10.18), we can obtain the following property of $\varepsilon_n^{\mathrm{q}}$.

Lemma 10.2: $\varepsilon_n^{\mathrm{q}}$ strictly decreases with the queueing delay D_n^{q} when λ_n and E_n^{B} are given.

Proof. According to (10.17), we have

$$\ln(\varepsilon_n^{\mathrm{q}}) = D_n^{\mathrm{q}} \phi(\lambda_n, E_n^{\mathrm{B}}). \tag{10.25}$$

Since $\varepsilon_n^{\mathrm{q}}$ is in the order of 10^{-5} to 10^{-8} and $D_n^{\mathrm{q}} > 0$, $\ln(\varepsilon_n^{\mathrm{q}}) < 0$, and thus $\phi(\lambda_n, E_n^{\mathrm{B}}) < 0$. As such, $\varepsilon_n^{\mathrm{q}}$ decreases with D_n^{q} in (10.17) when $\phi(\lambda_n, E_n^{\mathrm{B}})$ is given. The proof follows.

Lemma 10.2 indicates that with the same packet arrival process and service process, the queueing system with a smaller queueing delay bound requirement has a larger queueing delay violation probability. The intuition is that for a given CDF of the steady state queueing delay, the queueing delay violation probability decreases with the queueing delay bound.

10.4.4 Packet Loss Probability in Transmissions

With predictions, the communication delay can be longer than the required delay bound D_{max} (e.g. 1 ms). As such, retransmissions or repetitions become possible. To avoid feedback delay caused by retransmissions, we apply K-repetitions to reduce the packet loss probability in the communication system, i.e. the device sends K copies of each coding block no matter whether the first few copies are successfully decoded or not [1]. The transmission duration of each copy is denoted as D_n^{τ}. Then, we have $D_n^{\tau} = D_n^{\mathrm{t}}/K_n$. Some time and frequency resources are reserved for channel estimation at the AP. The fraction of time and frequency resources for data transmission is denoted as $\eta < 1$. To avoid overhead and extra delay caused by channel estimation at the device, we assume the device does not have Channel State Information (CSI). The impacts of CSI and training pilots on the achievable rate have been studied in the short blocklength regime [10, 23, 25, 28]. If more resource blocks are occupied by pilots, the accuracy of the estimated CSI can be improved. However, the remaining resource blocks for data transmission reduces. How to allocate radio resources for pilots and data transmissions is a complicated problem and deserves further study. By assuming CSI is not available at the transmitters, our approach can serve as a benchmark for future research.

For the transmission of each copy, we assume that the transmission duration is smaller than the channel coherence time and the bandwidth is smaller than the coherence bandwidth. This assumption is reasonable for short packet transmissions in URLLC. Then, the achievable rate in the short blocklength regime over a quasi-static SIMO channel can be accurately approximated by the following normal approximation [40][5],

$$b_n \approx \frac{\eta D_n^{\tau} T_s B_n}{\ln 2} \left[\ln(1+\gamma_n) - \sqrt{\frac{V_n}{\eta D_n^{\tau} T_s B_n}} f_{\mathrm{Q}}^{-1}(\varepsilon_n^{\tau}) \right] \quad \text{(bits/block)}, \tag{10.26}$$

where B_n is the bandwidth, γ_n represents the received SNR, $V_n = 1 - [1+\gamma_n]^{-2}$ [40], $f_{\mathrm{Q}}^{-1}(\cdot)$ is the inverse function of the Q-function, and ε_n^{τ} is the decoding error probability. The blocklength of the channel codes is $\eta D_n^{\tau} T_s B_n$. When the blocklength is large, (10.26) approaches the Shannon capacity [6].

5 The bounds of the decoding error probability can be obtained using the saddlepoint method [18], which is very accurate but has no closed-form expression. Since the gap between the normal approximation and practical coding schemes is around 0.1 dB [34], it is accurate enough for our framework.

6 The results in [27] indicate that if the Shannon capacity is used in the analyses, the delay bound and delay bound violation probability will be underestimated. Thus, the requirements of URLLC cannot be satisfied.

According to (10.26), the expected decoding error probability of each transmission over the SIMO channel is given by [40]

$$\bar{\varepsilon}_n^{\tau} = \int_0^{\infty} f_{\mathrm{Q}} \left\{ \sqrt{\frac{\eta D_n^{\tau} T_s B_n}{V_n}} \left[\ln\left(1 + \frac{a_n g_n P_n^{\mathrm{t}}}{\vartheta N_0 B_n}\right) - \frac{b_n \ln 2}{\eta D_n^{\tau} T_s B_n} \right] \right\} \cdot f_g(x) dx, \tag{10.27}$$

where $\gamma_n = \frac{a_n g_n P_n^{\mathrm{t}}}{\vartheta N_0 B_n}$ is applied, a_n denotes the large-scale channel gain, g_n is the small-scale channel gain, P_n^{t} represents the transmit power, $\vartheta > 1$ is the SNR loss due to inaccurate channel estimation, N_0 denotes the noise power spectral density, and $f_g(x)$ is the distribution of the instantaneous channel gain. For the Rayleigh fading channel, we have $f_g(x) = \frac{1}{(N_{\mathrm{r}}-1)!} x^{N_{\mathrm{r}}-1} e^{-x}$, where N_{r} is the number of antennas at the AP. From the approximation in [31][7], $\bar{\varepsilon}_n^{\tau}$ can be accurately approximated by

$$\bar{\varepsilon}_n^{\tau} = \frac{\omega_n a_n P_n^{\mathrm{t}} \sqrt{\eta D_n^{\tau} T_s B_n}}{\vartheta N_0 B_n} \left[(g_n^{\mathrm{U}} - g_n^{\mathrm{L}}) - \sum_{i=0}^{N_{\mathrm{r}}} (N_{\mathrm{r}} - i) A_n^{\mathrm{i}} \right], \tag{10.28}$$

where $\omega_n = \frac{1}{2\pi\sqrt{2^{2r_n^{\mathrm{c}}}-1}}$, $r_n^{\mathrm{c}} = \frac{b_n}{\eta D_n^{\tau} T_s B_n}$ is the number bits in each coding block, $g_n^{\mathrm{U}} = \frac{\vartheta N_0 B_n \xi_n}{a_n P_n^{\mathrm{t}}}$, $g_n^{\mathrm{L}} = \frac{\vartheta N_0 B_n \zeta_n}{a_n P_n^{\mathrm{t}}}$, $A_n^{\mathrm{i}} = \frac{(g_n^{\mathrm{L}})^i}{i!} e^{-g_n^{\mathrm{L}}} - \frac{(g_n^{\mathrm{U}})^i}{i!} e^{-g_n^{\mathrm{U}}}$, $\xi_n = \theta_n + \frac{1}{2\omega_n \sqrt{\eta D_n^{\tau} T_s B_n}}$, $\zeta_n = \theta_n - \frac{1}{2\omega_n \sqrt{\eta D_n^{\tau} T_s B_n}}$, and $\theta_n = 2^{r_n^{\mathrm{c}}-1}$.

After K repetitions, the packet loss probability in the communication system is given by

$$\varepsilon_n^{\mathrm{t}} = (\bar{\varepsilon}_n^{\tau})^{K_n}. \tag{10.29}$$

From (10.29), we can obtain the following property of $\varepsilon_n^{\mathrm{t}}$.

Lemma 10.3: When D_n^{τ} is given, $\varepsilon_n^{\mathrm{t}}$ strictly decreases with the repetition time K_n.

Proof. When D_n^{τ} is given, $\bar{\varepsilon}_n^{\tau}$ is fixed. According to (10.29), $\varepsilon_n^{\mathrm{t}}$ decreases with K_n since $\bar{\varepsilon}_n^{\tau} < 1$.

Lemma 10.3 indicates that there is a trade-off between the transmission delay and the reliability in communications. K-Repetition can be used to improve the transmission reliability at the cost of increasing the transmission delay.

---

7 As validated in [31], the approximation in (10.28) is accurate, especially when the number of antennas is large or the packet loss probability is small.

10.5 Prediction and Communication Co-design

In the above trade-off analyses, we obtained closed-form relations between each delay component (or prediction horizon) and its corresponding packet loss factor in terms of prediction, queueing, and wireless transmission, respectively. Based on the above analyses, the trade-off between the overall reliability and prediction horizon is revealed. As such, we could formulate the optimization problem in the following subsection.

10.5.1 Problem Formulation

To maximize the number of devices that can be supported by the system, we optimize the delay components, prediction horizon, and bandwidth allocation of wireless networks. The optimization problem can be formulated as follows,

$$\max_{\substack{D_n^{\mathrm{q}}, D_n^{\mathrm{t}}, T_n^{p}, B_n,\\ n=1,\dots,N,}} \quad N \tag{10.30}$$

$$\text{s.t.} \quad \sum_{n=1}^{N} B_n \leq B_{\max}, \tag{10.30a}$$

$$D_n^{\mathrm{q}} + D_n^{\mathrm{t}} + D_n^{\mathrm{d}} + D_n^{\mathrm{r}} - T_n^{p} \leq D_{\max}, \tag{10.30b}$$

$$\varepsilon_n^{\mathrm{q}} + \varepsilon_n^{\mathrm{t}} + \varepsilon_n^{\mathrm{p}} \leq \varepsilon_{max}, \tag{10.30c}$$

$$\varepsilon_n^{\mathrm{q}} = \exp\left\{ D_n^{\mathrm{q}} \left[\frac{\mathbb{W}_{-1}(-\lambda_n D_n^{\mathrm{t}} e^{-\lambda_n D_n^{\mathrm{t}}})}{D_n^{\mathrm{t}}} + \lambda_n \right] \right\}, \tag{10.30d}$$

$$\varepsilon_n^{\mathrm{t}} = \left\{ \frac{\omega_n a_n P_n^{\mathrm{t}} \sqrt{\eta D_n^{\tau} T_s B_n}}{\vartheta N_0 B_n} \cdot \left[(g_n^{\mathrm{U}} - g_n^{\mathrm{L}}) - \sum_{i=0}^{N_{\mathrm{r}}} (N_{\mathrm{r}} - i) A_n^i \right] \right\}^{K_n}, K_n D_n^{\tau} = D_n^{\mathrm{t}} \tag{10.30e}$$

$$\varepsilon_n^{\mathrm{p}} = 1 - \prod_{j=1}^{F} \left[1 - \psi\left(\frac{-\delta_j}{\sqrt{\sigma_j^2 + \sum_{i=1}^{T_n^{p}-1} \sum_{m=1}^{F} (\phi_{n,j,m,T_n^{p}-i})^2 \sigma_m^2}} \right) \right], \tag{10.30f}$$

$$n = 1, 2, 3, \cdots, N, \tag{10.30g}$$

where (10.30a) is the constraint on total bandwidth, (10.30b) is the constraint on user experienced delay, (10.30c) is the constraint on reliability. (10.30d) is obtained by substituting (10.18) and (10.16) into (10.17), (10.30e) is obtained from (10.28) and (10.29), and (10.30f) is obtained by substituting (10.12) into (10.14).

Problem (10.30) is not a deterministic optimization problem since the numbers of optimization variables and constraints depend on the number of users, which is not given. In addition, some optimization variables are integers and the constraints in (10.30c), (10.30d), and (10.30e) are non-convex. Thus, it is very challenging to solve this problem.

10.5.2 Algorithm for Solving Problem (10.30)

To solve the problem (10.30), we first find the minimal bandwidth B_n required for each user to ensure its delay and reliability requirements, i.e. $(D_{max}, \varepsilon_{max})$. By minimizing the bandwidth allocated to each user, the total number of users that can be supported with a given amount of total bandwidth can be maximized. Without the constraint on total bandwidth, the problem (10.30) can be decomposed into multiple single-user problems:

$$\min_{D_n^{\mathrm{q}}, D_n^{\mathrm{t}}, T_n^p} \quad B_n \tag{10.31}$$

$$\text{s.t.} \quad (10.30\text{b}), (10.30\text{c}), (10.30\text{d}), (10.30\text{e}) \text{ and } (10.30\text{f}). \tag{10.32}$$

To solve the above problem, we need the minimal bandwidth that is required to ensure a certain overall reliability. We denote it as $B_n^{min}(\varepsilon_n^{\mathrm{o}})$. However, deriving the expression of $B_n^{min}(\varepsilon_n^{\mathrm{o}})$ is very difficult. To overcome this difficulty, we first minimize $\varepsilon_n^{\mathrm{o}}$ for a given B_n. Then, we find the minimal required bandwidth that can satisfy $\varepsilon_n^{\mathrm{o}} \leq \varepsilon_{\max}$ via binary search.

When B_n is given, the minimal overall error probability can be obtained by optimizing T_n^p in solving the following problem,

$$\varepsilon_n^{\mathrm{o},min}(B_n) = \min_{D_n^{\mathrm{q}}, D_n^{\mathrm{t}}, T_n^p} \quad \varepsilon_n^{\mathrm{q}} + \varepsilon_n^{\mathrm{t}} + \varepsilon_n^{\mathrm{p}} \tag{10.33}$$

$$\text{s.t.} \quad (10.30\text{b}), (10.30\text{d}), (10.30\text{e}) \text{ and } (10.30\text{f}),$$

For mathematical tractability, we set $\varepsilon_n^{\mathrm{q}} = \varepsilon_n^{\mathrm{t}}$. According to [32], this simplification leads to negligible performance loss. We will first prove $\varepsilon_n^{\mathrm{q}}$ and $\varepsilon_n^{\mathrm{t}}$ decreases with T_n^p in the Lemma 10.4 when $\varepsilon_n^{\mathrm{q}} = \varepsilon_n^{\mathrm{t}}$.

Lemma 10.4: $\varepsilon_n^{\mathrm{q}}$ and $\varepsilon_n^{\mathrm{t}}$ decrease with T_n^p when $\varepsilon_n^{\mathrm{q}} = \varepsilon_n^{\mathrm{t}}$.

Proof. According to (10.36), we have $D_n^{\mathrm{q}} + D_n^{\mathrm{t}} = D_{max} + T_n^p - D_n^{\mathrm{r}}$. To prove this lemma, we need to prove that $\varepsilon_n^{\mathrm{q}}$ or $\varepsilon_n^{\mathrm{t}}$ decreases with $D_n^{\mathrm{q}} + D_n^{\mathrm{t}}$.

Next, we will prove D_n^{q} increases with D_n^{t}, and thus $D_n^{\mathrm{q}} + D_n^{\mathrm{t}}$ increases with D_n^{t}. According to (10.17) and (10.18), we have

$$D_n^{\mathrm{q}} = \frac{\ln(\varepsilon_n^{\mathrm{q}})}{\phi(\lambda_n, E_n^{\mathrm{B}})},$$

where

$$\phi(\lambda_n, E_n^{\mathrm{B}}) = \frac{\mathbb{W}_{-1}\left(-\lambda_n D_n^{\mathrm{t}} e^{-\lambda_n D_n^{\mathrm{t}}}\right)}{D_n^{\mathrm{t}}} + \lambda_n.$$

To check the monotonicity of D_n^{q} in terms of $\varepsilon_n^{\mathrm{q}}$ and D_n^{t}, we have the following partial derivatives,

$$\frac{\partial D_n^{\mathrm{q}}}{\partial \varepsilon_n^{\mathrm{q}}} = \frac{1}{\varepsilon_n^{\mathrm{q}} \phi(\lambda_n, E_n^{\mathrm{B}})} < 0, \tag{10.34}$$

and

$$\begin{aligned} \frac{\partial D_n^{\mathrm{q}}}{\partial D_n^{\mathrm{t}}} &= \frac{\ln(\varepsilon_n^{\mathrm{q}})}{\left[\mathbb{W}_{-1}(-\lambda_n D_n^{\mathrm{t}} e^{-\lambda_n D_n^{\mathrm{t}}}) + 1\right]} \\ &\cdot \frac{\mathbb{W}_{-1}(-\lambda_n D_n^{\mathrm{t}} e^{-\lambda_n D_n^{\mathrm{t}}})}{\left[\mathbb{W}_{-1}(-\lambda_n D_n^{\mathrm{t}} e^{-\lambda_n D_n^{\mathrm{t}}}) + \lambda_n D_n^{\mathrm{t}}\right]} > 0. \end{aligned} \tag{10.35}$$

As such, we prove D_n^{q} increases with D_n^{t} when $\varepsilon_n^{\mathrm{q}}$ is given. According to Lemma 10.3, $\varepsilon_n^{\mathrm{t}}$ strictly decreases with the transmission delay D_n^{t}. Since $\varepsilon_n^{\mathrm{q}} = \varepsilon_n^{\mathrm{t}}$, $\varepsilon_n^{\mathrm{q}}$ also strictly decreases with the transmission delay D_n^{t}. According to (10.34), D_n^{q} increases with a smaller $\varepsilon_n^{\mathrm{q}}$. So D_n^{q} increases with D_n^{t} when $\varepsilon_n^{\mathrm{q}}$ is determined by D_n^{t}.

In summary, $\varepsilon_n^{\mathrm{q}}$ or $\varepsilon_n^{\mathrm{t}}$ decreases with D_n^{t} and $D_n^{\mathrm{q}} + D_n^{\mathrm{t}}$, and thus decreases with T_n^p. This completes the proof.

Lemma 10.4 reveals the relation between the reliability of the queueing system (or the reliability of the wireless link) and the prediction horizon. With this relation, the number of independent optimization variables can be reduced.

It can be recalled that $\varepsilon_n^{\mathrm{p}}$ increases with T_n^p. Thus, together with Lemma 1.1, the optimal solution is obtained when the equality in (10.36) holds, which is

$$D_n^{\mathrm{q}} + D_n^{\mathrm{t}} + D_n^{\mathrm{r}} - T_n^p = D_{max}. \tag{10.36}$$

Moreover, for a given value of T_n^p, the values of D_n^{q} and D_n^{t} that satisfies $\varepsilon_n^{\mathrm{q}} = \varepsilon_n^{\mathrm{t}}$ and (10.36) can be obtained via a binary search. Therefore, we only need to optimize T_n^p in problem (10.33). The optimal solution and the minimal overall reliability in this simplified scenario are denoted as T_n^{p*} and $\varepsilon_n^{\mathrm{o},min*}(B_n)$, respectively.

Unfortunately, the simplified problem is still non-convex. As such, we will propose an approximated solution as follows. According to Lemma 10.1, $\varepsilon_n^{\mathrm{p}}$ increases with T_n^p, and we have proved $\varepsilon_n^{\mathrm{q}}$ and $\varepsilon_n^{\mathrm{t}}$ decreases with T_n^p in Lemma 10.4. A near optimal solution can be obtained when $\varepsilon_n^{\mathrm{q}} + \varepsilon_n^{\mathrm{t}} = \varepsilon_n^{\mathrm{p}}$. Since the optimization variables are not integers, $\varepsilon_n^{\mathrm{q}} + \varepsilon_n^{\mathrm{t}} = \varepsilon_n^{\mathrm{p}}$ may not hold strictly. To address this issue, we can use a binary search to find $\tilde{T}_n^p$ that satisfies $\varepsilon_n^{\mathrm{p}} \leq 2\varepsilon_n^{\mathrm{t}}$ when $T_n^p \leq \tilde{T}_n^p$, and $\varepsilon_n^{\mathrm{p}} > 2\varepsilon_n^{\mathrm{t}}$ when $T_n^p > \tilde{T}_n^p$. The corresponding reliability is denoted as $\hat{\varepsilon}_n^{\mathrm{o,min}}(B_n)$. The overall reliability achieved by this near optimal solution is denoted as $\hat{\varepsilon}_n^{\mathrm{o,min}}(B_n)$.

The performance gap between the near optimal solution and optimal one is analyzed in the following Lemma 10.5.

Lemma 10.5: The gap between $\hat{\varepsilon}_n^{\mathrm{o,min}}(B_n)$ and $\varepsilon_n^{\mathrm{o,min*}}(B_n)$ is less than $\varepsilon_n^{\mathrm{o,min*}}(B_n)$, where $\varepsilon_n^{\mathrm{o,min*}}(B_n)$ is the reliability achieved by the optimal solution.

Proof. In this proof, we use the notation $\varepsilon_n^{\mathrm{o}}(T_n^p, B_n)$, (or $\varepsilon_n^{\mathrm{t}}(T_n^p, B_n)$ or $\varepsilon_n^{\mathrm{p}}(T_n^p, B_n)$) to represent the relationship between the prediction horizon and the overall reliability (or the decoding error probability or the prediction error probability). For notational simplicity, we first omit B_n.

To prove this lemma, we first introduce an upper bound of $\varepsilon_n^{\mathrm{o}}(T_n^p) = 2\varepsilon_n^{\mathrm{t}}(T_n^p + D_{\max}) + \varepsilon_n^{\mathrm{p}}(T_n^p)$, i.e. $\varepsilon_{\mathrm{o},n}^{\mathrm{ub}}(T_n^p) = 2\max\{2\varepsilon_n^{\mathrm{t}}(T_n^p + D_{\max}), \varepsilon_n^{\mathrm{p}}(T_n^p)\}$.

Suppose $\tilde{T}_n^p$ is the maximal prediction horizon that satisfies $2\varepsilon_n^{\mathrm{t}}(T_n^p + D_{\max}) - \varepsilon_n^{\mathrm{p}}(T_n^p) > 0$ for all $0 \leq T_n^p \leq \tilde{T}_n^p$, and hence $\varepsilon_n^{\mathrm{o,ub}}(T_n^p) = 4\varepsilon_n^{\mathrm{t}}(T_n^p + D_{\max})$, which strictly decreases with T_n^p. On the other hand, when $T_n^p > \tilde{T}_n^p$, $2\varepsilon_n^{\mathrm{t}}(T_n^p + D_{\max}) - \varepsilon_n^{\mathrm{p}}(T_n^p) < 0$, and hence $\varepsilon_n^{\mathrm{o,ub}}(T_n^p) = 2\varepsilon_n^{\mathrm{p}}(T_n^p)$, which strictly increases with T_n^p. In other words, $\varepsilon_n^{\mathrm{o,ub}}(T_n^p)$ strictly decreases with T_n^p when $T_n^p \leq \tilde{T}_n^p$ and strictly increases with T_n^p when $T_n^p > \tilde{T}_n^p$. Therefore, the upper bound $\varepsilon_n^{\mathrm{o,ub}}(T_n^p)$ is minimized at $\hat{T}_n^p = \tilde{T}_n^p$ or $\hat{T}_n^p = \tilde{T}_n^p + 1$.

Let $2\varepsilon_n^{\mathrm{t}}(\hat{T}_n^p + D_{\max}) - \varepsilon_n^{\mathrm{p}}(\hat{T}_n^p) = \Delta$, where Δ is the small gap between $2\varepsilon_n^{\mathrm{t}}$ and $\varepsilon_n^{\mathrm{p}}$ at $\hat{T}_n^p$, which is every closed to zero. We have

$$\varepsilon_n^{\mathrm{o}}(\hat{T}_n^p) \approx \varepsilon_n^{\mathrm{o,ub}}(\hat{T}_n^p). \tag{10.37}$$

Besides, $\varepsilon_n^{\mathrm{o,ub}}(\hat{T}_n^p)$ is the minimum of $\varepsilon_n^{\mathrm{o,ub}}(T_n^p), \forall n \in [0, \infty)$, and hence

$$\varepsilon_n^{\mathrm{o,ub}}(\hat{T}_n^p) \leq \varepsilon_n^{\mathrm{o,ub}}(T_n^{p*}), \tag{10.38}$$

where T_n^{p*} is the optimal prediction horizon that minimizes $\varepsilon_n^{\mathrm{o}}(T_n^p)$. According to the definition of $\varepsilon_n^{\mathrm{o,ub}}(T_n^p)$, we have

$$\begin{aligned}\varepsilon_n^{\mathrm{o,ub}}(T_n^{p*}) &= 2\max\{2\varepsilon_n^{\mathrm{t}}(T_n^{p*} + D_{\max}), \varepsilon_n^{\mathrm{p}}(T_n^{p*})\}\\ &< 2\left[2\varepsilon_n^{\mathrm{t}}(T_n^{p*} + D_{\max}) + \varepsilon_n^{\mathrm{p}}(T_n^{p*})\right]\\ &= 2\varepsilon_n^{\mathrm{o}}(T_n^{p*}).\end{aligned} \tag{10.39}$$

From (10.37), (10.38), and (10.39), we have $\varepsilon_n^{\mathrm{o}}(\hat{T}_n^p) < 2\varepsilon_n^{\mathrm{o}}(T_n^{p*})$, i.e. $\varepsilon_n^{\mathrm{o}}(\hat{T}_n^p) - \varepsilon_n^{\mathrm{o}}(T_n^{p*}) < \varepsilon_n^{\mathrm{o}}(T_n^{p*})$.

Since $\varepsilon_n^{\rm o}(\hat{T}_n^p)$ and $\varepsilon_n^{\rm o}(T_n^{p*})$ are defined as $\hat{\varepsilon}_n^{\rm o,min}(\hat{T}_n^p, B_n)$ and $\varepsilon_{{\rm o},n}^{\rm min*}(T_n^{p*}, B_n)$, respectively, we have $\hat{\varepsilon}_n^{{\rm o},min}(\hat{T}_n^p, B_n) - \varepsilon_n^{{\rm o},min*}(T_n^{p*}, B_n) < \varepsilon_n^{{\rm o},min*}(T_n^{p*}, B_n)$. The proof follows.

Lemma 10.5 shows that the gap between the near optimal overall reliability and the optimal one is bounded by the value of the optimal overall reliability. Since the optimal overall reliability is of the order of 10^{-5}, the gap is very small.

The required minimal bandwidth to guarantee the overall reliability can be obtained from the following optimization problem,

$$\min_{B_n} B_n \tag{10.40}$$

$$\text{s.t. } \hat{\varepsilon}_n^{{\rm o},min}(B_n) \leq \varepsilon_{\max}. \tag{10.40a}$$

Since the packet loss in the communication system decreases with bandwidth, the optimal solution of problem (10.40) is achieved when the equality in (10.40a) holds. Thus, the minimal bandwidth can be obtained via a binary search.

10.5.3 Discussions on Implementation Complexity and Optimality

The original optimization problem is decomposed into N single-user problems. To solve each single-user problem, we search the required bandwidth and optimal prediction horizon in the regions $[0, \overline{B}]$ and $[0, \overline{T}^p]$, respectively, where $\overline{B}$ and $\overline{T}^p$ are the upper bounds of bandwidth and prediction horizon. Therefore, the complexity of the proposed algorithm is around $\mathcal{O}\left(N \log_2(\bar{B}) \log_2(\bar{T}^p)\right)$.

The performance loss of the near optimal solution relative to the global optimal solution results from simplification $\varepsilon_n^{\rm q} = \varepsilon_n^{\rm t}$ and the differences between $\hat{\varepsilon}_n^{{\rm o},min}(B_n)$ and $\varepsilon_n^{{\rm o},min*}(B_n)$. According to the analysis in [32] and Lemma 10.5, the performance loss is minor. We will further validate the performance loss with numerical results.

10.6 Performance Evaluation

In this section, we evaluate the effectiveness of the proposed co-design method via simulations and experiments.

10.6.1 Simulations

In the simulations, we consider a one-dimensional movement as an example to evaluate the proposed co-design method. With this example, we show how the proposed method helps improve the trade-offs among latency, reliability, and resource utilizations (i.e. bandwidth and antenna). For comparison, the performance achieved by the traditional transmission scheme with no prediction is provided. The simulation parameters are listed in Table 10.1. In all simulations, SNRs are computed according to $\gamma_n = \frac{a_n g_n P_n^t}{\vartheta N_0 B_n}$. The path loss model is $10\log_{10}(a_n) = 35.3 + 37.6\log_{10}(d_n) + S_n$, where d_n is the distance from the nth device to the AP and S_n is the shadowing. The shadowing S_n follows log normal distribution with a zero mean and a standard deviation of 8. To ensure the reliability and latency requirements, we consider the worst case of shadowing $S_{\rm w} = -34.1$ dB (i.e. $\Pr\{S_n \leq S_{\rm w}\} = 10^{-5}$), which is defined as the probability that the delay and reliability of a device can be satisfied [31].

For the one-dimensional movement, the state transition function in (10.6) can be simplified as follows [16],

$$\begin{bmatrix} r(k+1) \\ v(k+1) \\ a(k+1) \end{bmatrix} = \begin{bmatrix} 1 & T_s & \frac{T_s^2}{2} \\ 0 & 1 & T_s \\ 0 & 0 & 1 \end{bmatrix} \begin{bmatrix} r(k) \\ v(k) \\ a(k) \end{bmatrix} + \begin{bmatrix} 0 \\ 0 \\ w(k) \end{bmatrix}.$$

where $r(k)$, $v(k)$, and $a(k)$ represent the location, velocity, and acceleration in the kth slot, respectively, $w(k)$ is the Gaussian noise on acceleration, and Φ is given by

$$\Phi = \begin{bmatrix} 1 & T_s & \frac{T_s^2}{2} \\ 0 & 1 & T_s \\ 0 & 0 & 1 \end{bmatrix}, \tag{10.41}$$

which follows Newton's laws of motion. In predictions, the standard deviation of the transition noise of acceleration is $\sigma_w = 0.01$ m s^{-2}, and the required threshold is δ_l=0.1 m. The standard derivatives of the initial errors of location, velocity, and

Table 10.1 Simulation Parameters [2].

Parameters	Values
Maximal transmit power of a user $P_{\rm t}$	23 dBm
Single-sided noise spectral density N_0	-174 dBm Hz^{-1}
Information load per block b s^{-1}	160 bits
Slot duration $T_{\rm s}$	0.1 ms
Transmission duration D_τ	0.5 ms
Delay of core network and backhaul $D_{\rm r}$	10 ms

acceleration are set to be 0.01 m, 0.2 m s^{-1}, and 0.1 m s^{-2}, respectively. In practice, the values of initial errors depend on the accuracy of observation and residual filter errors [16].

10.6.1.1 Single-user Scenarios

In single-user scenarios, the distance between the user and the AP is set to be 200 m. To evaluate the proposed co-design method, the prediction horizon T_n^{p} is optimized to obtain the minimal overall error probability.

Under the given delay requirement (i.e. $D_{\max} = 0$ ms), the packet loss probability in communications $\varepsilon_n^{\mathrm{c}}$, the prediction error probability $\varepsilon_n^{\mathrm{p}}$, and the overall error probability $\varepsilon_n^{\mathrm{o}}$ are shown in Figure 10.3. To achieve target reliability, the bandwidth B is set as $B = 440$ KHz and the number of antennas at the AP is set to be $N_{\mathrm{r}} = 32$. It should be noted that the reliability depends on the amount of bandwidth and the number of antennas, but the trend of the overall reliability does not change.

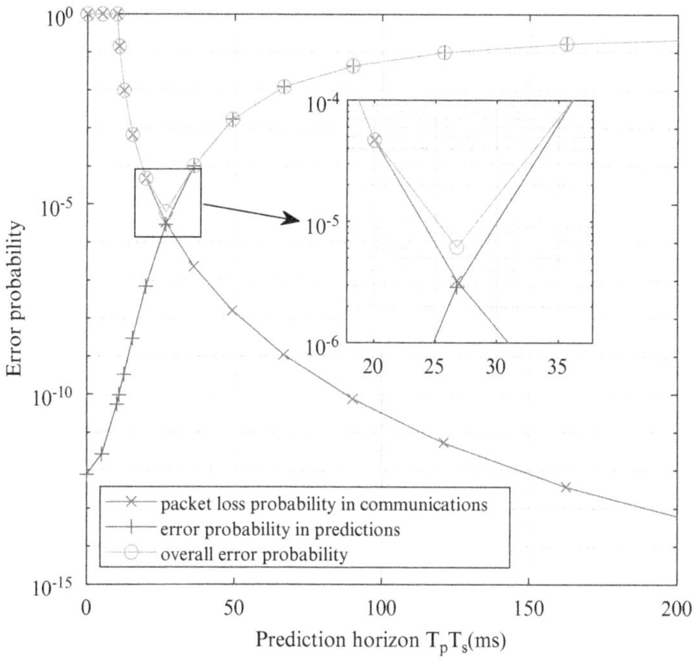

Figure 10.3 Joint optimization of predictions and communications: the packet loss probability ε_c in communications, the prediction error probability ε_p, and the over error probability ε_o are drawn as functions of prediction horizon T_pT_s. ©IEEE 2020. Reprinted with permission from [12].

In Figure 10.3, the communication delay and prediction horizon are set to be equal, i.e. $D_n^{\rm q}+D_n^{\rm t} = T_n^{\rm p}$. In this case, user experienced delay is zero. The results in Figure 10.3 show that when the E2E communication delay $D_n^{\rm q}+D_n^{\rm t} = T_n^{\rm p} < 10$ ms, i.e. less than the delays in the core network and the backhaul $D_n^{\rm r}$, it is impossible to achieve zero latency without prediction. When $D_n^{\rm q} + D_n^{\rm t} = T_n^{\rm p} > 10$ ms, the required transmission duration KD_n^{τ} increases with prediction horizon $T_n^{\rm p}$. As a result, the overall error probability, $\varepsilon_n^{\rm o}$, is first dominated by $\varepsilon_n^{\rm c}$ and then by $\varepsilon_n^{\rm p}$. As such, $\varepsilon_n^{\rm o}$ first decreases and then increases with $T_n^{\rm p}$. The results in Figure 10.3 indicate that the reliability achieved by the proposed method is 6.52×10^{-6} with $T_n^{\rm p} = 26.8$ ms, $D_n^{\rm t*} = 12.5$ ms, $D_n^{\rm q*} = 14.3$ ms and $K_n^* = 5$. The optimal solution obtained by exhaustive search is 6.15×10^{-6}. The gap between above two solutions is 3.7×10^{-7}, which is very small.

In Figure 10.4, the proposed co-design method is compared with a baseline method without prediction. When there is no prediction, the user experienced delay is equals to the communication delay. The results in Figure 10.4 show that when the requirement on user experienced delay is less than 10 ms, it cannot be

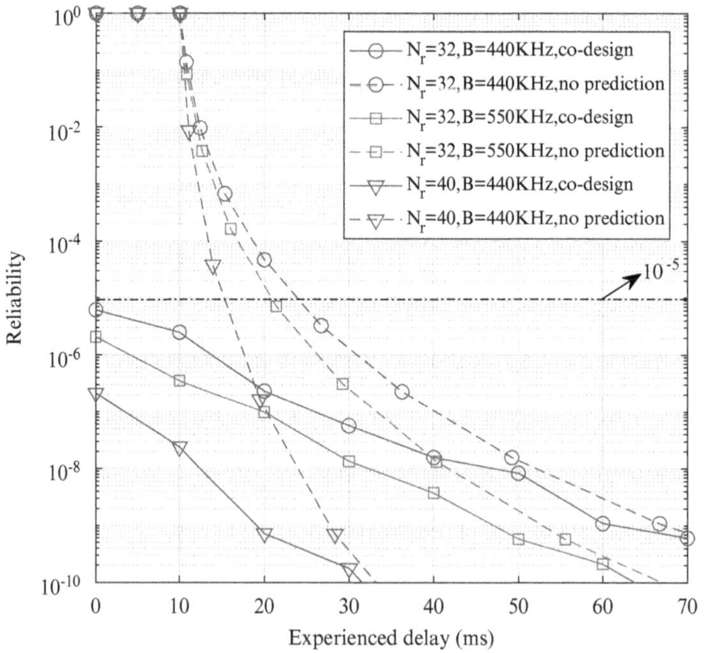

Figure 10.4 Comparison of reliability-delay trade-off curves between co-design and no predictions with different bandwidth B and numbers of received antennas N_r. ©IEEE 2020. Reprinted with permission from [12].

satisfied without prediction. When the required user experienced delay is larger than 10 ms, the reliability achieved by the co-design method is much better than the baseline method. In other words, by prediction and communication co-design, the trade-off between user experienced delay and overall reliability can be improved remarkably. Particularly, in the case $N_r = 32$ and $B = 440$ KHz, to ensure the same reliability 10^{-5}, the user experienced delay can be reduced by 23 ms and zero-latency can be achieved by the proposed co-design method.

10.6.1.2 Multiple-user Scenarios

In multiple-user scenarios, we will consider two scenarios: the distribution of large-scale fading of the mobile devices is available/unavaibale. In the first scenario, the distances from devices to the AP are uniformly distributed in the region [50, 200] m. In the second scenario, the worst case is considered in the optimization, i.e. the distances from all the devices to the AP are 200 m.

Since the large-scale fading of devices is a random variable in the first scenario, the sum of the required bandwidth is also a random variable. In Figure 10.5, we illustrated the probability that the sum of the required bandwidth is smaller than

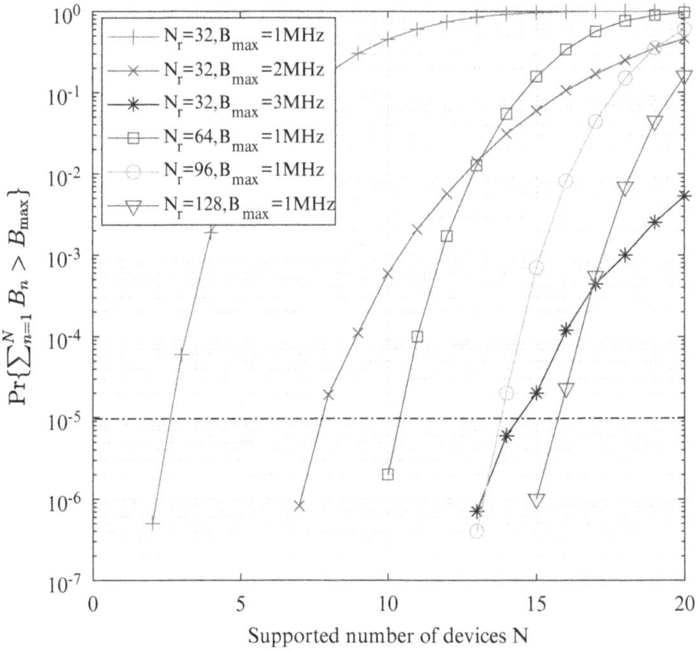

Figure 10.5 $\Pr\{\sum_{n=1}^{N} B_n > B_{\max}\}$ versus the number of devices when the distribution of large-scale fading of devices is known. ©IEEE 2020. Reprinted with permission from [12].

B_{max}. For URLLC services, we need to guarantee the delay and reliability requirements with high probability, e.g. 99.999 %. The results in Figure 10.5 show that when $B_{max} = 1$ MHz and $N_r = 32$, the system can only support 2 devices. By doubling the number of antennas (or the total bandwidth), 10 (or 7) devices can be supported. This implies that increasing the number of antennas at the AP is an efficient way to increase the number of devices that can be supported by the system. This is because SNR increases with the number of antennas due to array gain. To achieve the same reliability, i.e. packet loss probability, higher order modulation schemes can be used if more antennas are deployed at the AP. Since the spectrum efficiency increases with the order of the modulation scheme, more URLLC devices can be supported with a given amount of bandwidth.

If the distribution of large-scale fading of devices is unknown, the worst case is considered. Then, the total bandwidth that is required to support a given number of devices is deterministic. The results in Figure 10.6 show that the required total bandwidth linearly increases with the number of devices. This is because

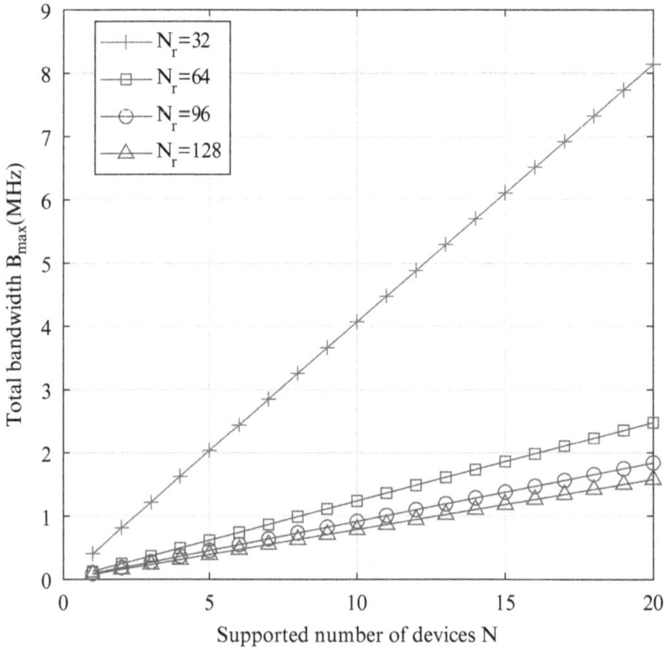

Figure 10.6 Required total bandwidth versus the number of devices when the distribution of large-scale fading of devices is unknown. ©IEEE 2020. Reprinted with permission from [12].

the required bandwidth for different devices is the same since the worst case is considered for all the devices. In addition, by increasing the number of antennas from 32 to 64, we can save 75 % of the bandwidth. This implies that increasing the number of antennas is an efficient way to improve spectrum efficiency of URLLC.

10.6.2 Experiments

To validate whether mobility prediction works for URLLC in practice, we record the real movement data from the experiment shown in Figure 10.7. In this experiment, a typical application of Tactile Internet is implemented in a virtual environment, where a box of hazardous chemicals or radioactive substances is dragged along the floor by a virtual slave device. A tactile hardware device named the 3D System Touch (previously named Phantom Omni, or Geomagic) is used as a master device, which sends real time location information to the virtual slave device. A cable is used to connect the master device to the virtual slave device in a virtual environment. The slave device in the virtual environment receives the locations from the master device, so it can move synchronously with the master device.

Human operators are invited to drag the virtual box from one corner to another corner of the floor in the virtual environment. In this experiment, we are mainly interested in the motion prediction, so the location information on the x-axis produced by the tactile hardware device is recorded and used to verify the predictions. A general linear prediction method in (10.9) is used to predict the future state system. Since only location information is available from the hardware, the velocity and acceleration are obtained from the first and the second order differences of

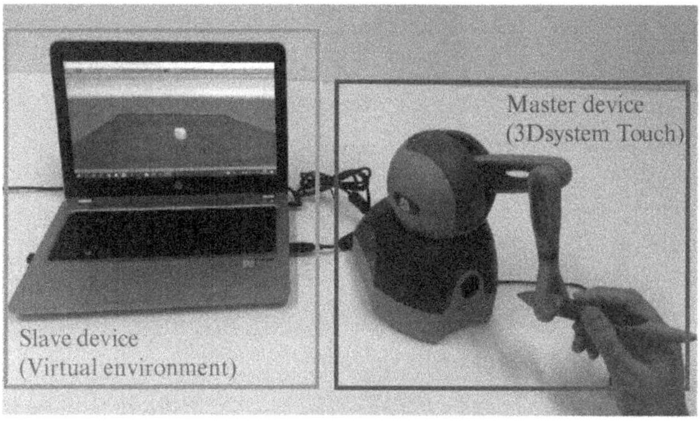

Figure 10.7 Experiment to obtain real movement data in the Tactile Internet. ©IEEE 2020. Reprinted with permission from [12].

locations [22]. Moreover, due to the limitation of the hardware, the duration of each slot is $T_s = 1$ ms.

The prediction error probabilities, ε_n^p, with different thresholds, δ, are shown in Table 10.2, where the prediction horizon, nT_s, is fixed. The results in Table 10.2 show that for the constant prediction horizon $nT_s = 5$ ms or $nT_s = 20$ ms, ε_n^p decreases with the required threshold δ.

The relation between the prediction error probability and the prediction horizon is shown in Table 10.3, where the required threshold is fixed. The results indicate that ε_n^p increases with nT_s. This observation is consistent with Lemma 10.1.

The results in Tables 10.2 and 10.3 imply that prediction and communication co-design has the potential to achieve zero-latency in practice. It should be noted that the results from the experiment are generally worse than those of the simulations. This is because we only have the location information of the device, and extra estimation errors are introduced during the estimations of the velocity and acceleration.

Table 10.2 Prediction error probability with fixed nT_s. ©IEEE 2020. Reprinted with permission from [12].

δ(m)	ε_n^p ($nT_s = 5$ ms)	ε_n^p ($nT_s = 20$ ms)
0.002	2.95×10^{-4}	0.45
0.01	1.62×10^{-5}	0.42
0.02	6.61×10^{-6}	3.2×10^{-3}
0.1	2.40×10^{-6}	2.40×10^{-5}
0.2	1.80×10^{-6}	7.82×10^{-6}

Table 10.3 Prediction error probability with given δ. ©IEEE 2020. Reprinted with permission from [12].

nT_s(ms)	ε_n^p ($\delta = 0.002$ m)	nT_s(ms)	ε_n^p ($\delta = 0.2$ m)
1	3.00×10^{-6}	10	1.80×10^{-6}
2	1.62×10^{-5}	20	7.82×10^{-6}
3	3.49×10^{-5}	30	2.71×10^{-5}
4	5.77×10^{-5}	40	4.45×10^{-5}
5	2.95×10^{-4}	50	1.69×10^{-4}

10.7 Conclusions

In this chapter, we studied how to achieve URLLC by prediction and communication co-design. We first derived the decoding error probability, the queueing delay violation probability, and the prediction error probability in closed-form expressions. Then, we established an optimization framework for maximizing the number of devices that can be supported in a system by optimizing time and frequency resources in the communication system and the prediction horizon in the prediction system. Simulation results showed that by prediction and communication co-design the trade-off between delay and reliability can be improved remarkably, i.e. the spectrum efficiency is improved subject to the delay and reliability constraints. In addition, an experiment was carried out to validate the accuracy of prediction in a remote-control system. The results showed that the proposed concept on prediction and communication co-design works well in the practical remote-control system.

Bibliography

[1] 3GPP TR 38.802 V2.0.0. Study on new radio (NR) access technology; physical layer aspects (release 14). 2017.

[2] 3GPP TSG RAN TR38.913 R14. Study on scenarios and requirements for next generation access technologies, Jun. 2017.

[3] Hussein Al-Zubaidy, Jörg Liebeherr, and Almut Burchard. Network-layer performance analysis of multihop fading channels. *IEEE/ACM Trans. Netw.*, 24 (1):204–217, Feb. 2016.

[4] Mehdi Bennis, Mérouane Debbah, and H Vincent Poor. Ultra-reliable and low-latency wireless communication: Tail, risk and scale. *arXiv preprint arXiv:1801.01270*, 2018.

[5] Gilberto Berardinelli, Nurul Huda Mahmood, Renato Abreu, Thomas Jacobsen, Klaus Pedersen, István Z Kovács, and Preben Mogensen. Reliability analysis of uplink grant-free transmission over shared resources. *IEEE Access*, 6: 23602–23611, 2018.

[6] Chen-Shang Chang and Joy A. Thomas. Effective bandwidth in high-speed digital networks. *IEEE J. Sel. Areas Commun.*, 13(6):1091–1100, 1995.

[7] M. Condoluci, T. Mahmoodi, E. Steinbach, and M. Dohler. Soft resource reservation for low-delayed teleoperation over mobile networks. *IEEE Access*, 5: 10445–10455, May 2017.

[8] Gerhard P. Fettweis. The Tactile Internet: applications & challenges. *IEEE Veh. Technol. Mag.*, 9(1):64–70, Mar. 2014.

[9] Seyedshams Feyzabadi, Sirko Straube, Michele Folgheraiter, et al. Human force discrimination during active arm motion for force feedback design. *IEEE Trans. Haptics*, 6(3):309–319, 2013.

[10] Babak Hassibi and Bertrand M Hochwald. How much training is needed in multiple-antenna wireless links? *IEEE Trans. Inf. Theory*, 49(4):951–963, 2003.

[11] Z. Hou, C. She, Y. Li, et al. Burstiness aware bandwidth reservation for ultra-reliable and low-latency communications (URLLC) in Tactile Internet. *IEEE J. Sel. Areas Commun.*, 36(11):2401–2410, 2018.

[12] Zhanwei Hou, Changyang She, Yonghui Li, Li Zhuo, and Branka Vucetic. Prediction and communication co-design for ultra-reliable and low-latency communications. *IEEE Tran. Wireless Commun.*, 19(2):1196–1209, 2020.

[13] Yulin Hu, Anke Schmeink, and James Gross. Blocklength-limited performance of relaying under quasi-static Rayleigh channels. *IEEE Wireless Commun.*, 15 (7):4548–4558, Jul. 2016.

[14] Thomas Jacobsen, Renato Abreu, Gilberto Berardinelli, et al. System level analysis of uplink grant-free transmission for URLLC. In *Proc. IEEE Globecom Workshops*, pages 1–6, 2017.

[15] Lei Kang, Wei Zhao, Bozhao Qi, and Suman Banerjee. Augmenting self-driving with remote control: Challenges and directions. In *Proc. ACM Mobile Computing Systems & Applications*, pages 19–24, 2018.

[16] Steven M Kay. *Fundamentals of statistical signal processing, volume I: estimation theory*. Prentice Hall, 1993.

[17] Gregor Klančar and Igor Škrjanc. Tracking-error model-based predictive control for mobile robots in real time. *Robotics and autonomous systems*, 55(6): 460–469, 2007.

[18] Alejandro Lancho, Jöhan Ostman, Giuseppe Durisi, et al. Saddlepoint approximations for Rayleigh block-fading channels. 2019. URL https://arxiv.org/abs/1904.10442.

[19] Mingyan Li, Xinping Guan, Cunqing Hua, Cailian Chen, and Ling Lyu. Predictive pre-allocation for low-latency uplink access in industrial wireless networks. In *Proc. IEEE INFOCOM*, pages 306–314, 2018.

[20] X. Lin, J. Li, R. Baldemair, et al. 5G New Radio: unveiling the essentials of the next generation wireless access technology. 2018. URL https://arxiv.org/abs/1806.06898.

[21] Behrooz Makki, Tommy Svensson, Giuseppe Caire, and Michele Zorzi. Fast HARQ over finite blocklength codes: A technique for low-latency reliable communication. *IEEE Wireless Commun.*, 18(1):194–209, 2018.

[22] John Mathews and Kurtis Fink. *Numerical methods using MATLAB*. Pearson Prentice Hall, NJ, 2004.

[23] Mohammadreza Mousaei and Besma Smida. Optimizing pilot overhead for ultra-reliable short-packet transmission. In *Proc. IEEE ICC*, pages 1–5, 2017.

[24] J. J. Nielsen, R. Liu, and P. Popovski. Ultra-reliable low latency communication using interface diversity. *IEEE Trans. on Commun.*, 66(3):1322–1334, Mar. 2018.

[25] Johan Östman, Giuseppe Durisi, Erik G Ström, Mustafa C Coşkun, and Gianluigi Liva. Short packets over block-memoryless fading channels: Pilot-assisted or noncoherent transmission? *IEEE Trans. Commun.*, 67(2): 1521–1536, 2018.

[26] Joachim Sachs, Gustav Wikstrom, Torsten Dudda, et al. 5G radio network design for ultra-reliable low-latency communication. *IEEE Network*, 32(2): 24–31, 2018.

[27] Sebastian Schiessl, James Gross, and Hussein Al-Zubaidy. Delay analysis for wireless fading channels with finite blocklength channel coding. In *Proc. ACM MSWiM*, pages 13–22, 2015.

[28] Sebastian Schiessl, Hussein Al-Zubaidy, Mikael Skoglund, and James Gross. Delay performance of wireless communications with imperfect CSI and finite-length coding. *IEEE Trans. Commun.*, 66(12):6527–6541, 2018.

[29] Philipp Schulz, Maximilian Matthe, Henrik Klessig, et al. Latency critical IoT applications in 5G: Perspective on the design of radio interface and network architecture. *IEEE Commun. Mag.*, 55(2):70–78, 2017.

[30] Changyang She, Chenyang Yang, and Tony Q. S. Quek. Radio resource management for ultra-reliable and low-latency communications. *IEEE Commun. Mag.*, 55(6):72–78, 2017.

[31] Changyang She, Zhengchuan Chen, Chenyang Yang, et al. Improving network availability of ultra-reliable and low-latency communications with multi-connectivity. *IEEE Trans. Commun.*, 66(11):5482–5496, Nov. 2018.

[32] Changyang She, Chenyang Yang, and Tony Q. S. Quek. Cross-layer optimization for ultra-reliable and low-latency radio access networks. *IEEE Wireless Commun.*, 17(1):127–141, 2018.

[33] Changyang She, Yifan Duan, Guodong Zhao, Tony Q. S. Quek, Yonghui Li, and Branka Vucetic. Cross-layer design for mission-critical IoT in mobile edge computing systems. *IEEE Internet of Things J., early access*, 2019.

[34] Mahyar Shirvanimoghaddam, Mohammad Sadegh Mohammadi, Rana Abbas, et al. Short block-length codes for ultra-reliable low latency communications. *IEEE Commun. Mag.*, 57(2):130–137, 2018.

[35] Meryem Simsek, Adnan Aijaz, Mischa Dohler, Joachim Sachs, and Gerhard Fettweis. 5G-enabled tactile internet. *IEEE J. Sel. Areas Commun.*, 34(3): 460–473, Mar. 2016.

[36] Beatriz Soret, Preben Mogensen, Klaus I Pedersen, and Mari Carmen Aguayo-Torres. Fundamental tradeoffs among reliability, latency and throughput in cellular networks. In *Proc. IEEE Globecom Workshops*, pages 1391–1396, 2014.

[37] Nils Strodthoff, Barış Göktepe, Thomas Schierl, et al. Enhanced machine learning techniques for early HARQ feedback prediction in 5G. *arXiv preprint arXiv:1807.10495*, 2018.

[38] Xin Tong, Guodong Zhao, Muhammad Ali Imran, et al. Minimizing wireless resource consumption for packetized predictive control in real-time cyber physical systems. In *Proc. IEEE ICC Workshops*, May 2018.

[39] Trung Kien Vu, Chen-Feng Liu, Mehdi Bennis, et al. Ultra-reliable and low latency communication in mmwave-enabled massive MIMO networks. *IEEE Commun. Letters*, 21(9):2041–2044, Sep. 2017.

[40] Wei Yang, Giuseppe Durisi, Tobias Koch, and Yury Polyanskiy. Quasi-static multiple-antenna fading channels at finite blocklength. *IEEE Trans. Inf. Theory*, 60(7):4232–4265, Jul. 2014.

[41] Guodong Zhao, Muhammad Ali Imran, Zhibo Pang, Zhi Chen, and Liying Li. Toward real-time control in future wireless networks: communication-control co-design. *IEEE Commun. Mag.*, 57(2):138–144, 2019.

[42] Zhongyuan Zhao, Mugen Peng, Zhiguo Ding, Wenbo Wang, and H Vincent Poor. Cluster content caching: An energy-efficient approach to improve quality of service in cloud radio access networks. *IEEE J. Sel. Areas Commun.*, 34(5): 1207–1221, May 2016.

11

Relay Robot-Aided URLLC in 5G Factory Automation with Industrial IoTs

Dang Van Huynh[1], Saeed R. Khosravirad[2], Yuexing Peng[3], Antonino Masaracchia[1], and Trung Q. Duong[1,]*

[1] *School of Electronics Electrical Engineering and Computer Science, Queen's University Belfast, BT7 1NN, Northern Ireland, Belfast, University Road, United Kingdom.*
[2] *Nokia Bell Labs, NJ 07974-0636, Murray Hill, New Jersey, Mountain Avenue, USA.*
[3] *Beijing University of Posts and Telecommunications, 100876, Haidian District, Beijing, Xitucheng Rd, China.*
[*] *Corresponding Author*

11.1 Introduction

Ultra-Reliable Low-Latency Communications (URLLC) is one of three main services of 5G New Radio (5G NR). According to 3GPP Release 15 [1], URLLC is targeted to enable mission-critical applications, where the reliability requirement for transmitting a packet is $1 - 10^{-5}$ for 32 bytes with a user plane latency of 1 ms. Release 16 enhances URLLC to more stringent requirements with $1 - 10^{-6}$ as the reliability target and the latency down to a range of 0.5 to 1 ms [2]. Therefore, URLLC brings many new kinds of services and applications that require URLLC as a prerequisite. For instance, URLLC can be applied in smart healthcare systems with telesurgery applications, or in entertainment with a wide range of Augmented Reality (AR) and Virtual Reality (VR) utilities [3]. Importantly, URLLC opens opportunities to transform current industrial automation to the next generation, which entails wireless control and real-time monitoring for autonomous factories [11, 13]. The ongoing 3GPP Release 17 targets further URLLC enhancements with the Industrial Internet-of-Things (IIoT). These situations pose many challenges yet introduce open opportunities for URLLC studies, which has attracted many researchers in academia and industry alike.

Ultra-Reliable and Low-Latency Communications (URLLC) Theory and Practice: Advances in 5G and Beyond, First Edition. Edited by Trung Q. Duong, Saeed R. Khosravirad, Changyang She, Petar Popovski, Mehdi Bennis and Tony Q.S. Quek.
© 2023 John Wiley & Sons Ltd. Published 2023 by John Wiley & Sons Ltd.

Despite its huge potential in the industrial revolution, research into URLLC in industry automation is still in an early stage. Specifically, to minimize the decoding error probability of URLLC systems, several studies considered scenarios such as point-to-point communication and single relay-assisted transmission [15, 18, 20]. In [15], the author considered joint power allocation and transmission blocklength optimization for industrial automation scenarios, which was based on a path-following algorithm. This work only examined the direct point-to-point transmission between the central controller and machine devices, while the relay-assisted transmission scheme showed significant improvement in reliability for URLLC systems [8, 12, 20]. In [20], the authors introduced joint power and blocklength optimization for URLLC in a factory automation scenario, where the reliability optimization of URLLC in different transmission schemes such as Orthogonal Multiple Access (OMA), Non-Orthogonal Multiple Access (NOMA), relay-assisted transmission, and Cooperative-NOMA (C-NOMA), was formulated. Their simulations showed that the relay-assisted scheme significantly outperformed other schemes. However, this study only explored general multiple devices with the OMA scheme, while the extensions to other schemes remain for future work.

URLLC has unravelled wireless intelligent control and operation in factory automation. However, the severe shadowing effects in highly dense factories render the control signal's reception unreliable, which poses a huge challenge in meeting stringent Quality-Of-Service (QoS) in industry automation. Motivated by the aforementioned discussion, to overcome the severe shadowing and blockage effect in the factories, we propose a multiple relay robots-assisted URLLC scheme for industrial automation [9]. Our study aims to minimize the decoding error probability of IoT devices in a factory automation scenario with the support of multiple relay robots. Due to the stringent constraints of URLLC, the optimization between blocklength and power allocation to minimize the error probability is non-convex and mathematically troublesome. In this chapter, we take a step further to propose a new algorithm for jointly optimizing the blocklength and power allocation where the two parameters are simultaneously considered in the optimization process. With this optimization scheme, the URLLC performance can be significantly increased. The contributions of this study are summarized as follows.

- To deal with the optimal relay robot deployment problem, we propose an enhanced K-means algorithm to form groups of IIoT devices and the relay robot efficiently. The proposed solution allows IIoT clusters to be created with the maximum number of IIoT devices in a particular group, which provides flexibility for industrial control and communication.
- Aiming to minimizing the decoding error probability of URLLC system, an effective iterative optimization algorithm is presented for solving the reliability

maximization of the considered URLLC system under the joint blocklength and power allocation optimization.

- Various numerical results illustrate the effectiveness of the proposed scheme over the conventional approach and prove that relay-assisted transmission significantly improves the reliability of URLLC systems in industrial automation.

11.2 System Model and Problem formulation

11.2.1 System model

We consider the downlink communication in an industry automation scenario where a central controller acting as a Base Station (BS) sends control signals to the IIoT devices (or machines) in a factory. Due to the huge shadowing effect within the heavily compact factory's environment, the transmission scheme is assisted by a group of mobile robots acting as relays. The BS is located at $\mathbf{q}_0 = [x_0, y_0]^T$ and there are K movable robots working as relay nodes, denoted as a set of $\mathcal{K} = \{1, 2, ..., K\}$. The relay robots also work as the cluster heads in the G_k clusters of IIoT devices $\mathcal{G} = \{G_1, G_2, ...G_K\}$. There are N IIoT devices working as User Equipment (UE) in the network, which are denoted as the set of $\mathcal{N} = \{1, 2, ...N\}$. The kth cluster can serve a finite number N_k of IIoT devices, $\mathcal{N}_k = \{1, 2, ...N_k\}$. All the IIoT devices are randomly distributed, and the location of the nth IIoT device is denoted by $\mathbf{q}_n = [x_n, y_n]^T$. The kth relay is positioned at $\mathbf{q}_k = [x_k, y_k]^T$. Each BS, relay, and IIoT device is equipped with a single antenna.

11.2.2 Achievable Rate and Decoding Error in an URLLC Regime

The coding rate R at a finite blocklength (m) in URLLC is approximated as [19]:

$$R \approx \log_2 [1 + \gamma(p)] - \sqrt{\frac{V(p)}{m}} \frac{Q^{-1}(\varepsilon)}{\ln 2}, \tag{11.1}$$

where p is the transmit power, $\gamma(p)$ is the Signal-to-Noise Ratio (SNR) at the receiver, $\varepsilon(p)$ is the decoding error probability, Q^{-1} is the inverse function of $Q(x) = \frac{1}{\sqrt{2\pi}} \int_x^\infty \exp^{\left(\frac{-t^2}{2}\right)} dt$, called the Q-function, and V given by $V(p) = 1 - [1 + \gamma(p)]^{-2}$. From (11.1), the decoding error probability can be presented as follows:

$$\varepsilon = Q[f(\gamma, m, p, D)], \tag{11.2}$$

where

$$f(\gamma, m, p, D) = \ln 2 \sqrt{\frac{m}{V(p)}} \left\{ \log_2 [1 + \gamma(p)] - \frac{D}{m} \right\}, \tag{11.3}$$

and D is the size of the transmitted packet.

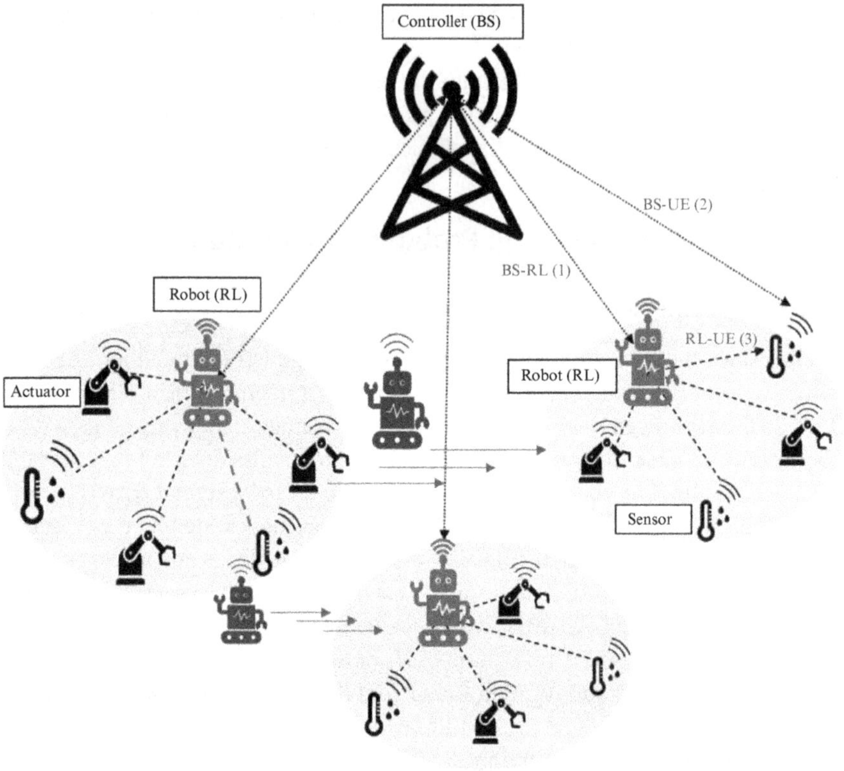

Figure 11.1 The system model of relay-assisted enabling URLLC in industry automation.

Since the decoding error probability with Q function is highly complex, it is profoundly difficult to solve the reliability optimization directly with the original form. Therefore, we apply the linear approximation $\varepsilon \approx \mathbb{E}[E(\gamma)]$ [21, 22] as

$$E(\gamma) = \begin{cases} 1, & \gamma < \alpha \\ 1 - \upsilon\sqrt{m}(\gamma - \theta), & \alpha < \gamma < \beta \\ 0, & \gamma \geq \beta, \end{cases} \tag{11.4}$$

where $\upsilon = \frac{1}{2\pi\sqrt{2^{\frac{2k}{m}}-1}}$, $\theta = \frac{2k}{m} - 1$, $\alpha = \theta - \frac{1}{2\upsilon\sqrt{m}}$, $\beta = \theta + \frac{1}{2\upsilon\sqrt{m}}$. Then, by applying the above approximation, the decoding error probability can be evaluated as

$$\varepsilon \approx \int_0^\infty E(x) f_\gamma(x)\, dx = \upsilon\sqrt{m} \int_\alpha^\beta F_\gamma(x)\, dx. \tag{11.5}$$

From [22], the cumulative distribution function is defined as

$$F_\gamma(x) = 1 - \exp\left(-\frac{x}{\gamma}\right) \overset{\text{high SNR}}{\approx} \frac{x}{\gamma}. \tag{11.6}$$

Finally, combining (11.5) and (11.6) together, the point-to-point decoding error probability is written as

$$\varepsilon \approx \frac{1}{\gamma}\left(2^{\frac{D}{m}} - 1\right). \tag{11.7}$$

11.2.3 Relay-Robot Assisted Transmissions Model

In this subsection, we will utilize the approximated decoding error probability derived in (11.7) in the relay-assisted URLLC scenario. In industrial scenarios, the communications between central controller to other devices (e.g. robots, Automated Guided Vehicles (AGV), actuators, etc.) require stringent reliability and end-to-end delay, while transmission links suffer severe shadowing due to many obstacles within the factory setting. Current research in URLLC has revealed that relay-assisted transmission outperforms other schemes in terms of satisfying ultra-reliability in factory automation [20]. Additionally, in industrial scenarios, the performance improvement of applying relaying is significant in URLLC networks operating with short blocklength [8]. Therefore, in this chapter, we explore the advantages of multiple relay robots to assist URLLC in industrial automation.

We examine the two phases of transmission of URLLC in industry automation. In the first phase, the BS broadcasts signals to all relay robots and IIoT devices. In the second phase, the robots, acting as Decode-and-Forward (DF) relays, assist the transmission for IIoT devices. The packet ID is assumed already defined in packet header for device identification. The whole blocklength is divided into two phases, m_b for the broadcasting phase, and $m_{k,n}$ for the relaying phase, where the kth relay and nth UE belong to the kth cluster.

For the first phase of the relay-assisted URLLC transmission, we assume that the BS broadcasts a combined packet of size $2D$ to the relays and IIoT devices. The probability that the kth relay fails to decode the information from the BS can be expressed as

$$\varepsilon_{s,k} = Q\left[f\left(\gamma_{s,k}, m_b, p_b, 2D\right)\right] \approx \frac{1}{\gamma_{s,k}}\left(2^{\frac{2D}{m_b}} - 1\right), \tag{11.8}$$

where p_b is the transmit power of the BS, $\gamma_{s,k} = p_b g_k / N_0 B$ is the SNR at the kth relay, B and N_0 are the bandwidth of the system and the one-side power spectral density of additive white Gaussian noise, $g_k = \beta_0 / d_{s,k}^2$ is the channel gain for the link between the BS and the kth relay, $d_{s,k} = \|\mathbf{q}_0 - \mathbf{q}_k\|$ is the distance from the BS to the kth relay.

Similarly, the probability that the nth IoT device fails to decode the combined packet from the BS is expressed as

$$\varepsilon_{s,n} = Q\left[f\left(\gamma_{s,n}, m_b, p_b, 2D\right)\right] \approx \frac{1}{\gamma_{s,n}}\left(2^{\frac{2D}{m_b}} - 1\right), \tag{11.9}$$

where $\gamma_{s,n} = p_b g_{s,n} / N_0 B$, $g_{s,n} = \beta_0 / d_{s,n}^2$ is the channel gain, $d_{s,n} = \|\mathbf{q}_0 - \mathbf{q}_n\|$.

For the second phase of the relay-assisted URLLC transmission, if the relay successfully decodes the combined packet, it will forward a packet of size D to the IIoT devices in its cluster with blocklength $m_{k,n}$. The probability that the nth IIoT device in the kth cluster fails to decode the information from the kth relay can be given by

$$\varepsilon_{k,n} = Q\left[f\left(\gamma_{k,n}, m_{k,n}, p_{k,n}, D\right)\right] \approx \frac{1}{\gamma_{k,n}}\left(2^{\frac{D}{m_{k,n}}} - 1\right), \tag{11.10}$$

where $p_{k,n}$ is the transmit power of the kth relay to nth UE, $\gamma_{k,n} = p_{k,n} g_{k,n} / N_0 B$ is the SNR at the nth IIoT device, $g_{k,n} = \beta_0 / d_{k,n}^2$ is the channel gain for the link between the kth relay and the nth IIoT device, $d_{k,n} = \|\mathbf{q}_k - \mathbf{q}_n\|$ is the distance from the kth relay to the nth IIoT device.

The final decoding error probability of the nth IIoT device can be built from the three components as defined in (11.8), (11.9) and (11.10):

$$\xi_n = \varepsilon_{s,n}\left[\varepsilon_{s,k} + (1 - \varepsilon_{s,k}).\varepsilon_{k,n}\right] \approx \varepsilon_{s,n}.\varepsilon_{s,k} + \varepsilon_{s,n}.\varepsilon_{k,n}, \tag{11.11}$$

In (11.11), $\varepsilon_{s,k}$ and $\varepsilon_{s,n}$ are the probability that the BS fails to transmit a packet to the kth relay and nth device at the broadcasting phase, meanwhile, $(1 - \varepsilon_{s,k}).\varepsilon_{k,n}$ indicates that the BS sends the packet successfully to the kth relay but the relay fails to transmit this packet to the nth device. Based on (11.6) and (11.7), the decoding error probability (11.11) can be calculated as

$$\xi_n \approx \frac{1}{\gamma_{s,n}}\left(2^{\frac{2D}{m_b}} - 1\right)\left[\frac{1}{\gamma_{s,k}}\left(2^{\frac{2D}{m_b}} - 1\right) + \frac{1}{\gamma_{k,n}}\left(2^{\frac{D}{m_{k,n}}} - 1\right)\right]. \tag{11.12}$$

11.2.4 Optimization Problem Formulation

In this subsection, we focus on optimizing the URLLC reliability of signal transmission from the BS to IIoT devices with the help of multiple relay robots. In this work, we propose a joint optimization of power and blocklength allocation instead of the fix-and-optimize method as in our previous work [9] for solving the reliability optimization problem. In [9], our approach is to iteratively solve the decomposition subproblems based on separated power and blocklength allocation problems. This method cannot guarantee the global optimal solution for both power and blocklength allocation. In our current approach, two types of variable will be jointly optimized based on new mathematical models and approximations for achieving a much more efficient solution in the decoding error probability minimization. We now formulate the reliability optimization problem as follows:

$$\min_{\mathbf{m},\mathbf{p},\mathbf{q}} \max_{n=1,\dots,N} \quad \xi_n \tag{11.13a}$$

$$\text{s.t.}\ m_b \le M_{\text{BS}}^{\max}, \tag{11.13b}$$

$$\sum_{n\in\mathcal{N}_k} m_{k,n} \le M_{\text{RL}}^{\max}, \forall n \in \mathcal{N}_k, \forall k \in \mathcal{K}, \tag{11.13c}$$

$$m_b \frac{p_b}{M_{\text{BS}}^{\max}} \le P_{\text{BS}}^{\max}, \tag{11.13d}$$

$$\sum_{n\in\mathcal{N}_k} \frac{m_{k,n} p_{k,n}}{M_{\text{RL}}^{\max}} \le P_{\text{RL}}^{\max}, \forall k \in \mathcal{K}, \tag{11.13e}$$

$$m_b, m_{k,n} \in \mathbb{Z}^+, \forall n \in \mathcal{N}_k, \forall k \in \mathcal{K}, \tag{11.13f}$$

$$(k,n) \in G_k, n \in \mathcal{N}_k, \forall k \in \mathcal{K}, \tag{11.13g}$$

$$N_k \in \mathbb{Z}^+, N_k \le N_k^{\max}, \forall k \in \mathcal{K}, \tag{11.13h}$$

$$\mathbf{q}_k \in [\mathbf{q}_k^{min}, \mathbf{q}_k^{\max}], \forall k \in \mathcal{K}. \tag{11.13i}$$

where $\mathbf{m} = \left[m_b, [m_{k,n}]_{k=1,n=1}^{K,N_k}\right]^T$, $\mathbf{p} = \left[p_b, [p_{k,n}]_{k=1,n=1}^{K,N_k}\right]^T$ and $\mathbf{q} = \left[\mathbf{q}_0, [\mathbf{q}_k]_{k=1}^{K}, [\mathbf{q}_n]_{n=1}^{N}\right]^T$. $M_{\text{BS}}^{\max}$, $M_{\text{RL}}^{\max}$, $P_{\text{BS}}^{\max}$ and $P_{\text{RL}}^{\max}$ are the blocklength and power budget of BS and relays, respectively. Here, (11.13b), (11.13c) are respectively the latency constraints of the broadcasting phase and relaying forwarding phase in terms of blocklength. The constraints (11.13d), (11.13e) refer to the power allocation of the BS and relays, respectively, while the constraints (11.13f) ensures that the blocklength allocated to each terminal is a positive integer ($\mathbb{Z}^+$). Finally, the constraints (11.13g), (11.13h) and (11.13i) indicate the forming of K clusters of IIoT devices under K relays. The constraint (11.13h) refers to the number of IIoT devices in each cluster that needs to be limited for the requirement of low latency. The problem in (11.13) is very difficult to handle due to its highly complex non-convex objective function (11.13a) and the integer constraint (11.13f). Therefore, we first propose to separately deal with the optimal relay deployment and the resource allocation in the next sections. In particular, we employ deep neural networks for fast optimal deployment of relays in Subsection 11.3.1, and enhanced K-means clustering in Subsection 11.3.2. Then, the optimal resource allocation under the joint blocklength and power allocation for minimizing the decoding error probability is discussed in Section 11.4.2.

11.3 Relay Robot Deployment

11.3.1 Optimizing the Position of Relay Robots with DNN-based Solutions

In this section, we aim to optimize the position of relay nodes to provide a best-effort transmission for the clustered networks in each group. We form K groups of IIoT devices under K relays based on their distances, which means $(k, n) \in G_k$ if the Euclidean distance between the kth relay and the nth IIoT device is less than the coverage distance $D_{k,cov}$ $\left(d_{k,n} \leq D^{\max}_{k,cov}, \forall k \in \mathcal{K}, n \in \mathcal{N}\right)$. We define an indicator function for clustering as follows:

$$\phi_{kn} = \begin{cases} 1, & \text{if } (k, n) \text{ belongs to the same cluster} \\ 0, & \text{otherwise.} \end{cases} \tag{11.14}$$

The optimization problem of relay deployment can be given by

$$\max_{\mathbf{q}_k, \phi_{kn}} \sum_{k=1}^{K} \sum_{n=1}^{N_k} \phi_{kn} \tag{11.15a}$$

$$\text{s.t. } q_k \in [q_k^{min}, q_k^{\max}], \forall k \in \mathcal{K}, \tag{11.15b}$$

$$d^2_{k,n} \leq \left(\mathcal{D}^{\max}_{k,cov}\right)^2 + \lambda_k(1 - \phi_{k,n}), \tag{11.15c}$$

$$k \in \mathcal{K}, n \in \mathcal{N}_k, \tag{11.15d}$$

where s_k is the position of the kth relay, λ_k is chosen as a specific value corresponding to the maximum network coverage area of the kth relay node (i.e. $\lambda_k > \left(\mathcal{D}^{\max}_{k,cov}\right)^2$) and $\mathcal{D}^{\max}_{k,cov}$ is the maximum coverage range of the kth relay. The problem (11.15) aims to optimize the position of the kth relay robot so that it can serve as many IIoT devices as possible in the cluster G_k. This is a mixed-integer

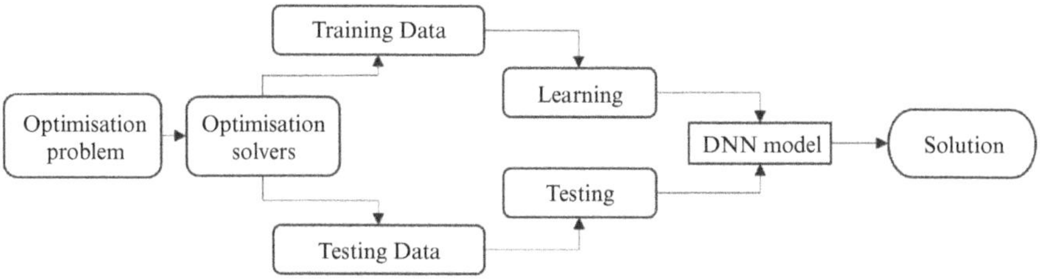

Figure 11.2 Learning-based optimization for rapid relays deployment in factory.

(binary) quadratic programming problem, which is a non-convex problem. Solving this kind of problem leads to combinatorial (or discrete) optimization, which is often very difficult and has a long processing time. To solve this problem, there are several well-known tools such as CVX in Matlab [6] or CVXPY [4] in Python. However, these methods result in a significant amount of processing time to find the optimal solution.

Therefore, we propose an effective learning-based method to solve the problem (11.15) rapidly. The learning-based optimization model based on DNN is conducted as in [5]. The first step of this learning-based model is to generate the training dataset by repeatedly solving the optimization problem (11.15) in order to create sufficient "labelled data" for the training process. With the training dataset, we design the DNN model to predict the optimal positions of the relay robots. The designed DNN model consists of the input, the output and three hidden layers. The first hidden layer has 200 nodes while the second and the third layers each have 80 nodes. The sigmoid activation function is used at the input and hidden layers while linear function is used at the output layer. We use Stochastic Gradient Descent (SGD) for optimizing our DNN model and the mean squared error to evaluate the learning process.

11.3.2 Optimizing the Position of Relay Robots with a K-means Clustering Algorithm

In this section, we aim to optimize the position of relay robots to provide a best-effort transmission for the clustered networks in each group. The proposed clustering algorithm addresses the constraints (11.13g), (11.13h), (11.13i) before solving the reliability optimization problem. The detailed description of the clustering algorithm is presented in Algorithm 1. Our clustering solution is based on the well-known K-means algorithm with enhancements for limiting the number of IIoT device in a cluster to satisfy the constraint (11.13h). In particular, at the

ith iteration, the nth IIoT device chooses the nearest kth relay robot based on the distance between the IIoT device and the relay robots. Then the kth relay robot checks the number of IIoT devices which is currently in its cluster and decides whether to allow the nth IIoT device to join its cluster. If the number of IIoT devices in the kth cluster reaches the maximum value (N_k^{max}), the nth IIoT device has to find the next nearest relay robot which is available. After successfully forming current clusters of relay robots and IIoT devices, the location of relay robots are updated equal to the average locations of all IIoT devices in their cluster. The procedure repeatedly executes until there is no change in cluster members or reaches the maximum number of iterations I_{max}. In our simulations, we ran the clustering algorithm in both constrained and unconstrained numbers of IIoT devices to investigate the effectiveness of robot relays in assisting URLLC transmissions. Figure 11.3 illustrates two examples of the clustering algorithm to form relay robots and IIoT devices network in factory automation using the proposed K-means clustering algorithms.

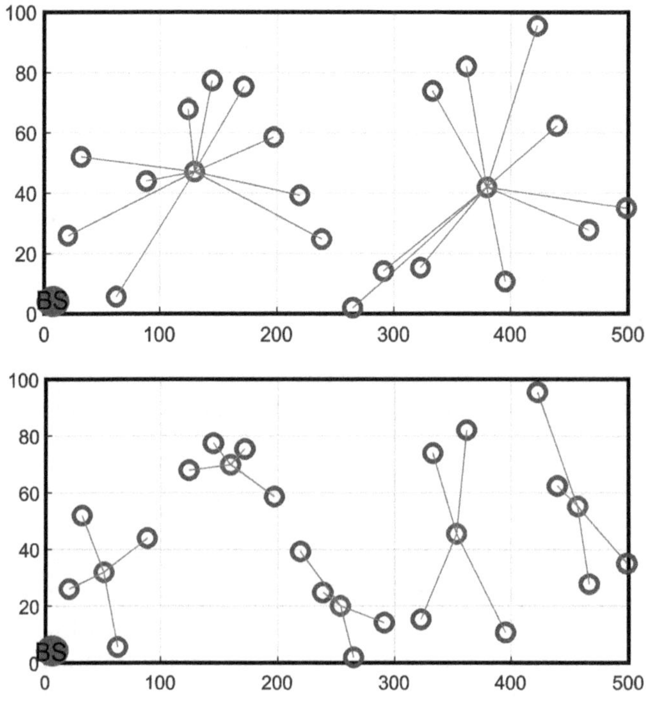

Figure 11.3 Examples of industrial IoTs clustering in the scenarios of $N = 20$ IIoT devices with $K = 2$ and $K = 5$ relay robots.

Algorithm 1 Constrained K-means clustering of IIoT devices

Input: The IoT devices' location ($\mathbf{q}_n$). The locations of the cluster centroid are initialized as $\{\theta_j = \mathbf{q}_k\}, J = K$. The maximum number of iterations is set to $I_{\max}$.
Set $J^{real} = \oslash$ and $\mathcal{N} = \{1, ..., N\}$. The maximum number of IIoT devices in a cluster $N_k^{\max}$

Repeat
  Update index of clusters:
    Compute the distance $\text{dist}(\mathbf{q}_n, \theta_j), \forall j, n$.
    Set $j = 0$.
    while $j \leq J$ **do**
      Set $j = j + 1, \hat{n} = 1$.
      for $n = \hat{n}$ to N **do**
        if $\left(\text{dist}(\mathbf{q}_{\hat{n}}, \theta_j) \leq D_{j,cov} \; \&\& \; \mathcal{N}_j < N_k^{\max}\right)$, assign the IoT device nth to cluster jth. Set $\mathcal{N}_j = \{n\}$.
        else find the next available cluster head;
        end if
      end for
      Set $J^{real} = \{j\}$
    end while
  Update the number of IoT devices in each cluster
  Update centroids:
    Update the location of the jth cluster:
    $\theta_j = \frac{1}{N_j} \sum_{n \in \mathcal{N}_j} \mathbf{q}_N, \forall j \in J$
Until There is no change in cluster members, or the iteration number reaches $I_{\max}$.

Output: $\{J^{real}\}, \{\theta_j\}$ and $\{\mathcal{N}_j = \{1, ..., N_j\}\}, \forall j$.

11.4 Resource Allocation Solution

11.4.1 Divide and Conquer Approach for Decoding Error Optimization (DnC-PABO)

In this section, we solve the joint blocklength and the power allocation optimization of the BS and relay robots in (11.13) with the "divide and conquer" approach. In particular, we consider optimizing for each cluster first and then for the whole network. Firstly, for the G_k cluster, we consider sub-problem optimization of minimizing the decoding error probability at the kth relay/nth terminal is expressed as (11.16)

To solve the problem (11.16), we first relax the integer constraint (11.16d) and find the optimal transmit power of the problem (11.16) with an initial value of fixed blocklength. Then we use the current optimal transmit power to find the current optimal blocklength at the ith iteration. After finding the current optimal blocklength, we initialize the integer solution as a floor-function of m [15] to use for the next iteration. The algorithm runs until convergence or the maximum number of iterations is reached. This optimization procedure is repeated among the K clusters before the optimal value of blocklength and transmit power of the BS can be found. Then the overall decoding error probability of the nth IIoT device can be obtained by following (11.7). The full description of the optimization algorithm is presented in Algorithm 2.

Algorithm 2 DnC-PABO: Divide and Conquer optimization algorithm

Input: The number of relays (K), IIoT devices (N).
The power budget of the BS and relays, ($P_{\max}^{BS}$), ($P_{\max}^{RL}$).
The blocklength budget of BS and relays, ($M_{\max}^{BS}$), ($M_{\max}^{RL}$).
The maximum number of iterations is set to $N_{\max}^{iter}$.

for $k = 1$ **to** K **do**
    Repeat
        Optimize decoding error probability in K **clusters**
            Find the optimal transmit power (p_{kn})
                Solve the optimization problem (11.16) with fixed blocklength(m_{kn}^{ini})
                Obtain the current optimal power (p_{kn}^{ref})
            Find optimal blocklength ($m_k n$) by solving (11.16) with (p_{kn}^{ref})
    Until
    Convergence, or the iteration number reaches $N_{\max}^{iter}$
end for
Optimize decoding error probability from BS
        Find the optimal value of BS blocklength (m_b^{ref}) based on (11.13b)
Calculate overall error
    Calculate current optimal error probability of the nth device by following (11.12)

Output: the optimal value of worst-case decoding error probability

$$\min_{\mathbf{m},\mathbf{p}} \max \frac{N_0 B d_{k,n}^2}{p_{kn}\beta_0}\left(2^{\frac{D}{m_{kn}}}-1\right), \tag{11.16a}$$

$$\text{s.t.} \sum_{n=1}^{N_k} m_{kn} \le M_{\max}^{\mathrm{RL}}, \tag{11.16b}$$

$$\sum_{n=1}^{N_k} \frac{m_{kn}p_{kn}}{M_{\max}^{\mathrm{RL}}} \le P_{\max}^{\mathrm{RL}}, \tag{11.16c}$$

$$\forall m_{kn} \in \mathbb{Z}^+, \forall k \in \mathcal{K}, n \in \mathcal{N}, (k,n) \in G_k. \tag{11.16d}$$

11.4.2 Joint Power Allocation and Blocklength Optimization (JPABO)

We provide the solution of joint blocklength and the power allocation optimization of the BS and relay robots. The original problem (11.13) will be derived as

$$\min_{\mathbf{m},\mathbf{p}} \max_{n=1,\dots,N} \quad \xi_n(\mathbf{m},\mathbf{p}) = \varepsilon_{s,n}.\varepsilon_{s,k} + \varepsilon_{s,n}.\varepsilon_{k,n} \tag{11.17a}$$

$$\text{s.t.} \quad (11.13b), (11.13c), (11.13d), (11.13e).(11.13f) \tag{11.17b}$$

However, due to the highly complex non-convex optimization problem in (11.17), we develop the solution by using convex approximations to transform this problem into a low-complexity convex optimization problem [7, 10, 16, 17]. We first handle the objective function (11.17a) and then address other non-convex constraints step-by-step in the rest of this section.

The objective function (11.17a) is described as (11.18)

$$\begin{aligned}
\xi_n(\mathbf{m},\mathbf{p}) &= \varepsilon_{s,n}.\varepsilon_{s,k} + \varepsilon_{s,n}.\varepsilon_{k,n} \\
&= (N_0B)^2 \frac{1}{p_b g_{s,n}}\left(2^{\frac{2D}{m_b}}-1\right)\left[\frac{1}{p_b g_{s,k}}\left(2^{\frac{2D}{m_b}}-1\right)+\frac{1}{p_{k,n}g_{k,n}}\left(2^{\frac{D}{m_{k,n}}}-1\right)\right] \\
&= (N_0B)^2\left[\frac{1}{g_{s,n}g_{s,k}}\left(\frac{1}{p_b}\left(2^{\frac{2D}{m_b}}-1\right)\right)^2+\frac{1}{g_{s,n}g_{k,n}}\frac{1}{p_b p_{k,n}}\left(2^{\frac{2D}{m_b}}-1\right)\left(2^{\frac{D}{m_{k,n}}}-1\right)\right].
\end{aligned} \tag{11.18}$$

Firstly, by introducing a new variable t, we rewrite the problem (11.17) as follow:

$$\min_{\mathbf{m},\mathbf{p}} \quad t \tag{11.19a}$$

$$\begin{aligned}
\text{s.t.} \quad & (11.13b), (11.13c), (11.13d), (11.13e), (11.13f) \\
& t \ge \xi_n(\mathbf{m},\mathbf{p}), \forall n \in \mathcal{N}.
\end{aligned} \tag{11.19b}$$

To handle the non-convex constraint (11.19b), we follow the inequality for approximating the function $\xi_n(\mathbf{m}, \mathbf{p})$ as below

$$a^2 \geq 2\bar{a}a - \bar{a}^2 \tag{11.20}$$

for $a > 0$. Then, applying (11.20) with

$$a = \frac{1}{p_b}\left(2^{2D/m_b} - 1\right), \bar{a} = \frac{1}{\bar{p}_b}\left(2^{2D/\bar{m}_b} - 1\right)$$

we have

$$\left(\frac{1}{p_b}\left(2^{2D/m_b}-1\right)\right)^2 \geq 2\frac{1}{\bar{p}_b}\left(2^{2D/\bar{m}_b}-1\right)\frac{1}{p_b}\left(2^{2D/m_b}-1\right)-\left(\frac{1}{\bar{p}_b}\left(2^{2D/\bar{m}_b}-1\right)\right)^2. \tag{11.21}$$

Next, we follow the inequality as below

$$zt \leq \frac{1}{2}\left(\frac{\bar{t}}{\bar{z}}z^2 + \frac{\bar{z}}{\bar{t}}t^2\right), \tag{11.22}$$

for $z > 0, t > 0$. Then, applying (11.22) with

$$z = \frac{1}{p_b}, \bar{z} = \frac{1}{\bar{p}_b}, t = \left(2^{2D/m_b} - 1\right), \bar{t} = \left(2^{2D/\bar{m}_b} - 1\right)$$

we have

$$\frac{1}{p_b}\left(2^{2D/m_b} - 1\right) \leq \frac{1}{2}\left(\bar{p}_p\left(2^{2D/\bar{m}_b} - 1\right)\frac{1}{p_b^2} + \frac{1}{\bar{p}_p\left(2^{2D/\bar{m}_b} - 1\right)}\left(2^{2D/m_b} - 1\right)^2\right). \tag{11.23}$$

Thus, the first term of $\xi_n(\mathbf{m}, \mathbf{p})$ can be approximated as

$$\left(\frac{1}{p_b}\left(2^{2D/m_b}-1\right)\right)^2 \geq \left(2^{2D/\bar{m}_b}-1\right)^2\frac{1}{p_b^2} + \frac{1}{\bar{p}_p^2}\left(2^{2D/m_b}-1\right)^2 - \left(\frac{1}{\bar{p}_b}\left(2^{2D/\bar{m}_b}-1\right)\right)^2. \tag{11.24}$$

Based on (11.20) and (11.22), we follow the inequality as below

$$\begin{aligned} xyzt &\leq \frac{1}{2}\left(\bar{z}\bar{t}(2xy - \bar{x}\bar{y}) + \bar{x}\bar{y}(2zt - \bar{z}\bar{t})\right) \\ &\leq \frac{1}{2}\left(\bar{z}\bar{t}\left[\left(\frac{\bar{y}}{\bar{x}}x^2 + \frac{\bar{x}}{\bar{y}}y^2\right) - \bar{x}\bar{y}\right] + \bar{x}\bar{y}\left[\left(\frac{\bar{t}}{\bar{z}}z^2 + \frac{\bar{z}}{\bar{t}}t^2\right) - \bar{z}\bar{t}\right]\right) \\ &= \frac{1}{2}\bar{x}\bar{y}\bar{z}\bar{t}\left(\frac{1}{\bar{x}^2}x^2 + \frac{1}{\bar{y}^2}y^2 + \frac{1}{\bar{z}^2}z^2 + \frac{1}{\bar{t}^2}t^2 - 2\right) \end{aligned} \tag{11.25}$$

for $x, y, z, t, \bar{x}, \bar{y}, \bar{z}, \bar{t} > 0$. Then, applying (11.25) with

$$x = \frac{1}{p_b}, \bar{x} = \frac{1}{\bar{p}_b}, y = \frac{1}{p_{k,n}}, \bar{y} = \frac{1}{\bar{p}_{k,n}}, z = 2^{2D/m_b} - 1,$$

$$\bar{z} = 2^{2D/\bar{m}_b} - 1, t = 2^{D/m_{k,n}} - 1, \bar{t} = 2^{D/\bar{m}_{k,n}} - 1$$

the last term of $\xi_n(\mathbf{m}, \mathbf{p})$ can be approximated as

$$\frac{1}{p_b p_{k,n}}\left(2^{\frac{2D}{m_b}} - 1\right)\left(2^{\frac{D}{m_{k,n}}} - 1\right)$$
$$\leq \frac{1}{2}\frac{1}{\bar{p}_b \bar{p}_{k,n}}\left(2^{2D/\bar{m}_b} - 1\right)\left(2^{2D/\bar{m}_{k,n}} - 1\right)$$
$$\left(\frac{\bar{p}_b^2}{p_b^2} + \frac{\bar{p}_{k,n}^2}{p_{k,n}^2} + \frac{(2^{2D/m_b} - 1)^2}{(2^{2D/\bar{m}_b} - 1)^2} + \frac{(2^{D/m_{k,n}} - 1)^2}{(2^{D/\bar{m}_{k,n}} - 1)^2} - 2\right). \tag{11.26}$$

Moreover, we define

$$\hat{g}_n(\mathbf{m}, \mathbf{p}) \geq \xi_n(\mathbf{m}, \mathbf{p}), \forall n \in \mathcal{N}, \tag{11.27}$$

where $\hat{g}_n(\mathbf{m}, \mathbf{p})$ defined as (11.28)

$$\hat{g}_n(\mathbf{m}, \mathbf{p}) = (N_0 B)^2 \left[\frac{1}{g_{s,n} g_{s,k}}\left(\left(2^{\frac{2D}{\bar{m}_b}} - 1\right)^2 \frac{1}{p_b^2}\right) + \frac{1}{\bar{p}_p^2}\left(2^{\frac{2D}{m_b}} - 1\right)^2 - \left(\frac{1}{\bar{p}_b}\left(2^{\frac{2D}{\bar{m}_b}} - 1\right)\right)^2\right.$$
$$\left. + \frac{1}{g_{s,n} g_{k,n}} \frac{1}{2} \frac{1}{\bar{p}_b \bar{p}_{k,n}}\left(2^{\frac{2D}{\bar{m}_b}} - 1\right)\left(2^{\frac{D}{\bar{m}_{k,n}}} - 1\right)\left(\frac{\bar{p}_b^2}{p_b^2} + \frac{\bar{p}_{k,n}^2}{p_{k,n}^2} + \frac{(2^{\frac{2D}{m_b}} - 1)^2}{(2^{\frac{2D}{\bar{m}_b}} - 1)^2} + \frac{(2^{\frac{D}{m_{k,n}}} - 1)^2}{(2^{\frac{D}{\bar{m}_{k,n}}} - 1)^2} - 2\right)\right]. \tag{11.28}$$

And thus, the constraint (11.19b) can be rewritten as

$$t \geq \hat{g}_n(\mathbf{m}, \mathbf{p}), \forall n \in \mathcal{N}. \tag{11.29}$$

To handle the non-convex constraints (11.13d) and (11.13e), we follow (11.22) as below

$$\frac{1}{2}\left(\frac{\bar{p}_b}{\bar{m}_b} m_b^2 + \frac{\bar{m}_b}{\bar{p}_b} p_b^2\right) \leq M_{\text{BS}}^{\max} P_{\text{BS}}^{\max}, \tag{11.30}$$

$$\sum_{n \in \mathcal{N}_k} \frac{1}{M_{\text{RL}}^{\max}} \frac{1}{2}\left(\frac{\bar{p}_{k,n}}{\bar{m}_{k,n}} m_{k,n}^2 + \frac{\bar{m}_{k,n}}{\bar{p}_{k,n}} p_{k,n}^2\right) \leq P_{\text{RL}}^{\max}. \tag{11.31}$$

One simple approach is to handle the non-convex constraints (11.13f), we linearly approximate $m_b, m_{k,n}$ into

$$m_b \in [0, M_{\text{BS}}^{\max}], \sum_{n \in \mathcal{N}_k} m_{k,n} \in [0, M_{\text{RL}}^{\max}]. \tag{11.32}$$

Then, after implementation of the optimization algorithm, we choose $m_b, m_{k,n}$ as the nearest integer values from their solutions.

At the ith iteration, given feasible points $(\mathbf{m}^{(i)}, \mathbf{p}^{(i)})$, the following convex program is solved to generate the next feasible point as follows:

$$\min_{\mathbf{m},\mathbf{p}} \quad t \tag{11.33a}$$

$$\text{s.t.} \quad (11.13b), (11.13c), (11.29), (11.30), (11.31), (11.32). \tag{11.33b}$$

We now proceed by proposing Algorithm 3 to solve the reliability optimization problem (11.33).

Algorithm 3 : Joint blocklength and power allocation procedure for solving problem (11.33)

Input: Set $i = 0$ and a feasible point $(\mathbf{m}^{(0)}, \mathbf{p}^{(0)})$. Set the tolerance $\varepsilon = 10^{-3}$ and the maximum number of iterations $I_{\max} = 20$.
Repeat
Solve problem (11.33) for the next optimal solution $(\mathbf{m}^{(i+1)}, \mathbf{p}^{(i+1)})$.
Set $i := i + 1$
Until Convergence of the objective function in (11.33) or $i > I_{\max}$.
Output: $\{\mathbf{m}^*, \mathbf{p}^*\}$ and $\{t^*\}$.

Note that the optimal solution of $\{t^*\}$ in (11.33a) will be a suboptimal choice to minimize the decoding error probability for the network.

The initial point $(\mathbf{m}^{(0)}, \mathbf{p}^{(0)})$ for (11.33a) can be set up by randomness approaches for satisfying constraints (11.13b), (11.13c), (11.13d), (11.13e), (11.13f) in the optimization problem.

11.5 Simulation Results

To demonstrate the effectiveness of our proposed optimization scheme and the use of multiple relay robots in assisting the URLLC, in this section, several representative numerical results are provided.

11.5.1 Parameter Setup

In our simulations, the noise power spectral density is set to -173 dBm Hz^{-1} [15] with communication bandwidth of the channel 200 KHz [18]. The channel gain at reference distance is set to $\beta_0 = -50$ dB [18]. The size of transmission packet is from $D = 100$ bits to $D = 125$ [15, 20]. The transmit power budget of the relay robots is examined in the range 14–20 dBm [15], while the transmit power of the BS is set to 30 dBm [14]. The autonomous factory is assumed in the scale of 100m × 500m [20].

11.5.2 Effectiveness of Multi-relay Robot-aided Transmission

To demonstrate the effectiveness of relay-assisted transmission in URLLC systems, we conducted simulations to examine the worst-case decoding error probability with and without relays. The results are presented in Table 11.1. According to this table, it can be seen that relay-assisted transmissions provide significant improvement in reliability by reducing the decoding rate from 10^{-5} to 10^{-10}, which satisfies the stringent URLLC's requirements for industrial automation proposed in 3GPP releases.

The effectiveness of multiple relay robots in enhancing the reliability is clearly presented in Figure 11.4 and Figure 11.5. As can be observed from the figures, in the scenario of 20 IIoT devices, the system with five relays provides better performance than that with two relays, especially when the transmit power of the relays is small. Similarly, Figure 11.5 shows that with a fixed number of relays, the fewer IIoT devices the system has, the more the reliability level can be guaranteed for the system.

11.5.3 Optimal Blocklength and Power Allocation for Minimizing Decoding Error Probability

Figure 11.6 illustrates the error probability versus blocklength in different numbers of relays and IoT devices. This figure also demonstrates the effectiveness of the joint blocklength and power optimization in terms of finding the optimal allocation. It is clearly seen that the proposed optimization algorithm effectively reduces the probability error rate of users approximately five times with the same budget of symbols compared with equal allocation.

Table 11.1 The effectiveness of relay-aided transmission in enabling URLLC systems.

Power (dBm)	15	16	17	18	19	20
K=2; N=20	3.25E-09	3.18E-09	2.61E-09	1.97E-09	1.54E-09	1.37E-09
K=0; N=20	7.81E-04	6.21E-04	4.93E-04	3.92E-04	3.11E-04	2.47E-04
K=5; N=40	2.32E-09	1.97E-09	1.73E-09	1.38E-09	1.15E-09	9.98E-10
K=0, K=40	7.86E-04	6.24E-04	4.96E-04	3.94E-04	3.13E-04	2.49E-04

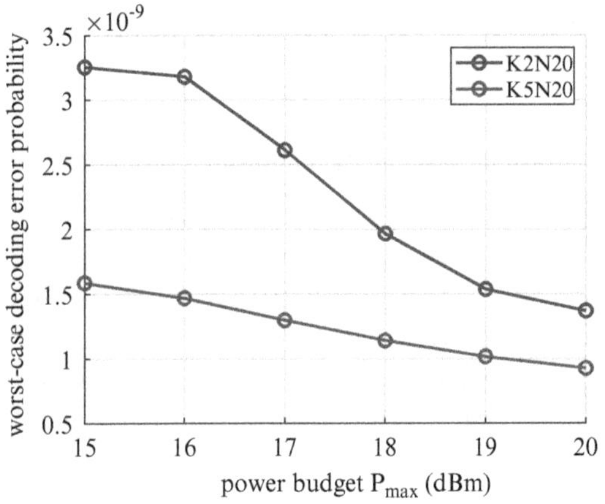

Figure 11.4 The worst-case decoding error probability in a range of powers in scenarios of $N = 20$ IIoT devices with $K = 2$ versus $K = 5$ relay robots.

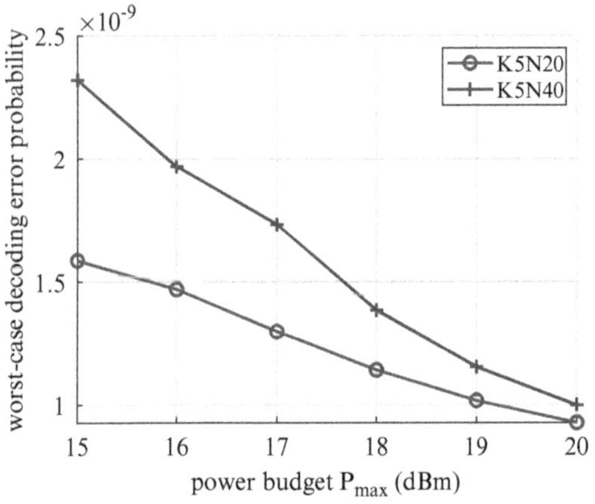

Figure 11.5 The worst-case decoding error probability in range of powers in scenarios of $K = 5$ relay robots with $N = 20$ versus $N = 40$ IIoT devices.

11.5.4 Decoding Error Probability Versus the Packet Size Under Two Clustering Approaches

Figure 11.7 plots the impact of packet size parameter (D bits) on the reliability of URLLC-based communications. As expected, in the scenario consisting of $N = 20$ IIoT devices with the support of $K = 5$ relay robots, a larger packet size leads to a higher decoding error probability for both clustering approaches. In addition, the cluster-free algorithm results in a poorer performance than the constrained-clustering because the distance between the relay robot and the IIoT devices in the constrained-clustering is smaller than that in the cluster-free approach. Under the same number of IIoT devices, the constrained-clustering naturally needs more relay robots than the cluster-free solution so that the distance between relay robots and IIoT devices becomes smaller.

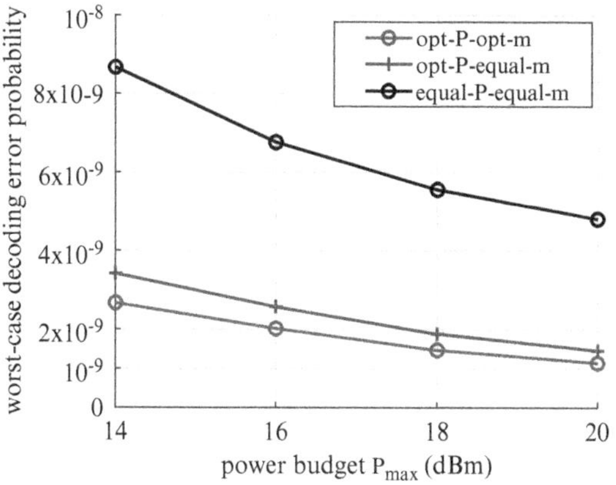

Figure 11.6 The comparison of worst-case decoding error probability for optimal blocklength, power allocation, and equal allocation in the scenarios of $K = 5$ relay robots, $N = 20$ IIoT devices.

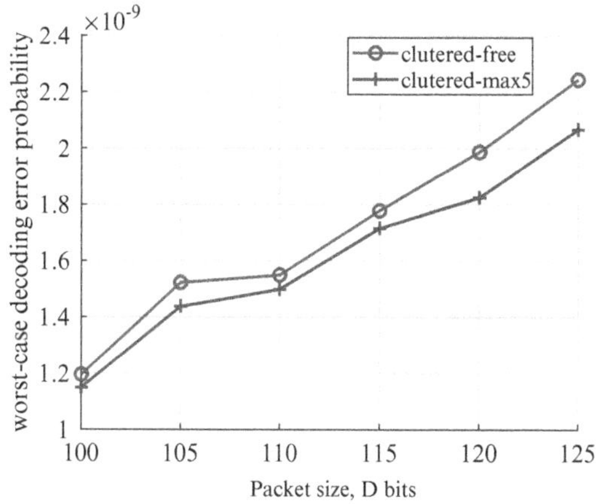

Figure 11.7 The comparison of worst-case decoding error probability for different packet sizes under two clustering approaches in the scenarios of $K = 5$ relay robots, $N = 20$ IIoT devices.

11.5.5 Decoding Error Probability Versus the Range of IIoT Devices

Figure 11.8 demonstrates that under the same level of transmit power and the same number of supporting relay robots, the decoding error probability of IIoT devices decreases when there are more IIoT devices in the system. In addition, the results illustrate that the more transmit power the relay provides, the higher the reliability that the IIoT devices can achieve, especially in small-scale scenarios.

11.6 Conclusion

In this chapter, we have investigated the practical problem of industrial automation in clustering, joint blocklength, and power optimization for multiple relay

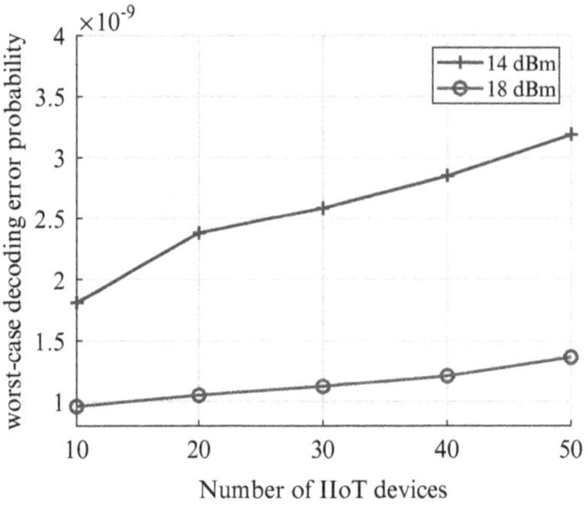

Figure 11.8 The comparison of worst-case decoding error probability for different numbers of IIoT devices in scenarios which have the same numbers of relay robots at two transmit power levels $P_{RL}^{max} = 14$ dBm and $P_{RL}^{max} = 18$ dBm.

robot-aided URRLC systems. We have formulated the decoding error probability minimization problem under stringent constraints of URLLC in terms of low latency and ultra-high reliability. To deal with this problem, the enhanced K-means clustering solution is introduced to effectively form the networks of relay robots and IIoT devices. Then we propose the joint blocklength and power optimization to minimize the worst-case decoding error probability of IIoT devices. The numerical results of various experiments demonstrate the effectiveness of our proposed solutions. As future work, we will explore some promising directions of combining URLLC with other emerging technologies such as multi-access edge computing, Reconfigurable Intelligent Surface(RIS)-assisted communications for industrial automation.

Bibliography

[1] 3GPP. Study on scenarios and requirements for next generation access technologies. Technical Report (TR) 38.913, 3rd Generation Partnership Project (3GPP), 2018. Version 15.0.0.

[2] 3GPP. Release 16 description. Technical Report (TR) 21.916, 3rd Generation Partnership Project (3GPP), 2020. Version 1.0.0.

[3] Mehdi Bennis, Merouane Debbah, and H. Vincent Poor. Ultrareliable and low-latency wireless communication: Tail, risk, and scale. *Proc. IEEE*, 106(10): 1834–1853, 2018.

[4] Steven Diamond and Stephen Boyd. CVXPY: A Python-embedded modeling language for convex optimization. *Journal of Machine Learning Research*, 17 (83):1–5, 2016.

[5] Trung Q. Duong, Long D. Nguyen, Hoang Duong Tuan, and Lajos Hanzo. Learning-aided realtime performance optimisation of cognitive UAV-assisted disaster communication. In *Proc. IEEE Glob. Commun. Conf. 2019*, pages 1–6, Waikoloa, HI, December 2019.

[6] Michael Grant and Stephen Boyd. CVX: MATLAB software for disciplined convex programming, version 2.1. http://cvxr.com/cvx, March 2014.

[7] Tuy Hoang. *Convex Analysis and Global Optimization*. Springer, second edition, 2016.

[8] Yulin Hu, M. Cenk Gursoy, and Anke Schmeink. Relaying-enabled ultra-reliable low-latency communications in 5G. *IEEE Netw.*, 32(2):62–68, March 2018.

[9] Dang Van Huynh, Saeed R. Khosravirad, Long D. Nguyen, and Trung Q. Duong. Multiple relay robots-assisted URLLC for industrial automation with deep neural networks. In *Proc. IEEE Global Communications Conference (GLOBECOM'21)*, Madrid, Spain, December7–11 2021.

[10] H.H. Kha, H. D. Tuan, and Ha H. Nguyen. Fast global optimal power allocation in wireless networks by local D.C. programming. *IEEE Trans. Wireless Commun.*, 11(2):510–515, 2012.

[11] Saeed R. Khosravirad, Harish Viswanathan, and Wei Yu. Exploiting diversity for ultra-reliable and low-latency wireless control. *IEEE Trans. Wireless Commun.*, 20(1):316–331, 2021.

[12] Liang Liu and Wei Yu. A D2D-based protocol for ultra-reliable wireless communications for industrial automation. *IEEE Trans. Wireless Commun.*, 17 (8):5045–5058, August 2018.

[13] Jakub Mazgula, Jakub Sapis, Umair Sajid Hashmi, and Harish Viswanathan. Ultra reliable low latency communications in mmwave for factory floor automation. *J. Indian Inst. Sci.*, 100(2):303–314, 2020.

[14] A. A. Nasir, H. D. Tuan, H. Q. Ngo, T. Q. Duong, and H. V. Poor. Cell-free massive mimo in the short blocklength regime for URLLC. *IEEE Transactions on Wireless Communications*, April 2021.

[15] Ali A. Nasir. Min-max decoding-error probability-based resource allocation for a urllc system. *IEEE Commun. Lett.*, 24(12):2864–2867, December 2020.

[16] Long D. Nguyen, Hoang Duong Tuan, Trung Q. Duong, Octavia A. Dobre, and H. Vincent Poor. Downlink beamforming for energy-efficient heterogeneous networks with massive MIMO and small cells. *IEEE Trans. Wireless Commun.*, 17(5):3386–3400, May 2018.

[17] Long D. Nguyen, Hoang Duong Tuan, Trung Q. Duong, and H. Vincent Poor. Multi-user regularized zero-forcing beamforming. *IEEE Trans. Signal Process.*, 67(11):2839–2853, 2019.

[18] Weichen Ning, Ying Wang, Yuanbin Chen, and Man Liu. Resource allocation in relay-assisted mission-critical industrial internet of things. In *Proc. IEEE Wirel. Commun. Netw. Conf. Workshop*, pages 1–6, Seoul, Korea (South), April 2020.

[19] Yury Polyanskiy, H. Vincent Poor, and Sergio Verdu. Channel coding rate in the finite blocklength regime. *IEEE Trans. Inf. Theory*, 56(5):2307–2359, May 2010.

[20] Hong Ren, Cunhua Pan, Yansha Deng, Maged Elkashlan, and Arumugam Nallanathan. Joint power and blocklength optimization for URLLC in a factory automation scenario. *IEEE Trans. Wireless Commun.*, 19(3):1786–1801, March 2020.

[21] Wei Yang, Giuseppe Durisi, Tobias Koch, and Yury Polyanskiy. Quasi-static multiple-antenna fading channels at finite blocklength. *IEEE Trans. Inf. Theory*, 60(7):4232–4265, 2014.

[22] Gu Yifan, Chen He, Li Yonghui, and Vucetic Branka. Ultra-reliable short-packet communications: Half-duplex or full-duplex relaying? *IEEE Wirel. Commun. Lett.*, 7(3):348–351, June 2018.

Index

Ultra-Reliable and Low-Latency Communications (URLLC) Theory and Practice: Advances in 5G and Beyond, First Edition. Edited by Trung Q. Duong, Saeed R. Khosravirad, Changyang She, Petar Popovski, Mehdi Bennis and Tony Q.S. Quek.
© 2023 John Wiley & Sons Ltd. Published 2023 by John Wiley & Sons Ltd.